OT 39
Operator Theory: Advances and Applications
Vol. 39

Birkhäuser Verlag
Basel · Boston · Berlin

Mircea Martin
Mihai Putinar

Lectures on Hyponormal Operators

1989

Birkhäuser Verlag
Basel · Boston · Berlin

Authors' addresses:

Prof. Mircea Martin
Prof. Mihai Putinar
Department of Mathematics
National Institute for Scientific
and Technical Creation
B-dul Pacii 220
79622 Bucharest, Romania

CIP-Titelaufnahme der Deutschen Bibliothek

Martin, Mircea:
Lectures on hyponormal operators / Mircea Martin ; Mihai Putinar. –
Basel ; Boston ; Berlin ; Birkhäuser, 1989
(Operator theory ; Vol. 39)
ISBN 3-7643-2329-9 (Basel ...) Pb.
ISBN 0-8176-2329-9 (Boston) Pb.
NE: Putinar, Mihai:; GT

Printed in Germany on acid-free paper
ISBN 3-7643-2329-9
ISBN 0-8176-2329-9

CB-S+K
SCIMON

FOREWORD

The present lectures are based on a course delivered by the authors at the University of Bucharest, in the winter semester 1985-1986.

Without aiming at completeness, the topics selected cover all the major questions concerning hyponormal operators. Our main purpose is to provide the reader with a straightforward access to an active field of research which is strongly related to the spectral and perturbation theories of Hilbert space operators, singular integral equations and scattering theory.

We have in view an audience composed especially of experts in operator theory or integral equations, mathematical physicists and graduate students. The book is intended as a reference for the basic results on hyponormal operators, but has the structure of a textbook. Parts of it can also be used as a second year graduate course.

As prerequisites the reader is supposed to be acquainted with the basic principles of functional analysis and operator theory as covered for instance by Reed and Simon [1].

At several stages of preparation of the manuscript we were pleased to benefit from proper comments made by our colleagues: Grigore Arsene, Tiberiu Constantinescu, Raúl Curto, Jan Janas, Bebe Prunaru, Florin Rădulescu, Khrysztof Rudol, Konrad Schmüdgen, Florian-Horia Vasilescu. We warmly thank them all.

We are indebted to Professor Israel Gohberg, the editor of this series, for his constant encouragement and his valuable mathematical advice.

We wish to thank Mr. Benno Zimmermann, the Mathematics Editor at Birkhäuser Verlag, for cooperation and assistance during the preparation of the manuscript.

An editing group whose principal member was Mrs. Irén Némethi dealt with the difficult task of implementing the text by means of a word processor. We thank them for their high standard job.

We are already aware of certain limits of our lectures. Any inaccuracies concerning the paternity of a result or an idea, as well as any omission, are non-intentional. With regard to this or to anything concerning the book we would be glad to know the personal opinions of our readers.

Bucharest — The Authors

February 20, 1989

CONTENTS

INTRODUCTION

> *Dans les champs élyséens de la pensée, toute satisfaction correspond à un rajeunissement et à un développement, et rien n'est plus sain pour l'esprit que les ivresses et les débauches de la curiosité, de la compréhension et de l'admiration.*
>
> *Maurice Maeterlinck: "Le Temple Enseveli"*

In the modern theory of linear operators two different tendencies can be perceived. One oriented towards an abstract setting based on the general principles of functional analysis and the second dealing with special classes of operators and relying usually on classical analysis, mathematical physics or other external fields. Fortunately, the theory of hyponormal operators grew up in the last three decades as a byproduct of both tendencies. In many respects this theory repeats that of self-adjoint, normal, dissipative or accretive operators. Quite specifically, all these objects defined by an abstract condition and fitted for an axiomatic treatment, turned out to be rather concrete due to the existence of some functional models carrying a rich additional structure.

As a matter of fact the theory of hyponormal operators is equivalent to the study of the commutation relation $i[X, Y] \geq 0$, where X and Y are bounded self-adjoint operators acting on a complex Hilbert space. A landmark contribution to the field was the discovery in the fifties of absolute continuity results for concrete pairs of self-adjoint operators satisfying similar commutator restrictions, cf. **C.R. Putnam** [1]-[4] and **M. Rosenblum** [2]. While investigating these phenomena, **C.R. Putnam** [4] and **T. Kato** [3] proved among other things that every irreducible pair (X,Y) as above contains only purely absolutely continuous self-adjoint operators. The book by **C.R. Putnam** [5] crowned these researches.

Later on, using the absolute continuity theorem, **Daoxing Xia** [1], **J.D. Pincus** [1] and **T. Kato** [5] independently obtained a functional model with one dimensional singular integral operators for the pair (X,Y). Further, increasingly more elaborated models appeared for more general pairs of operators, cf. the book by **D. Xia** [3].

The importance of the singular integral model resides in that it brings into the theory the well established methods of singular integral equations and the perturbation and scattering theories. These methods allow an invariant interpretation of the parameters describing the singular integral model and the classification of pairs (X,Y) in terms of these invariants. The individual work of **J.D. Pincus** and then his joint work with **R.W. Carey** marked the culminating points of this direction of research. In the seventies the two authors introduced a complete unitary invariant for (X,Y) - the mosaic -, and its trace - the principal function. Specifically, these invariants exist under the additional but not very restrictive assumption, that the commutator [X,Y] is trace-class.

From another viewpoint, the operator $T = X + iY$ associated to a pair (X,Y) as above satisfies $[T^*, T] \geq 0$. Such an operator was called hyponormal. In the sixties several authors among which we cite **S. Berberian** [1], **J. Stampfli** [1], **T. Ando** [1] and **P.R. Halmos** [4] discovered that hyponormal operators inherit some sharp spectral properties from normal operators (defined by the commutator identity $[T^*, T] = 0$). The abstract spectral theory of hyponormal operators was settled in the seventies by **J. Stampfli** [2]-[9], **K. Clancey** [1]-[4], **S. Clary** [1], **R. Howe** [1], **C.R. Putnam** [6]-[11], **M. Radjabalipour** [2], [3] and others. An intermediate class between hyponormal and normal operators is that of subnormal operators. These are modelled as multiplication operators on spaces of analytic functions and their theory is better understood, cf. the book by **J.B. Conway** [2].

In the case of hyponormal operators T with trace-class self-commutator $[T^*, T]$, **J.W. Helton** and **R. Howe** [1], [2] obtained a remarkable trace formula. Their contribution combined with the work of **R.W. Carey** and **J.D. Pincus** [4]-[10] has carried out the theory of the principal function. Further contributors to this key branch of the theory of hyponormal operators were **C.A. Berger** [1], [2], **C.A. Berger** and **B.L. Shaw** [1], [2], **K. Clancey** [4]-[9], **D. Xia** [3]-[7], **D. Voiculescu** [5], [6]. From the perspective of the theory of C^*-algebras, the trace formula of Helton and Howe lies at the heart of some new and very active theories, cf. **R.G. Douglas** and **D. Voiculescu** [1], **A. Connes** [1] and others. The book by **K. Clancey** [5] presents in a concise and unifying manner the essentials of the theory of the principal function.

In recent times other alternative functional models for hyponormal operators have appeared: the Toeplitz model of **C. Foiaş** [2], **B.Sz.-Nagy** and **C. Foiaş** [2], [3] and also the two-dimensional models investigated by **K. Clancey** [9], **D. Xia** [4], [7], [8], **J.D. Pincus**, **J. Xia** and **D. Xia** [1], and by the authors of these lectures.

Last but not least we should mention the essential contribution of **S. Brown** [1], [2], followed by **E. Albrecht** and **B. Chevreau** [1] and many other authors, to the investigation of the lattice of invariant subspaces for large classes of hyponormal operators.

Some very recent contributions have deliberately not been included in our text. Such are the joint hyponormality for n-tuples of commuting operators, unbounded hyponormal operators, C^*-algebras techniques applied to hyponormal operators, semi-hyponormal operators. At a certain point the theory of the commutation relation $i[X,Y] \geq 0$ becomes parallel to that of the relation **trace**$|[X, Y]| < \infty$. Also, we have not pursuited the latter direction, but the reader can easily fill this gap by consulting the papers **Helton** and **Howe** [1], [2], **Carey** and **Pincus** [5]-[9] and the book by **Clancey** [5].

Part of the most important results contained in the book are presented with different proofs than the original ones. This allowed us to arrange the material in a new and, we hope, conceptually simple form.

The order of the chapters is not linear, and the cross-references between chapters are reduced to a minimum. For instance chapters III-VI, VIII-IX are rather technical so the reader may postpone their reading until he has got a feeling of the theory by means of the other chapters.

Each chapter begins with an informal introduction, contains several sections and ends with some bibliographical notes and exercises. The cross-references are marked as follows: Theorem 1.8 inside the same chpater and Theorem VII.1.8 between different chapters. The exercises are of several types: exercises pertaining to the text, theorems which are not proved in the text but are stated for the information of the reader, usually mentioned with the name of the author, difficult exercises of one of the previous types, marked by a star, and open problems.

Chapter I, III, VI, VII and IX contain also some supplements. Their role is to isolate without going into complete details some external theoretical or technical tools. The rest of the book, except §VII.4 which is isolated, is self-contained.

The book ends up with an index and a list of bibliographical references, the last far from being exhaustive. In the text these are referred to as, e.g. Pincus [2], [3], Howe [1], Helton and Howe [1] (or Helton-Howe [1] in the case of several authors).

A list of the most frequently used notational symbols precedes the index.

Chapter I

SUBNORMAL OPERATORS

The purpose of this first chapter is to recall some basic properties of subnormal operators. They anticipate those which will be resumed later in the case of hyponormal operators. Many other topics related to subnormal operators are omitted or only mentioned among the exercises. An optimal reference for the theory of subnormal operators up to 1981 is Conway's monograph [2].

1. Elementary properties and examples

Let K be a Hilbert space.

Recall that $N \in L(K)$ is said to be a *normal operator* on the Hilbert space K, if it satisfies the commutator condition:

(1) $$[N^*, N] = N^*N - NN^* = 0,$$

or equivalently

$$\|N\xi\| = \|N^*\xi\|, \quad \xi \in K.$$

The classification (and implicitly the structure) of normal operators reduces generically to that of finite Borel measures, compactly supported on the complex plane **C**. A comprehensive reference for this classical subject is Dunford-Schwartz [2].

1.1. Definition. The restriction of a normal operator to a closed invariant subspace is called a *subnormal operator.*

More precisely, if $N \in L(K)$ is normal and $H \subset K$ is a closed subspace of K such that $NH \subset H$, then $S = N|H$ is called subnormal. If we decompose K into the direct orthogonal sum $K = H \oplus H^{\perp}$, then N has a matrix representation:

(2) $$N = \begin{bmatrix} S & A \\ 0 & B \end{bmatrix} : \begin{matrix} H \\ \oplus \\ H^{\perp} \end{matrix} \rightarrow \begin{matrix} H \\ \oplus \\ H^{\perp} \end{matrix}.$$

With the above notations, we say that N is a *normal extension* of the operator S. Notice that N, as a normal extension of S, is not necessarily unique.

The self-commutator of a subnormal operator S, and later of a hyponormal operator, will play on important role in our approach, so we adopt the following simple notation:

$$D = [S^*, S].$$

Sometimes, in order to avoid ambiguities, we shall put D_S instead of D.

By writing the adjoint N^* in matrix representation, and by using (1), we infer

(3) $$[S^*, S] = AA^*,$$

hence the *self-commutator* $[S^*, S]$ of S is a positive operator.

An alternative proof of this fact runs as follows: let P denote the orthogonal projection of K onto H. Then, for any vector $\xi \in H$,

$$\|S\xi\| = \|N\xi\| = \|N^*\xi\| \geq \|PN^*\xi\| = \|S^*\xi\|,$$

since $S^* = PN^*|H$. But

$$\langle [S^*, S]\xi, \xi\rangle = \langle S^*S\xi, \xi\rangle - \langle SS^*\xi, \xi\rangle = \|S\xi\|^2 - \|S^*\xi\|^2.$$

1.2. Proposition. *Let* S *be a subnormal operator with self-commutator* D. *Then*
(i) **Ker** D *is invariant under* S;
(ii) (**Ran** D)$^-$ *is invariant under* S^*.

Proof. From the normality condition $NN^* = N^*N$, written for the normal extension (2) of S, we infer $BA^* = A^*S$, hence $S(\textbf{Ker}\, A^*) \subset \textbf{Ker}\, A^*$. As $D = AA^*$, we have **Ker** D = **Ker** A^*, and the proof of (i) is complete.

The second statement is obtained from (i) as a consequence of the duality between the invariant subspaces of S and S^*.

Because we are mainly interested in specific properties of subnormal operators, it will be necessary to avoid any (direct summand) normal part of such an operator. The next simple but important theorem asserts the existence of the largest normal part of a subnormal operator. First, a definition.

1.3. Definition. A subnormal operator S is called *pure* or *completely non-normal* if there is no nontrivial reducing subspace for S on which S is normal.

1.4. Theorem. *Let* $S \in L(H)$ *be a subnormal operator with self-commutator* D. *Then there is a unique orthogonal decomposition* $H = H_p(S) \oplus H_n(S)$, *where* $H_p(S)$ *and* $H_n(S)$ *are reducing subspaces for* S, *such that:*

(i) $S_p = S|H_p(S)$ *is pure;*

(ii) $S_n = S|H_n(S)$ *is normal.*

Moreover, one has

(4) $$H_p(S) = \bigvee\{S^k D\xi;\ k \in \mathbf{N},\quad \xi \in H\},$$

(5) $$H_n(S) = \{\eta \in H;\ DS^{*k}\eta = 0,\quad k \in \mathbf{N}\}.$$

Proof. Let us define for every natural k the subspace

$$L_k = \bigvee\{S^i D\xi;\ 0 \leq i \leq k,\quad \xi \in H\}.$$

By Proposition 1.2 one finds $S^*L_o \subseteq L_o$, and by an induction argument, based on the relation $S^*S = D + SS^*$, we infer

$$S^*L_k \subseteq L_k, \qquad \text{for every } k \in \mathbf{N}.$$

Let us denote $H_o = (\bigcup_{k\geq 0} L_k)^-$. The subspace H_o is clearly invariant under S and S^*. Its complement $H_1 = H_o^\perp$ is given by the relation:

$$H_1 = \{\eta \in H\ ;\ DS^{*k}\eta = 0,\ k \in \mathbf{N}\}.$$

Both subspaces H_o and H_1 reduce S, and for an arbitrary vector $\eta \in H_1$ we have

$$\|S\eta\|^2 - \|S^*\eta\|^2 = \langle D\eta, \eta\rangle = 0.$$

This shows that the operator $S|H_1$ is normal.

Let L be a reducing subspace for S, such that $S|L$ is normal. Then, as we have remarked above, $\langle D\eta, \eta\rangle = 0$ for any $\eta \in L$, and consequently $D\eta = 0$. After replacing η by $S^{*k}\eta \in L$, one obtains $DS^{*k}\eta = 0$ for $k \in \mathbf{N}$, whence $L \subset H_1$.

Concluding, we have proved that H_1 is the largest reducing subspace of S, on which S is normal, therefore $H_o = H_p(S)$ and $H_1 = H_n(S)$. The proof is complete.

The above subspace $H_p(S)$ (respectively $H_n(S)$) is called the *pure* (respectively, the *normal) subspace* of S. The operators S_p and S_n will be referred to as the *pure* part, and respectively, the *normal part*, of S.

By relation (4) above one gets the next result.

1.5. Corollary. *The pure subspace* $H_p(S)$ *is the smallest closed invariant subspace of* S, *which contains the range of the self-commutator.*

In order to give a pertinent application of Theorem 1.4 we need to discuss first the simplest class of examples of subnormal operators.

Example 1. *The isometries*

Recall that an operator $V \in L(H)$ is said to be an *isometry* if $\|V\xi\| = \|\xi\|$ for every $\xi \in H$, or equivalently $V^*V = I$. Then it turns out that the self-commutator D_V is an orthogonal projection. Let us denote $D = \mathbf{Ran}\, D_V$.

The operator V is subnormal because it extends to the normal operator acting on $H \oplus H$:

$$U = \begin{bmatrix} V & D_V \\ 0 & V^* \end{bmatrix}.$$

Actually U is a unitary operator, that is $UU^* = U^*U = I$.

By Theorem 1.4 we have the decomposition $V = V_p \oplus V_n$, known as the von Neumann-Wold decomposition. In fact the normal part V_n turns out to be unitary.

By (4) the pure subspace of V is

$$H_p(V) = \bigvee_{k=0}^{\infty} V^k D = D \oplus VD \oplus V^2 D \oplus \dots,$$

the last equality being a consequence of the assumption $V^*V = I$.

The pure part V_p of the isometry V (which in other terminology is called the completely non-unitary part of V) acts extremely simply on the direct sum decomposition of $H_p(V)$. Namely, it shifts parallelwisely a vector from one summand to the next one. When $\dim D = 1$ there is exactly one way of doing this. More precisely, in that case

$$H_p(V) = l^2(\mathbf{N}),$$

and

$$V(\xi_o, \xi_1, \xi_2, \dots) = (0, \xi_o, \xi_1, \dots)$$

for any vector $(\xi_o, \xi_1, \dots) \in l^2(\mathbf{N})$. The last operator is the familiar *unilateral shift* and it will be denoted by U_+.

In the general case one simply checks that V_p is unitary equivalent with $U_+ \otimes I_D$.

More details concerning the unilateral shift can be found in Halmos [4]. Some of them will also be discussed in the sequel.

Our next aim is to present an application of Theorem 1.4, which distinguishes the unilateral shift among the subnormal operators.

1.6. Proposition. (Morrel-Clancey). *Let* S *be a pure subnormal operator with one dimensional self-commutator. Then* S *is unitarily equivalent to* $\alpha U_+ + \beta$, *where* α, β *are complex numbers, and* $\alpha > 0$.

Proof. Take a unit vector e_o, such that

$$D\xi = \alpha^2 \langle \xi , e_o \rangle e_o, \quad \xi \in H,$$

with a suitable constant $\alpha > 0$. By Proposition 1.2 (ii) there is a constant $\beta \in \mathbf{C}$, so that

$$S^* e_o = \overline{\beta} e_o.$$

Further, we set $V = \alpha^{-1}(S - \beta)$ and $e_n = V^n e_o$, $n \in \mathbf{N}$. The operator V is obviously still subnormal,

$$(V^*V - VV^*)\xi = \langle \xi, e_o \rangle e_o, \quad \xi \in H,$$

and

$$V^* e_o = 0.$$

By Theorem 1.4 we infer $H = \bigvee\{e_n ; n \geq 0\}$, and a simple induction argument shows that

$$V^* e_{n+1} = e_n, \quad n \geq 0.$$

Therefore $V^*V = I$, $\{e_n\}$ is an orthonormal basis of H, $V = U_+$ with respect to this basis, and the proof is complete.

Example 2. *The Bergman operator*

Let Ω be a bounded open subset of $\mathbf{C}$ and let us denote by $L^p(\Omega)$, $1 \leq p \leq \infty$, the Lebesgue spaces relative to the planar Lebesgue measure $d\mu$.

For any function $\phi \in L^\infty(\Omega)$ we denote by M_ϕ the *multiplication operator:*

$$M_\phi f = \phi f, \quad f \in L^2(\Omega).$$

It is clear that M_ϕ is a bounded operator on $L^2(\Omega)$, with $\|M_\phi\| \leq \|\phi\|_\infty$. Moreover, since $(M_\phi)^* = M_{\bar{\phi}}$, it turns out that M_ϕ is a normal operator.

Of a particular interest will be the *position operator* M_z, which corresponds to the identity function $\phi(z) = z$, $z \in \Omega$.

Let $A^2(\Omega)$ denote the subspace of analytic functions on Ω, belonging to $L^2(\Omega)$. A standard application of the Cauchy formula shows that the uniform norm of an analytic function on a compact subset of Ω is majorized, up to a constant, by the $L^2(\Omega)$-norm. This shows that $A^2(\Omega)$ is a closed subspace of $L^2(\Omega)$, known as the *Bergman space* of the set Ω.

The operator M_z clearly leaves $A^2(\Omega)$ invariant, hence it defines a subnormal operator

$$B_\Omega = M_z \,|\, A^2(\Omega),$$

which will be called the *Bergman operator.*

In the case when Ω is the unit disc, a standard computation of a basis in $A^2(\Omega)$ shows that B_Ω is a compact perturbation of the shift U_+.

Some of the relevant properties of the Bergman operator are presented as exercises in the last part of this section.

Example 3. *Cyclic subnormal operators*

When the Lebesgue measure in the preceding example is replaced by an arbitrary Borel positive measure $d\nu$, compactly supported on **C**, then the operator M_z is still normal on $L^2(d\nu)$. Although the Bergman space has no meaning for $L^2(d\nu)$, we can replace it by some other subspaces arising from analytic functions, which are invariant under M_z. Thus we are led to a class of cyclic (respectively, rationally cyclic) subnormal operators. In Section 3 below we shall prove that these are all the cyclic (respectively, rationally cyclic) subnormal operators.

First we need a few notations. Let σ be a compact subset of **C**. We denote by **Pol**(σ) the set of all complex analytic polynomial functions on σ. Similarly **Rat**(σ) stands for the algebra of all rational functions with poles off σ. Let $d\nu$ be a Borel measure with **supp** $(d\nu) \subset \sigma$. We put

$$P^2(d\nu) = \mathbf{Pol}(\sigma)^-, \quad R^2_\sigma(d\nu) = \mathbf{Rat}(\sigma)^-,$$

where the closures are taken in $L^2(d\nu)$.

It is plain to check that both spaces are invariant under M_z. Let P_ν and $R_{\nu,\sigma}$ denote the corresponding subnormal operators.

If $\xi \in L^2(d\nu)$ stands for the class of the function **1** (identically equal to 1 on **C**), then it turns out from the definition that the vector spaces:

$$\{f(P_\nu)\xi \ ; \ f \in \mathbf{Pol}(\sigma)\}, \ \{f(R_{\nu,\sigma})\xi \ ; \ f \in \mathbf{Rat}(\sigma)\}$$

are dense in $P^2(d\nu)$, respectively in $R^2_\sigma(d\nu)$. Thus the operator P_ν is cyclic, and $R_{\nu,\sigma}$ is rationally cyclic. The operator P_ν is completely determined by the measure ν, while the operator $R_{\nu,\sigma}$ depends on ν and also on the compact set σ.

The following well known and important functional model for the shift U_+ fits into this scheme. Consider the one dimensional torus $\mathbf{T} = \{e^{i\theta} \ ; \ \theta \in \mathbf{R}\}$ endowed with its Haar measure $d\nu = (1/2\pi)d\theta$. Then $P^2(d\nu)$ coincides with the Hardy space $H^2(\mathbf{T})$, and the cyclic operator $P_{d\nu} = M_z | H^2(\mathbf{T})$ is unitarily equivalent to the unilateral shift U_+. For the convenience of the reader we recall that the Hardy space $H^2(\mathbf{T})$ may equally be defined as the subset of $L^2(\mathbf{T})$ consisting of those functions $f \in L^2(\mathbf{T})$ whose Fourier series

$$f(e^{i\theta}) = \sum_{n=-\infty}^{\infty} a_n e^{in\theta}$$

contain only non-negative index terms:

$$f \in H^2(\mathbf{T}) \Leftrightarrow a_n = 0, \text{ for every } n < 0.$$

2. Characterization of subnormality

In this section we establish a fundamental relationship between subnormal operators and some positive definite kernels. The main result is due to Halmos [4] and Bram [1]. As an application we present a solution to a two-dimensional moment problem, derived by Atzmon [1] from this result.

The terminology and basic facts concerning positive definite kernels are presented in Section 5 of this chapter.

2.1. Lemma. *Let H be a Hilbert space and let $K : \mathbf{N} \times \mathbf{N} \to L(H)$ be a positive definite kernel. The following conditions are equivalent:*

(i) *There are a Hilbert space K and operators $A \in L(K)$ and $V \in L(H,K)$, such that*

(6) $$K(i,j) = V^* A^{*i} A^j V, \quad i, j \in \mathbf{N};$$

(ii) *There is a constant $C > 0$, such that*

(7) $$\sum_{i,j=0}^{n} \langle K(i+1, j+1)\xi_j, \xi_i \rangle \le C \sum_{i,j=0}^{n} \langle K(i,j)\xi_j, \xi_i \rangle$$

for every finite system $(\xi_0, \xi_1, \ldots, \xi_n)$ *of vectors of* H.

Condition (ii) means that the shift operator S_1 by the element $1 \in \mathbf{N}$ is bounded relative to K (see Section 5).

Proof. (i) $\Rightarrow$(ii). If K is defined by (6), then

$$\sum_{i,j=0}^{n} \langle K(i+1,j+1)\xi_j, \xi_i\rangle = \|\sum_{j=0}^{n} A^{j+1}V\xi_j\|^2 \leq \|A\|^2 \|\sum_{j=0}^{n} A^j V\xi_j\|^2 =$$

$$= \|A\|^2 \sum_{i,j=0}^{n} \langle K(i,j)\xi_j, \xi_i\rangle.$$

(ii)$\Rightarrow$(i). By Kolmogorov's Theorem 5.1 there exist a Hilbert space K and a sequence of operators $A(n) \in L(H,K)$, $n \in \mathbf{N}$, such that

$$K(i,j) = A(i)^* A(j), \quad i,j \in \mathbf{N},$$

and

$$K = \bigvee_{i=0}^{\infty} \mathbf{Ran}\, A(i).$$

Condition (7) enables us to define the operator $A \in L(K)$, with the property:

$$AA(i) = A(i+1), \quad i \in \mathbf{N}.$$

Then we get

$$A(i) = A^i A(0),$$

and finally the desired form for the kernel K.

Let us assume that $S \in L(H)$ is a subnormal operator. Then the kernel $K : \mathbf{N} \times \mathbf{N} \to L(H)$, $K(i,j) = S^{*i}S^j$, is obviously positive definite. Let us denote by $K^\tau : \mathbf{N} \times \mathbf{N} \to L(H)$ the transposed kernel, that is $K^\tau(i,j) = K(j,i)$. Then K^τ is positive definite too. Indeed, let N be a normal extension of the operator S and let $(\xi_0, \ldots, \xi_n)$ be a finite system of elements of H. We have

$$\sum_{i,j=0}^{n} \langle K^\tau(i,j)\xi_j, \xi_i\rangle = \sum_{i,j=0}^{n} \langle K(j,i)\xi_j, \xi_i\rangle = \sum_{i,j=0}^{n} \langle S^i\xi_j, S^j\xi_i\rangle =$$

$$= \sum_{i,j=0}^{n} \langle N^i\xi_j, N^j\xi_i\rangle = \sum_{i,j=0}^{n} \langle N^{*j}N^i\xi_j, \xi_i\rangle = \|\sum_{j=0}^{n} N^{*j}\xi_j\|^2.$$

The interesting and rather unexpected fact is that the converse implication is still true. More precisely we can state the following.

2.2. Theorem. (Halmos-Bram). *Let* $S \in L(H)$ *and denote by* K^τ *the transpose of the kernel* $K(i,j) = S^{*i}S^j$.

The operator S *is subnormal if and only if the kernel* K^τ *is positive definite on* $\mathbf{N} \times \mathbf{N}$.

Proof. Suppose K^τ is positive definite. We may assume $\|S\| = \rho < 1$. Indeed, after the homotethy $S \to tS$, $t > 0$, the kernel K^τ becomes

$$K^\tau(i,j) \to t^{i+j}K^\tau(i,j),$$

and this transformation preserves the positivity.

We build a larger Hilbert space $H_\infty = \bigoplus_{k=0}^{\infty} H$, as a countable direct sum of copies of H. Let us consider the operator L on H_∞ with the matrix representation with respect to the above direct sum:

$$L = [K^\tau(i,j)]_{i,j\geq 0}.$$

First we have to prove that L is bounded on H_∞. Indeed, if $\xi = \bigoplus_{j=0}^{\infty} \xi_j$ is an element of the algebraic direct sum, then

$$\|\sum_{j=0}^{\infty} K^\tau(i,j)\xi_j\|^2 \leq (\sum_{j=0}^{\infty} \rho^{i+j}\|\xi_j\|)^2 \leq (\sum_{j=0}^{\infty} \rho^{2(i+j)})(\sum_{j=0}^{\infty} \|\xi_j\|^2) \leq$$
$$\leq \rho^{2i}(1 - \rho^2)^{-1}\|\xi\|^2.$$

Therefore

$$\|L\xi\|^2 = \sum_{i=0}^{\infty} \|\sum_{j=0}^{\infty} K^\tau(i,j)\xi_j\|^2 \leq (1 - \rho^2)^{-2}\|\xi\|^2.$$

Since the kernel K^τ is positive definite, the operator L turns out to be positive.

Analogously, the linear transform M with the matrix

$$M = [K^\tau(i+1, j+1)]_{i,j\geq 0}$$

extends continuously to a positive operator on H_∞.

Our next aim is to show that $M \leq L$. In order to prove this inequality it suffices to check (by a known device, see for instance Exercise 10) that $M^2 \leq L^2$. This, in turn, is equivalent with

$$\|M\xi\|^2 \leq \|L\xi\|^2,$$

for any algebraic direct sum vector $\xi = \bigoplus_{k=0}^{\infty} \xi_k$. By taking into account that $\|S\| = \|S^*\| = \rho < 1$, we find

$$\|M\xi\|^2 = \sum_{i=0}^{\infty} \|\sum_{j=0}^{\infty} K^{\tau}(i+1,j+1)\xi_j\|^2 = \sum_{i=0}^{\infty} \|\sum_{j=0}^{\infty} S^{*j+1}S^{i+1}\xi_j\|^2 \leq$$

$$\leq \sum_{i=0}^{\infty} \|S^*\|^2 \|\sum_{j=0}^{\infty} S^{*j}S^{i+1}\xi_j\|^2 = \rho^2 \sum_{i=1}^{\infty} \|\sum_{j=0}^{\infty} S^{*j}S^{i}\xi_j\|^2 \leq$$

$$\leq \rho^2 \sum_{i=0}^{\infty} \|\sum_{j=0}^{\infty} S^{*j}S^{i}\xi_j\|^2 = \rho^2 \|L\xi\|^2 \leq \|L\xi\|^2,$$

as desired.

By Lemma 2.1 applied to the kernel K^{τ}, we find a Hilbert space K and operators $V \in L(H,K)$, $A \in L(K)$, such that

$$K^{\tau}(i,j) = V^*A^{*i}A^jV, \quad i,j \in \mathbf{N}.$$

Moreover, recall from the proof of Lemma 2.1 that the auxiliary Hilbert space K is spanned by the vectors $A^jV\xi$, where $j \in \mathbf{N}$ and $\xi \in H$.

Since $V^*V = K^{\tau}(0,0) = I$, V is an isometry from H into K. By identifying H with the subspace VH, we shall prove that the operator $N = A^* \in L(K)$ is a normal extension of S, more precisely of the operator $VSV^* \in L(VH)$.

We assert that

$$NA^iV = A^iVS, \quad i \in \mathbf{N}. \tag{8}$$

Indeed, (8) is equivalent with

$$\langle NA^iV\eta, A^jV\xi\rangle = \langle A^iVS\eta, A^jV\xi\rangle$$

for any $\eta,\xi \in H$ and $i,j \in \mathbf{N}$. But

$$\langle NA^iV\eta, A^jV\xi\rangle = \langle V^*A^{*j}A^*A^iV\eta, \xi\rangle = \langle K^{\tau}(j+1,i)\eta, \xi\rangle = \langle S^{*i}S^{j+1}\eta, \xi\rangle =$$

$$= \langle K^{\tau}(j,i)S\eta, \xi\rangle = \langle V^*A^{*j}A^iVS\eta, \xi\rangle = \langle A^iVS\eta, A^jV\xi\rangle.$$

As a first consequence of (8), let us remark that $N|VH = VSV^*$. Secondly, again by (8), N is a normal operator because

$$(N^*N - NN^*)A^jV = ANA^jV - NA^{j+1}V = AA^jVS - A^{j+1}VS = 0.$$

This completes the proof of Theorem 2.2.

Note that the normal extension N which appears in the preceding proof has the additional property (by forgetting the isometry V):

$$K = \bigvee \{N^{*j}\xi;\ j \in \mathbf{N}, \xi \in H\}.$$

In the next section we shall prove that such a normal extension in unique up to a unitary equivalence. It is called the *minimal normal extension* of S.

The next simple characterization of subnormality uses Naimark's Dilation Theorem (see Section 5) rather than the general Kolmogorov factorization.

2.3. Proposition. *Let* $S \in L(H)$. *The following conditions are equivalent:*

(i) S *is subnormal;*

(ii) *There exists a positive* $L(H)$*-valued measure* dQ *on* $\mathbf{C}$, *with compact support, such that*

$$S^{*i}S^j = \int_{\mathbf{C}} \bar{z}^i z^j dQ(z), \quad i,j \in \mathbf{N}.$$

Proof. (i) $\Rightarrow$(ii). If $N \in L(K)$ is a normal extension of S, with spectral measure dE, then

$$S^{*i}S^j = PN^{*i}N^j \,|\, H = P\int_{\mathbf{C}} \bar{z}^i z^j dE(z)\cdot P; \quad i, j \in \mathbf{N},$$

where P denotes the orthogonal projection of K onto H. Consequently the desired measure is dQ = P(dE)P.

(ii) $\Rightarrow$(i). Let K be a Hilbert space and let dE be a spectral measure with values in the algebra $L(K)$, which dilates dQ, in virtue of Naimark's Theorem 5.2, that is :

$$Q(\beta) = PE(\beta) \,|\, H,$$

for every Borel subset $\beta \subset \mathbf{C}$. Again P is the orthogonal projection of K onto H.

Then $N = \int_{\mathbf{C}} z dE(z)$ is a normal operator on K, with the property :

$$PN^{*i}N^j \,|\, H = \int_{\mathbf{C}} \bar{z}^i z^j dQ(z) = S^{*i}S^j, \quad i,j \in \mathbf{N}.$$

Consequently

$$PN^*NP = S^*S = PN^*PNP,$$

whence $PN^*(I - P)NP = 0$, or equivalently $(I - P)NP = 0$, which shows that $N\,|\,H = S$.

The proof is complete.

As an illustration of the strength of the preceding results we close this section by presenting after Atzmon [1] a solution to a two dimensional moment problem.

The *moment problem* in two dimensions asks for characterization of the sequence of complex numbers $(a_{mn})_{m,n=0}^{\infty}$, which arises from a finite, positive, Borel measure $d\nu$ on $\mathbf{C}$, as follows:

(9) $$a_{mn} = \int_{\mathbf{C}} \bar{z}^m z^n d\nu(z); \quad m,n \in \mathbf{N}.$$

The integrals are naturally assumed to exist.

It is clear from the very beginning that certain relations must hold between the moments a_{mn}. Finding all these relations turned out be rather difficult. The solution of this problem goes back to the thirties, and it has a long history, see Akhiezer [1] and Krein-Nudelman [1].

We confine ourselves to present a characterization of the moments (9) only for compactly supported measures.

2.4. Theorem. *The sequence of complex numbers* $(a_{mn})_{m,n=0}^{\infty}$ *represents the moments* (9) *of a positive finite measure* $d\nu$*, compactly supported on* $\mathbf{C}$*, if and only if:*

(i) *The kernel* $K : \mathbf{N}^2 \times \mathbf{N}^2 \to \mathbf{C}$, $K(m,n;p,q) = a_{m+q,p+n}$ *is positive definite,*

(ii) *The shift* $S_{(1,0)}$ *is bounded with respect to* K.

For terminology see Section 5 below.

Proof. The necessity of both conditions is immediate. For a given measure $d\nu$ as in the statement, and for every polynomial in z and $\bar{z}$:

$$p(z,\bar{z}) = \sum_{k,\ell=0}^{n} \alpha_{k\ell} z^k \bar{z}^{\ell},$$

we have

$$\int_{\mathbf{C}} |p|^2 d\nu = \sum_{i,j,k,\ell} K(i,j;k,\ell)\alpha_{k\ell}\bar{\alpha}_{ij},$$

whence one obtains (i). In addition, there exists a constant $C > 0$ so that

$$\int_{\mathbf{C}} |\bar{z}p(z)|^2 d\nu(z) \leq C \int_{\mathbf{C}} |p|^2 d\nu,$$

and this inequality is an explicit form of condition (ii).

For the proof of the sufficiency implication, we assume (i) and (ii) and factorize the kernel K by Kolmogorov's theorem:

$$K(m,n;p,q) = A(m,n)^* A(p,q), \quad m, n, p, q \in \mathbf{N},$$

where $A(p,q) \in L(\mathbf{C},K)$, with a suitable Hilbert space K. We may identify $A(p,q)$ with a vector $\xi_{p,q} \in K$, so that

$$K(m,n\,;\,p,q) = \langle \xi_{p,q}, \xi_{m,n} \rangle, \quad m, n, p, q \in \mathbf{N}.$$

By condition (ii) and the minimality of K, there exists a unique bounded linear operator $T \in L(K)$, such that

$$T\xi_{p,q} = \xi_{p+1,q}, \quad p, q \in \mathbf{N}.$$

By taking into account condition (i), we obtain

$$K(m,n\,;\,p,q) = K(m+q,0\,;\,n+p,0) = \langle T^{n+p}\xi, T^{m+q}\xi \rangle,$$

where $\xi = \xi_{0,0}$.

Since the vectors $\xi_{p,q}$ span K, the operator T has ξ as a cyclic vector. By condition (i) we know that the kernel

$$\langle T^{*(m+q)}T^{n+p}\xi, \xi \rangle = K(m,n\,;\,p,q)$$

is positive definite. In particular this shows that the operator valued kernel

$$(n,q) \mapsto T^{*q}T^n$$

is positive definite on $\mathbf{N} \times \mathbf{N}$. Indeed, for any double sequence of complex numbers $(\alpha_{p,q})_{p,q=0}^{\infty}$, of finite support, we have

$$\sum_{q,n} \langle T^{*q}T^n(\sum_p \alpha_{pq}T^p\xi), \sum_m \alpha_{mn}T^m\xi \rangle = \sum_{p,q,m,n} \langle T^{*(m+q)}T^{n+p}\xi, \xi \rangle \alpha_{pq}\bar{\alpha}_{mn} \geq 0.$$

In view of the Halmos-Bram Theorem 2.2, T is a subnormal operator and Proposition 2.3 shows that there exists a positive measure with compact support $d\nu$, such that

$$K(m,n\,;\,p,q) = \int_{\mathbf{C}} \bar{z}^{m+q} z^{n+p} d\nu(z)$$

for any natural numbers m, n, p, q.

This completes the proof of Theorem 2.4.

3. The minimal normal extension

Some of the properties of the (unique) minimal normal extension of a subnormal operator are briefly recalled in this section.

3.1. Definition. Assume $S \in L(H)$ is a subnormal operator. A normal extension $N \in L(K)$ of S is called *minimal* if the smallest reducing subspace for N, which contains

H, is K itself.

A straightforward exercise shows that the normal extension $N \in L(K)$ is minimal if and only if

$$K = \bigvee \{N^{*i}\xi \,;\, i \in \mathbf{N}, \xi \in H\}.$$

3.2. Lemma. *Let* $S_i \in L(H)$, $i = 1, 2$, *be two subnormal operators with minimal normal extensions* $N_i \in L(K_i)$, $i = 1, 2$.

If $U \in L(H)$ *is a unitary operator, with* $US_1U^* = S_2$, *then there is a unitary operator* $V \in L(K_1, K_2)$ *such that* $V|H = U$ *and* $VN_1V^* = N_2$.

Proof. Define $V : K_1 \to K_2$ by the formula

$$V(N_1^{*n}\xi) = N_2^{*n}U\xi, \quad n \in \mathbf{N}, \xi \in H,$$

by remarking that

$$\langle N_1^{*n}\xi, N_1^{*m}\eta\rangle = \langle S_1^m\xi, S_1^n\eta\rangle = \langle S_2^m U\xi, S_2^n U\eta\rangle = \langle N_2^{*n}U\xi, N_2^{*m}U\eta\rangle.$$

3.3. Corollary. *Any two minimal normal extensions of a subnormal operator are unitarily equivalent.*

So we may unambiguously call one of such extensions *the minimal normal extension.* There exists, a now classical, simple relation between the spectra of a subnormal operator and its minimal normal extension, see Halmos [1], [4].

Recall that the *spectrum* $\sigma(T)$ of an operator $T \in L(H)$, is the set of those $z \in \mathbf{C}$, for which $T - z$ is not invertible. The *point spectrum* $\sigma_p(T)$ is the set of eigenvalues of T, that is of those $z \in \mathbf{C}$, with the property that $\mathbf{Ker}\,(T - z) \neq 0$. By the *approximate point spectrum* $\sigma_{ap}(T)$ we mean the set of those $z \in \mathbf{C}$ for which there is a sequence of unit vectors $(\xi_n)_{n=0}^{\infty}$, such that

$$\lim_n (T - z)\xi_n = 0.$$

3.4. Proposition. *If* N *is the minimal normal extension of a subnormal operator* S, *then*

(i) $\partial\sigma(S) \subset \sigma(N) \subset \sigma(S)$;

(ii) *If* Ω *is a bounded connected component of* $\mathbf{C} \setminus \sigma(N)$, *then either* $\Omega \subset \sigma(S)$ *or* $\Omega \cap \sigma(S) = \emptyset$.

Proof. (i) The inclusion of the boundary $\partial\sigma(S)$ into $\sigma_{ap}(S)$ is a general property. But $\sigma_{ap}(S) \subset \sigma_{ap}(N)$ by the definition of the approximate point spectrum.

In order to prove the inclusion $\sigma(N) \subset \sigma(S)$, it is a enough to check that N is invertible whenever S is invertible.

We may assume, after a homothety, that $\|S^{-1}\| = 1$. Let us consider the spectral measure dE of the normal operator N. Take $0 < \varepsilon < 1$ and the spectral subspace

$$L = E(\bar{D}(0,\varepsilon))H,$$

where $\bar{D}(0,\varepsilon) = \{z \in \mathbf{C}\,;\ |z| \leq \varepsilon\}$.

For any $\xi \in L$ and $k \in \mathbf{N}$ one finds

$$\|N^{*k}\xi\| = \|N^k\xi\| \leq \varepsilon^k \|\xi\|.$$

On the other hand, for any $\eta \in H$ we obtain

$$|\langle\xi,\eta\rangle| = |\langle\xi, S^kS^{-k}\eta\rangle| = |\langle\xi, N^kS^{-k}\eta\rangle| = |\langle N^{*k}\xi, S^{-k}\eta\rangle| \leq \varepsilon^k \|\xi\|\,\|\eta\|.$$

By letting k tend to infinity, one gets $\xi \perp H$.

Thus $H \subset L^{\perp}$. By the minimality of N it follows $L^{\perp} = H$, whence $L = 0$. Actually this means that $0 \notin \sigma(N)$, so N is invertible.

(ii) This assertion is a topological consequence of (i). Let Ω be a bounded connected component of $\mathbf{C} \setminus \sigma(N)$. Then $\Omega = [\Omega \setminus \sigma(S)] \cup [\Omega \cap \sigma(S)]$. The set $\Omega_1 = \Omega \setminus \sigma(S)$ is obviously open. But $\Omega_2 = \Omega \cap \sigma(S)$ is also open. Indeed, by (i) we find

$$\Omega_2 = \Omega \cap \sigma(S) = \Omega \cap [\partial\sigma(S) \cup \operatorname{int}\sigma(S)] = [\Omega \cap \partial\sigma(S)] \cup [\Omega \cap \operatorname{int}\sigma(S)] = \Omega \cap \operatorname{int}\sigma(S).$$

By the connectedness of Ω one of the open sets Ω_1 and Ω_2 must be empty. This proves assertion (ii).

3.5. Corollary. *Let* S *be a subnormal operator and let* $f \in \mathcal{O}(\sigma(S))$ *be an analytic function defined in a neighbourhood of* $\sigma(S)$.

The operator f(S) *is still subnormal, and*

$$\|f(S)\| = \|f\|_{\infty,\sigma(S)}.$$

Here f(S) is the Riesz-Dunford functional calculus, see Dunford-Schwartz [1].

Proof. By well-known properties of the functional calculus, the operator f(N) still extends f(S). Thus $\|f(S)\| \leq \|f(N)\|$. On the other hand, by the Spectral Mapping Theorem and inclusion (i), one obtains $\sigma(f(N)) \subset \sigma(f(S))$.

After evaluating the spectral radii of both operators, we find

$$\|f(N)\| = |f(N)|_{sp} \le |f(S)|_{sp} \le \|f(S)\|,$$

hence $\|f(S)\| = \|f(N)\|$. But

$$|f(N)|_{sp} := \max\{|z|;\ z \in \sigma(f(N))\} = \max\{|f(\lambda)|;\ \lambda \in \sigma(N)\} =: \|f\|_{\infty,\sigma(N)}.$$

Again by (i), (ii) and the maximum principle for analytic functions, we conclude

$$\|f(S)\| = \|f\|_{\infty,\sigma(S)},$$

and the proof is complete.

The above corollary, which in particular shows that the spectrum of a subnormal operator is a spectral set, was crucial for further developments of the theory, see Conway [2]. Below we mention only two direct applications of Corollary 3.5, which will be found again for hyponormal operators.

3.6. Corollary. *If* S *is a subnormal operator, then* $\|S^n\| = \|S\|^n$ *for every natural number* n.

3.7. Corollary. *For any point* $\lambda \in \mathbf{C} \setminus \sigma(S)$ *in the resolvent set of the subnormal operator* S, *we have*

$$\|(S - \lambda)^{-1}\| = 1/\mathrm{dist}(\lambda, \sigma(S)).$$

The reader will easily compute the minimal normal extensions of the subnormal operators presented in Examples 1, 2, 3 above. For instance the minimal normal extension of the unilateral shift U_+ is the bilateral shift (acting on $l^2(\mathbf{Z})$), which is a unitary operator.

We conclude this section with a discussion on the structure of (rationally) cyclic subnormal operators. First we recall some terminology and well known facts. Let K be a Hilbert space and let $T \in L(K)$. A vector $\xi \in K$ is said to be *$*$-cyclic* for the operator T, if the vector space $\{A\xi\ ;\ A \in A^*(T)\}$ is dense in K. Here $A^*(T)$ is the unital $*$-subalgebra of $L(K)$ generated by T.

Thus the vector $\xi \in K$ is $*$-cyclic for the normal operator $N \in L(K)$ if and only if

$$K = \bigvee\{N^{*i}N^j\xi\ ;\ i, j \in \mathbf{N}\}.$$

If dE is the spectral measure of N, then it is determined by the scalar measure $d\nu =$

$= \langle dE\xi, \xi \rangle$. Moreover, in this case, there is a unitary operator

$$U : K \to L^2(d\nu),$$

such that

$$U\xi = \mathbf{1},$$

$$UNU^* = M_z.$$

For details see Dunford - Schwartz [2]. In other words any $*$-cyclic normal operator can be represented as the position operator on a suitable space $L^2(d\nu)$. Note that $\sigma(N) = \mathbf{supp}(d\nu)$.

3.8. Proposition. *Let* $S \in L(H)$ *be a (rationally) cyclic subnormal operator. Then there is a positive finite measure* $d\nu$ *on* $\mathbf{C}$, *with* $\mathbf{supp}(d\nu)$ *compact, so that* S *is unitarily equivalent to the operator* P_ν *(respectively* $R_{\nu,\sigma}$ *where* $\sigma = \sigma(S)$ *).*

Proof. Let ξ be a cyclic vector for S, that is $H = \bigvee\{S^i\xi \; ; \; i \in \mathbf{N}\}$. Because $K = \bigvee\{N^{*j}\eta \; ; \; j \in \mathbf{N}, \; \eta \in H\}$, ξ turns out to be a $*$-cyclic vector for the minimal normal extension N of S.

If N is represented as M_z on $L^2(d\nu)$, with a suitable measure $d\nu$, then S becomes the operator M_z on the subspace $P^2(d\nu)$.

One proceeds similarly in the rationally cyclic case.

4. Putnam's inequality

In this section we reproduce after Axler and Shapiro [1] a short proof of Putnam's inequality, in the case of subnormal operators. Surprisingly, this new proof has been found a rather long while after the discovery of this inequality in the general case of hyponormal operators, cf. Putnam [6].

Recall that μ denotes the planar Lebesgue measure.

4.1. Theorem. (Putnam). *If* S *is a subnormal operator, then*

$$\|[S^*, S]\| \leq \pi^{-1}\mu(\sigma(S)).$$

The main tool in the proof of Theorem 4.1 will be a classical inequality from the theory of functions of two real variables :

4.2. Lemma. (Ahlfors - Beurling). *For any compact subset* σ *of* $\mathbf{C}$, *the following inequality*

$$\left|\int_\sigma d\mu(z)/(z-\zeta)\right| \leq [\pi\,\mu(\sigma)]^{\frac{1}{2}}$$

holds for every $\zeta \in \mathbf{C}$.

Proof. Recall that the function $1/z$ is locally integrable with respect to $d\mu$. After a linear change of coordinates we can assumme $\zeta = 0$ and

$$J = \int_\sigma d\mu(z)/z \geq 0.$$

The change to polar coordinates yields

$$J = \int_\sigma (\cos\theta - i\sin\theta)\,dr\,d\theta = \int_\sigma \cos\theta\,dr\,d\theta.$$

If we denote $\sigma_+ = \{z \in \sigma\,;\,\mathbf{Re}\,z \geq 0\}$ we have

$$J \leq \int_{\sigma_+} \cos\theta\,dr\,d\theta.$$

Let us denote $R = \mathbf{sup}\{|z|\,;\,z \in \sigma\}$ and, for $-\pi/2 \leq \theta \leq \pi/2$, $\sigma_+(\theta) = \sigma_+ \cap \{re^{i\theta}\,:\,: 0 \leq r \leq R\}$. The linear measure of the set $\sigma_+(\theta)$ is denoted by $\ell(\theta)$. Thus, by Cauchy-Schwarz inequality

$$\int_{\sigma_+} \cos\theta\,dr\,d\theta = \int_{-\pi/2}^{\pi/2} \ell(\theta)\cos\theta\,d\theta \leq \left((\pi/2)\int_{-\pi/2}^{\pi/2} \ell(\theta)^2 d\theta\right)^{\frac{1}{2}}.$$

If $\ell(t,\theta)$ is the linear measure of the set $\sigma_+ \cap \{re^{i\theta}\,;\,0 \leq r \leq t\}$, then one easily remarks that $\ell(\cdot,\theta)$ is an increasing function, with the property

$$\ell(t_1,\theta) - \ell(t_2,\theta) \leq t_1 - t_2,$$

whenever $t_1 \leq t_2$. Consequently, one obtains

$$\ell(\theta)^2/2 = \int_{\sigma_+(\theta)} \ell(t,\theta)\,d\ell(t,\theta) \leq \int_{\sigma_+(\theta)} \ell(t,\theta)\,dt \leq \int_{\sigma_+(\theta)} t\,dt.$$

In conclusion

$$(\pi/2)\int_{-\pi/2}^{\pi/2} \ell(\theta)^2 d\theta \leq \pi \int_{-\pi/2}^{\pi/2} \left(\int_{\sigma_+(\theta)} t\,dt\right) d\theta = \pi\,\mu(\sigma_+),$$

and the proof is complete.

Next we need some terminology. For a compact set $\sigma \subset \mathbf{C}$, $C(\sigma)$ is the Banach algebra of continuous functions on σ. By $R(\sigma)$ one denotes the closure of the algebra $\mathbf{Rat}(\sigma)$ (of rational functions with poles off σ) in $C(\sigma)$.

An immediate consequence of Ahlfors and Beurling's lemma is the next result.

4.3. Corollary. (Alexander). *If σ is a compact subset of $\mathbf{C}$, then*

$$\mathbf{dist}_{C(\sigma)}(\bar{z}, R(\sigma)) \leq (\mu(\sigma)/\pi)^{\frac{1}{2}}.$$

Proof. Let $\Omega \supset \sigma$ be a bounded domain with smooth boundary. By the Cauchy - Pompeiu formula we have

$$\bar{z} = (1/2\pi i) \int_{\partial\Omega} \bar{\zeta}/(\zeta - z) d\zeta - (1/\pi) \int_{\Omega} d\mu(\zeta)/(\zeta - z)$$

for every $z \in \sigma$. Since the curvilinear integral belongs to $R(\sigma)$ as a function of z, one obtains the desired inequality by letting Ω approach σ.

Alexander's inequality provides a quantitative version of the well-known Hartogs - Rosenthal approximation theorem.

Proof of Theorem 4.1. Let $S \in L(H)$ be a subnormal operator with a minimal normal extension $N \in L(K)$. The orthogonal projection of K onto H is denoted by P. Then

$$S^*S - SS^* = PN(I - P)N^*P,$$

so that

$$\| S^*S - SS^* \| = \| (I - P)N^*P \|^2.$$

Let $\xi \in H$ be a unit vector and let $f \in \mathbf{Rat}(\sigma(S))$ be arbitrary. Then

$$\| (I - P)N^*\xi \| = \mathrm{dist}_K(N^*\xi, H) \leq \| N^*\xi - f(S)\xi \| = \| (N^* - f(N))\xi \| \leq \| \bar{z} - f \|_{\infty,\sigma(N)} \leq$$

$$\leq \| \bar{z} - f \|_{\infty,\sigma(S)} = \mathbf{dist}_{C(\sigma(S))}(\bar{z}, R(\sigma(S))).$$

The proof finishes by using Corollary 4.3.

As on application of Putnam's inequality we mention at this point only the following fact.

4.4. Corollary. *A subnormal operator whose spectrum has Lebesgue measure zero is necessarily normal.*

Other relevant applications of Putnam's inequality will be presented later, starting with Chapter VI.

5. SUPPLEMENT : Positive definite kernels

Below we summarize some basic results related to positive definite kernels, which have been already used in Chapter I, or will be needed later.

Let H be a Hilbert space and Λ a nonempty set. By a *positive definite kernel* on $\Lambda \times \Lambda$ with values operators on H, we mean a map

$$K : \Lambda \times \Lambda \longrightarrow L(H),$$

with the property:

(10) $$\sum_{i,j=0}^{n} \langle K(\lambda_i,\lambda_j)\xi_j , \xi_i \rangle \geq 0$$

for every finite subset $\{\lambda_o,\lambda_1,\ldots,\lambda_n\} \subset \Lambda$ and every system of vectors $\{\xi_o,\ldots,\xi_n\} \subset H$.

The reader will easily check that a map K of the form

$$K(\lambda,\mu) = A(\lambda)^* A(\mu), \quad \lambda,\mu \in \Lambda,$$

where $A : \Lambda \longrightarrow L(H)$ is arbitrary, has property (10). The reverse statement, which has been established by Kolmogorov (and independently by others), is very useful.

5.1. Theorem. (Kolmogorov). *If* $K : \Lambda \times \Lambda \longrightarrow L(H)$ *is a positive definite kernel, then there exists a Hilbert space* K *and a map* $A : \Lambda \longrightarrow L(H,K)$*, such that:*

(i) $K(\lambda,\mu) = A(\lambda)^* A(\mu), \quad \lambda,\mu \in \Lambda,$

(ii) $K = \bigvee_{\lambda \in \Lambda} \mathbf{Ran}\, A(\lambda).$

Moreover, the pair (K,A) *with properties* (i) *and* (ii) *is unique, up to unitary equivalence.*

Proof. Let F be the vector space of all H-valued functions defined on Λ, and supported on a finite set. One defines a sesquilinear form on F by:

(11) $$[\phi,\psi] = \sum_{\lambda,\mu\in\Lambda} \langle K(\lambda,\mu)\phi(\mu), \psi(\lambda) \rangle ; \quad \phi,\psi \in F.$$

By (10) we have $[\phi,\phi] \geq 0$ for any $\phi \in F$. Take $F_o = \{\phi \in F \mid [\phi,\phi] = 0\}$. This is a linear subspace of F (by the Cauchy-Schwarz inequality), and

$$[\phi + F_o, \psi + F_o] = [\phi,\psi].$$

Hence the quotient space F/F_0 is endowed with a scalar product induced by $[\cdot,\cdot]$. Let K be its Hilbert space completion with the inner product again denoted by $[\cdot,\cdot]$.

For each $\lambda \in \Lambda$ and $\xi \in H$, let $\phi_{\lambda,\xi} \in K$ denote the class of the function

$$\phi_{\lambda,\xi}(\mu) = \begin{cases} \xi, \text{ if } \mu = \lambda \\ 0, \text{ otherwise.} \end{cases}$$

First, remark that for any $\lambda, \mu \in \Lambda$ and $\xi, \eta \in H$, we have by (11):

$$[\phi_{\mu,\eta}, \phi_{\lambda,\xi}] = \langle K(\lambda,\mu)\eta, \xi\rangle.$$

Furthermore, by our construction, the Hilbert space K is spanned by the vectors $\phi_{\lambda,\xi}$.

Next, we notice that, for a fixed $\lambda \in \Lambda$,

$$A(\lambda)\xi = \phi_{\lambda,\xi}$$

is a linear bounded operator with $\|A(\lambda)\| = \|K(\lambda,\lambda)\|^{\frac{1}{2}}$.

Finally, we find

$$K(\lambda,\mu) = A(\lambda)^* A(\mu),$$

as desired.

The uniqueness of the minimal factorization above of the kernel K is a standard matter and we leave it as an exercise to the reader.

As a direct consequence of (i), we mention that any positive definite kernel $K : \Lambda \times \Lambda \to L(H)$ must satisfy:

1) $K(\lambda,\lambda) \geq 0$

2) $K(\lambda,\mu) = K(\mu,\lambda)^*$, $\lambda, \mu \in \Lambda$.

The minimality condition (ii) assures that any continuity, smoothness or measurability property of the kernel function K is inherited by the map A. For instance, if Λ is an open subset of $\mathbf{C}$ and K is separately continuous in the strong operator topology, then $A : \Lambda \to L(H)$ is a continuous function. Or, if K is analytic in the second variable, then A is an operator-valued analytic function, and so on.

A result related to Kolmogorov's Theorem is Naimark's Dilation Theorem which we state without proof.

5.2. Theorem. (Naimark). *Let Q be a positive $L(H)$-valued measure on $\mathbf{R}^n$, with compact support and such that $Q(\mathbf{R}^n) = I$. Then there is a Hilbert space K containing H as a subspace, and a spectral measure E with values in $L(K)$, such that*

$$Q(\beta) = P_H^K E(\beta) | H,$$

for every Borel set $\beta \subset \mathbf{R}^n$.

Recall that by a *positive* $L(H)$ - *valued measure* Q on $\mathbf{R}^n$ we mean a map

$$Q : B(\mathbf{R}^n) \to L(H)_+$$

from the Borel sets of $\mathbf{R}^n$ into the positive operators on H, such that for every $\xi \in H$,

$$\beta \to \langle Q(\beta)\xi, \xi \rangle$$

is a Borel measure.

The positive measure Q is called *spectral* if, in addition,

$$Q(\alpha \cap \beta) = Q(\alpha) \cdot Q(\beta)$$

for any $\alpha, \beta \in B(\mathbf{R}^n)$.

The proof of Theorem 5.2 is reminiscent of the proof of Theorem 5.1, see Sz.-Nagy [1].

Of particular interest are the positive kernels defined on semigroups. For our purposes we may assume that Λ is the semigroup $\mathbf{N}^k$, with a natural k. When an element $\lambda \in \Lambda$ is fixed, we can associate to it the shift S_λ. This is defined on the space F appearing in the proof of Theorem 5.1, as follows,

$$(S_\lambda \phi)(\mu) = \begin{cases} \phi(\mu - \lambda), & \text{if } \mu \in \lambda + \Lambda, \\ 0, & \text{otherwise,} \end{cases}$$

where $\phi \in F$.

Thus, for a fixed kernel $K : \Lambda \times \Lambda \to L(H)$,

$$[S_\lambda \phi, S_\lambda \phi] = \sum_{\mu,\nu} \langle K(\lambda+\mu, \lambda+\nu)\phi(\nu), \phi(\mu) \rangle,$$

and the shift S_λ extends to a bounded operator on K, if and only if there is a constant $C > 0$, such that

$$[S_\lambda \phi, S_\lambda \phi] \leq C[\phi, \phi]$$

for every $\phi \in F$.

In Kolmogorov's and Naimark's Theorems the new Hilbert spaces are not necessarily separable. However, in all the concrete situations in which we shall use the theorems, these Hilbert spaces turn out to be separable. This is due to the special form of the positive definite kernels we shall meet.

Notes. The Halmos-Bram Theorem 2.2 and the spectral inclusion asserted by Proposition 3.4 can be found in Halmos' problem book [4]. Theorem 2.4 is presented after Atzmon [1]. Alexander's inequality stated in Corollary 4.3 was used in a new proof of the separate analiticity theorem of Hartogs, cf. Alexander [1]. For the origins of Kolmogorov's Theorem, consult Kolmogorov [1]. For a complete proof of Naimark's dilation theorem and some of its applications see for instance Sz.-Nagy [1].

Some new important advances in the theory of subnormal operators discovered after the publication of Conway's monograph are Axler [1], Miller - Olin - Thompson [1], Xia [7], [8].

EXERCISES

1. Let Ω denote a bounded domain in **C**, and $f \in O(\Omega)$ an analytic function on Ω.

a) Prove that for every disc $D(z,r)$ compactly contained in Ω, one has

$$|f(z)|^2 \leq (\pi r^2)^{-1} \int_{D(z,r)} |f(\zeta)|^2 d\mu(\zeta).$$

b) Show that $A^2(\Omega)$ is a closed subspace of $L^2(\Omega)$.

c) For every $\lambda \in \Omega$, there is a unique vector $k_\lambda \in A^2(\Omega)$, such that $f(\lambda) = \langle f, k_\lambda \rangle$ for every $f \in A^2(\Omega)$.

d) The kernel $K_\Omega(\lambda,\zeta) = \langle k_\zeta, k_\lambda \rangle$ is positive definite on $\Omega \times \Omega$ and has the property

$$f(\lambda) = \int_\Omega K_\Omega(\lambda,\zeta) f(\zeta) d\mu(\zeta), \quad \lambda \in \Omega,$$

for every function $f \in A^2(\Omega)$.

K_Ω is called *the Bergman kernel* of the domain Ω, see Bergman [1] for details.

e) Compute K_Ω in the case Ω is a disc.

2. Let B_Ω denote the Bergman shift associated to a bounded domain $\Omega \subset$ **C**.

a) The adjoint B_Ω^* is given by the formula:

$$(B_\Omega^* f)(\lambda) = \int_\Omega K_\Omega(\lambda,\zeta)\bar{\zeta} f(\zeta) d\mu(\zeta), \quad \lambda \in \Omega.$$

b) The spectrum of B_Ω is $\sigma(B_\Omega) = \overline{\Omega}$.

c) The point spectrum of B_Ω is empty.

d) If $\lambda \in \Omega$, then **Ran** $(B_\Omega - \lambda) = \{f \in A^2(\Omega) \,;\, f(\lambda) = 0\}$, hence it is a closed subspace.

3*. With the notations in Exercise 2, a point $\lambda \in \partial\Omega$ is called *removable* for Ω if there exists on open neighbourhood U of λ, such that any function $f \in A^2(\Omega)$ extends analitically over U. The set of all removable points of $\partial\Omega$ is denoted by $\partial_r\Omega$.

Prove that:

a) All isolated points of $\partial\Omega$ belong to $\partial_r\Omega$.

b) $\mu(\partial_r\Omega) = 0$.

c) $\partial(\overline{\Omega}) \subseteq \partial\Omega \setminus \partial_r\Omega$ and $\Omega \cup \partial_r\Omega$ is open.

4*. (Axler, Conway, McDonald, see Conway [2]).

a) The essential spectrum of the operator B_Ω is:

$$\sigma_{ess}(B_\Omega) = \overline{\Omega} \setminus (\Omega \cup \partial_r\Omega).$$

b) For every $\lambda \in \Omega \cup \partial_r\Omega$, index $(B_\Omega - \lambda) = -1$.

c) If $\partial(\overline{\Omega}) = \partial(\Omega)$, then $\sigma_{ess}(B_\Omega) = \partial\Omega$.

5. Let $A^\infty(\Omega)$ denote the algebra of all analytic functions on Ω, uniformly bounded on Ω.

Prove that the commutant of the Bergman operator B_Ω is

$$(B_\Omega)' = \{M_f \; ; \; f \in A^\infty(\Omega)\}.$$

6*. (Agler [1]). Let $S = B_{\mathbf{D}}$ denote the Bergman operator of the unit disc and put $S^{*(\infty)}$ for a countable direct sum of S^*.

An operator T is unitarily equivalent with the restriction of $S^{*(\infty)}$ to an invariant subspace if and only if $\|T\| \leq 1$, $1 - 2T^*T + T^{*2}T^2 \geq 0$ and $\operatorname{so-lim}_n T^n = 0$.

7*. (Halmos, Lumer, Schaffer [1]). The operator B_Ω has a square root if and only if the set $\{z \in \mathbf{C} \; ; \; z^2 \in \Omega\}$ is not connected.

8. There exist several characterizations of the subnormality among which we quote a few:

Let $S \in L(H)$ be an operator. The following conditions are equivalent:

a) S is subnormal;

b) (Bram [1]). There exists a positive operator valued measure Q, compactly supported on $\mathbf{C}$, such that

$$S = \int_{\mathbf{C}} z dQ(z) \quad \text{and} \quad S^*S = \int_{\mathbf{C}} |z|^2 dQ(z);$$

c) (Bram [1]). The function $T : \mathbf{C} \to L(H)$, $T(z) = \exp(-\bar{z}S^*)\exp(zS)$ is positive definite, that is the kernel

$$K : \mathbf{C}\times\mathbf{C} \to L(H),\ K(z,w) = T(w - z),$$

is positive definite;

d) (Embry [1]). For any finite sequence of vectors $\xi_o,\dots,\xi_n \in H$, one has

$$\sum_{j,k=0}^{n} \langle S^{j+k}\xi_j, S^{j+k}\xi_k\rangle \geq 0;$$

e) (Embry [1]). There is a positive operator valued measure M supported by the interval $[0, \|S\|]$, such that

$$S^{*n}S^n = \int_{\mathbf{R}} t^{2n} dM(t);$$

f) (Bunce - Deddens [1]). For any finite sequence of operators $A_o,\dots,A_n$, belonging to the C^*-algebra generated by S, one has

$$\sum_{j,k=0}^{n} A_j^* S^{*k} S^j A_k \geq 0;$$

g) (Agler [1]). For any natural number n,

$$\sum_{k=0}^{n} (-1)^k \binom{n}{k} \|S\|^{-2k} S^{*k}S^k \geq 0,$$

where $\binom{n}{k} = n!/k!(n-k)!$.

9. (Bishop [1]). The set of subnormal operators acting on a fixed Hilbert space is the closure in the strong operator topology of the set of normal operators.

Hint. Take $S \in L(H)$ a subnormal operator with minimal normal extension $N \in L(K)$. For an arbitrary orthogonal basis $\{e_n\}_o^\infty$ of H consider $H_n = \bigvee\{e_o,\dots,e_n, Se_o,\dots,Se_n\}$. Pick a unitary $U_n \in L(H,K)$ which acts as the identity on H_n, and then show that $\mathbf{so\text{-}lim}_n (U_n^* N U_n) = S$.

For the converse, let $S = \mathbf{so\text{-}lim}_n N_n$, where $N_n \in L(H)$ are normal operators. First prove that $\mathbf{so\text{-}lim}_n N_n^k = S^k$ for every natural number k and then use the Halmos-Bram Theorem.

10. If A and B are positive operators on a Hilbert space, then $A \leq B$ whenever $A^2 \leq B^2$.

Hint. One may assume B invertible. Then, $A^2 \leq B^2$ implies $B^{-1}A^2B^{-1} \leq I$, whence $\|AB^{-1}\| \leq 1$. By passing to spectral radii one finds $|AB^{-1}|_{sp} = |B^{-\frac{1}{2}}AB^{-\frac{1}{2}}|_{sp}$

and therefore $B^{-\frac{1}{2}}AB^{-\frac{1}{2}} \leq I$, as desired.

11. Characterize, by using Theorem 2.4, the sequence (not necessarily double) of moments of a finite, positive measure on the torus **T**, or compactly supported on **R**.

Chapter II

HYPONORMAL OPERATORS AND RELATED OBJECTS

As in the case of subnormal operators, the hyponormal operators uniquely decompose into a normal and a pure part. The pure subspace can be described analogously to Theorem I.1.4, and this description provides complete unitary invariants for pure hyponormal operators. The description of the internal structure and the parametrization of these unitary invariants, as well as of a few objects arising directly from the hyponormality assumption, are two of the leading motives of our approach.

In this chapter we introduce these objects of principal interest and we study their elementary properties. A generic example of hyponormal operator, which will turn out to be the skeleton of a functional model for any hyponormal operator, is also briefly discussed below.

1. Pure hyponormal operators

Throughout this chapter T is a linear bounded operator acting on the Hilbert space H.

1.1. Definition. The operator $T \in L(H)$ is said to be *hyponormal* if its self-commutator $[T^*, T]$ is positive, that is

$$\langle [T^*, T]\xi , \xi\rangle \geq 0,$$

or equivalently,

$$(1) \qquad \|T^*\xi\| \leq \|T\xi\|,$$

for every vector $\xi \in H$.

We have already seen in Chapter I that every subnormal operator satisfies condition (1). There are simple examples (see Section 2 below) which show that the converse is not always true.

As in Chapter I, our first goal is to isolate a maximal normal part of any hyponormal operator.

1.2. Definition. An operator T is said to be *pure hyponormal* or *completely non--normal hyponormal,* if there is no nontrivial normal operator, which is a direct orthogonal summand of T.

For instance every pure subnormal operator is pure hyponormal, too.

In the sequel we adopt for the sake of simplicity the notational convention:

$$D_T = [T^*, T], \text{ or simply } D = [T^*, T].$$

If we decompose T into its real and imaginary parts, $T = X + iY$, with

$$X = \mathbf{Re}\, T = (1/2)(T + T^*), \quad Y = \mathbf{Im}\, T = (1/2i)(T - T^*),$$

then

$$D = 2i[X, Y].$$

Thus, to give a hyponormal operator $T \in L(H)$ is equivalent with giving a pair $X, Y \in L(H)$ of self-adjoint operators, subject to the commutator condition

$$i[X, Y] \geq 0.$$

1.3. Theorem. *Let* $T \in L(H)$ *be a hyponormal operator with self-commutator* D.

There is a unique orthogonal decomposition $H = H_p(T) \oplus H_n(T)$ *where* $H_p(T)$ *and* $H_n(T)$ *are reducing subspaces for* T, *such that*

(i) $T_p = T|H_p(T)$ *is pure,*

(ii) $T_n = T|H_n(T)$ *is normal.*

Moreover,

$$H_p(T) = \bigvee \{T^{*k}T^{\ell}D\xi \;;\; \xi \in H, \quad k, \ell \in \mathbf{N}\},$$

and

$$H_n(T) = \{\eta \in H \;;\; DT^{*\ell}T^k\eta = 0 \quad \textit{for every } k, \ell \in \mathbf{N}\}.$$

Proof. Let $A = A^*(T)$ denote the unital $*$-algebra generated by T in $L(H)$, and put

$$H_o = \bigvee \{AD\xi \;;\; \xi \in H, \quad A \in A\},$$

$$H_1 = \bigcap_{A \in A} \mathbf{Ker}\,(DA).$$

It is clear that $H_1 = H_o^{\perp}$ and that H_o and H_1 are reducing subspaces for T.

If K is a reducing subspace of T, such that $T|K$ is a normal operator, then $DA|K = 0$ for every $A \in A$, whence $K \subset H_1$. This proves that H_1 is the largest reducing subspace of T, on which T is normal. Accordingly, $T|H_o$ is a pure hyponormal operator, $H_o = H_p(T)$, and $H_1 = H_n(T)$.

Moreover, by using the commutator relation $TT^* = T^*T - D$, one may rearrange the order of the factors in any monomial of the form

$$T^{*n_1}T^{m_1}T^{*n_2}T^{m_2}\dots T^{*n_k}T^{m_k}D$$

so that it becomes a linear combination of products of expressions like $T^{*k}T^{\ell}D$, $k, \ell \in \mathbf{N}$. It turns out that for any $A \in A$ the operator AD has a similar representation. This completes the proof of Theorem 1.3.

The operators T_p and T_n above are called the *pure*, respectively the *normal*, *parts* of T. Correspondingly, $H_p(T)$ and $H_n(T)$ are called the *pure*, and respectively the *normal*, *subspaces* of T.

Actually the rearrangement argument in the proof of Theorem 1.3 can be put in a precise quantitative form, which we isolate below for later use.

1.4. Lemma. *Let* A *and* B *be operators on a Hilbert space* H. *Then for every natural numbers* m *and* n *one has*

$$[A^n, B^m] = \sum_{p=0}^{n-1}\sum_{q=0}^{m-1} A^pB^q[A, B]B^{m-q-1}A^{n-p-1}.$$

Proof. It suffices to regard the bracket operations $[\cdot, B]$ and $[A, \cdot]$ as derivations on the algebra $L(H)$.

Theorem 1.3 has the following direct consequences.

1.5. Corollary. *The pure subspace* $H_p(T)$ *is the smallest reducing subspace of* T *which contains the range of the self-commutator* $[T^*, T]$.

Moreover, **dim** $H_p(T)$ *is* 0 *or* ∞.

1.6. Corollary. *The pure subspace* $H_p(T)$ *can be alternately described as follows:*

$$H_p(T) = \bigvee\{T^kT^{*\ell}D\xi : \xi \in H, \quad k, \ell \in \mathbf{N}\} =$$

$$= \bigvee\{X^kY^{\ell}D\xi : \xi \in H, \quad k, \ell \in \mathbf{N}\} =$$

$$= \bigvee\{Y^k X^\ell D\xi : \xi \in H, \quad k,\ell \in \mathbf{N}\},$$

where $X = \mathbf{Re}\, T$ *and* $Y = \mathbf{Im}\, T$.

For the proof it suffices to transform the expression of the space $H_p(T)$, as given in Theorem 1.3, by Lemma 1.4.

In the case of subnormal operators, the above description leads to an already proved result (see Theorem I.1.4). We insert it here again.

1.7. Corollary. *Let* $S \in L(H)$ *be a subnormal operator. Then*

$$H_p(S) = \bigvee\{S^k D\xi \; ; \; \xi \in H, \; k \in \mathbf{N}\}.$$

Proof. Since the range of D is invariant under S^* (Proposition I.1.2 (ii)) every vector like $S^k S^{*\ell} D\xi$ belongs to the right hand space.

1.8. Proposition. *Let* T *be a pure hyponormal operator and denote* $D = (\mathbf{Ran}\, D)^-$.

Each of the following sequences of operators on D *is a complete set of unitary invariants for* T:

(i) $\{D^{\frac{1}{2}} T^k T^{*\ell} T^m T^{*n} D^{\frac{1}{2}}\}$,

(ii) $\{D^{\frac{1}{2}} T^{*k} T^\ell T^{*m} T^n D^{\frac{1}{2}}\}$,

(iii) $\{D^{\frac{1}{2}} X^k Y^\ell X^m Y^n D^{\frac{1}{2}}\}$,

(iv) $\{D^{\frac{1}{2}} Y^k X^\ell Y^m X^n D^{\frac{1}{2}}\}$,

with $k, \ell, m, n \in \mathbf{N}$.

Proof. Let $T \in L(H)$ and $T' \in L(H')$ be pure hyponormal operators with self--commutators D and D', respectively. Denote $D' = (\mathbf{Ran}\, D')^-$.

Suppose $U : D \to D'$ is a unitary operator with the property

$$UD^{\frac{1}{2}} T^k T^{*\ell} T^m T^{*n} D^{\frac{1}{2}} U^* = D'^{\frac{1}{2}} T'^k T'^{*\ell} T'^m T'^{*n} D'^{\frac{1}{2}},$$

for any system of natural numbers k, ℓ, m, n.

In virtue of Corollary 1.6, there is a unitary operator $V : H \to H'$, defined by

$$VT^m T^{*n} D^{\frac{1}{2}} \xi = T'^m T'^{*m} D'^{\frac{1}{2}} U\xi, \quad \xi \in H, \quad m, n \in \mathbf{N}.$$

The definition of V shows that $VT = T'V$, hence the operators T and T' are unitarily equivalent.

This ends the proof in the case (i). Similar arguments apply to the other invariants, and the proposition is proved.

Let us note cursorily that the invariants (i) - (iv) correspond in fact to certain

$L(D)$-valued positive kernels, defined on $\mathbf{N}^2 \times \mathbf{N}^2$. For a subnormal operator, the simpler form of the pure subspace asserted in Corollary 1.7 prompts one to work with a kernel defined on $\mathbf{N} \times \mathbf{N}$. More precisely one finds the next result.

1.9. Corollary. *The positive definite kernel* $K : \mathbf{N} \times \mathbf{N} \to L(D)$

$$K(m,n) = D^{\frac{1}{2}} S^{*m} S^n D^{\frac{1}{2}}, \quad m, n \in \mathbf{N},$$

is a complete unitary invariant for the pure subnormal operator S.

By Lemma 1.4 one immediately obtains from Corollary 1.9 that the "reversed" kernel $L : \mathbf{N} \times \mathbf{N} \to L(D)$,

$$L(m,n) = D^{\frac{1}{2}} S^m S^{*n} D^{\frac{1}{2}}, \quad m, n \in \mathbf{N},$$

is also a complete unitary invariant for the subnormal operator S. More exactly, the relation

$$K(m,n) - L(m,n) = \sum_{p=0}^{n-1} \sum_{q=0}^{m-1} L(m-q-1,\ n-p-1) K(p,q)$$

follows from Lemma 1.4. Then an obvious induction shows that the kernel L can be determined from K, and conversely.

At this moment it is quite obvious that the sequences (i) - (iv) of invariants listed in Proposition 1.8, as well as the kernel K and L above, contain a certain redundancy in their definitions. In fact only half of the indices suffices to determine them completely. Next we perform this reduction in the case of subnormal operators. The structure of the invariants (i) - (iv) for hyponormal operators will be discussed in § 4, in a more enlightened context.

1.10. Lemma. *The sequence of operators* $\Theta(i) = D^{\frac{1}{2}} S^i D^{\frac{1}{2}}$, $i \in \mathbf{N}$, *forms a complete unitary invariant for the pure subnormal operator* S.

Proof. Roughly, one has to prove that expressions like $D^{\frac{1}{2}} S^m S^{*n} D^{\frac{1}{2}}$ can be recuperated from Θ.

Let us assume that the sequence Θ is known. If P denotes the orthogonal projection of the space H, where S acts, onto $D = (\mathbf{Ran}\ D)^-$, then the scalar products

$$\langle PS^i D^{\frac{1}{2}} \xi,\ D^{\frac{1}{2}} \eta \rangle, \quad i \in \mathbf{N};\ \xi, \eta \in D,$$

are also known. By them one recomposes the operators $T_i = PS^i | D \in L(D)$, $i \in \mathbf{N}$.

In virtue of Proposition I.1.2 (ii) one gets

$$T_j^* = PS^{*j}|D = S^{*j}|D,$$

whence the operators

$$D^{\frac{1}{2}}T_iT_j^*D^{\frac{1}{2}} = D^{\frac{1}{2}}S^iPS^{*j}D^{\frac{1}{2}} = D^{\frac{1}{2}}S^iS^{*j}D^{\frac{1}{2}} = L(i,j)$$

are perfectly detemined by Θ.

This completes the proof of the lemma.

1.11. Theorem. *The pair of operators* $D = [S^*, S]$ *and* $S^*|D$ *acting on the space* $D = (\mathbf{Ran}\,[S^*, S])^-$, *forms a complete system of unitary invariants for any pure subnormal operator* S.

Proof. Let $R = S^*|D \in L(D)$. Since for every natural number n, $R^n = S^{*n}|D$, it follows that $\Theta(n)^* = D^{\frac{1}{2}}R^nD^{\frac{1}{2}}$. Thus the invariant Θ can be reconstituted from the operators R and D. By Lemma 1.10, the proof of theorem is complete.

This theorem shows that the pure subnormal operators with self-commutator of a finite rank, say r, depend on at most r(2r + 1) real parameters. The classification in the case r = 1 goes back to Morrel [1], see Proposition I.1.5. Recently, Xia [8] has obtained the concrete form of all pure subnormal operators with finite rank self-commutator.

We close this section by recalling that an operator $T \in L(H)$ is called *irreducible* if there is no non-trivial decomposition $H = K \oplus K^{\perp}$ which reduces T (i.e. $TK \subset K$ and $T^*K \subset K$).

Obviously, any irreducible hyponormal operator is, by Theorem 1.3, either pure or normal. As every normal operator acting on a Hilbert space of dimension greater than 1 is reducible, we conclude with the following.

1.12. Lemma. *Any irreducible hyponormal operator acting on a Hilbert space of dimension greater than* 1 *is pure.*

As a partial converse, we note that any pure hyponormal operator with the self-commutator of rank 1 is irreducible. In connection with this remark see Example 3 below and Exercise 5.

2. Examples of hyponormal operators

We interrupt the main line of our presentation with a digression on a few illustrative examples of hyponormal operators.

Example 1. *A hyponormal operator which is not subnormal*

Recall that $U_+ \in L(l^2(\mathbf{N}))$ denoted the unilateral shift. Let us consider the operator $T = 2U_+ + U_+^*$. Since $[T^*, T] = 3[U_+^*, U_+]$ the operator T is hyponormal. If we denote by $(e_n)_{n=0}^{\infty}$ the standard basis of $l^2(\mathbf{N})$, then straightforward computations show that the operator $D = [T^{2*}, T^2]$ acts as follows:

$$De_0 = 15e_0 + 6e_2,$$

$$De_1 = 15e_1,$$

$$De_2 = 6e_0,$$

$$De_n = 0, \text{ for any } n \geq 3.$$

Thus the self-commuator D, compressed to the space spanned by the vectors e_0, e_1, e_2 has the matrix representation

$$\begin{bmatrix} 15 & 0 & 6 \\ 0 & 15 & 0 \\ 6 & 0 & 0 \end{bmatrix}.$$

But it is obvious that the above matrix is not positive definite.

In conclusion we got a hyponormal operator T such that T^2 is no longer hyponormal. In view of Corollary I.3.5 the operator T is not subnormal.

This example shows also that the functional calculus with analytic functions is not an internal operation in the class of hyponormal operators. There are, however, quite a lot of other internal operations whose systematic study is carried out in Chapter V.

Example 2. *Weighted shifts*

An operator $T \in L(l^2(\mathbf{N}))$ acting on the standard basis by the formulae

$$Te_n = \alpha_n e_{n+1}, \quad \alpha_n \in \mathbf{C}, \quad n \geq 0,$$

is called a *weighted shift*, with the sequence of weights $(\alpha_n)_{n=0}^{\infty}$. Since T is bounded, one gets

$$\sup_n |\alpha_n| < \infty.$$

The self-commutator $D = [T^*, T]$ turns out to be diagonal with respect to the basis $(e_n)_{n=0}^{\infty}$, with entries:

$$\langle De_0, e_0 \rangle = |\alpha_0|^2,$$

$$\langle De_n, e_n \rangle = |\alpha_n|^2 - |\alpha_{n-1}|^2, \quad n \geq 1.$$

Thus the weighted shift T is hyponormal if and only if the weights satisfy:

$$|\alpha_0| \leq |\alpha_1| \leq |\alpha_2| \leq \dots \ .$$

In order to produce a weighted shift T which is hyponormal but not subnormal we pick the weights:

$$(\alpha, \beta, 1, 1, \dots),$$

with $0 < \alpha < \beta < 1$.

By Halmos-Bram's Theorem I.2.2, we must investigate kernels like $\langle T^k \xi_\ell , T^\ell \xi_k \rangle_{k,\ell=0}^{\infty}$. An immediate computation shows that the matrix $\langle T^k e_\ell , T^\ell e_k \rangle_{k,\ell=0}^{2}$ is

$$\begin{bmatrix} 1 & \alpha & \alpha\beta \\ \alpha & \beta^2 & \beta \\ \alpha\beta & \beta & 1 \end{bmatrix} .$$

But it is obvious that this matrix is not positive definite, whence T is not subnormal.

In spite of all appearances weighted shifts are difficult to deal with. An extensive literature is devoted to them and their relationship with spaces of analytic functions, see Shields [1]. We shall not pursuit this direction.

Example 3. *Singular integral operators*

We now turn our attention to the most important example of hyponormal operator. It provided the source and the test field of every development of the general theory.

We have already remarked at the beginning of this chapter, that the hyponormality condition is equivalent to the positivity (up to the factor i) of the commutator of two self-adjoint operators X and Y.

Accordingly, take X to be one of the simplest self-adjoint operators, namely, the position operator M_t on the space $L^2[0,1]$ (with respect to the linear measure dt). Explicitly we have

$$(Xf)(t) = tf(t), \quad f \in L^2[0,1],\ t \in [0,1].$$

We seek for self-adjoint operators Y, acting also on $L^2[0,1]$, subject to the condition $D = 2i[X , Y] \geq 0$.

For instance, such an operator is given by a principal value integral $Y = -S$, as follows:

$$(Sf)(t) = (1/\pi i)\text{v.p.}\int_0^1 [f(s)/(s - t)]ds, \quad t \in [0,1].$$

Indeed

$$\langle Df,f\rangle = \langle 2i[X,Y]f,f\rangle = (2/\pi)\int_0^1\left(\text{v.p.}\int_0^1 sf(s)/(s - t)ds - \text{v.p.}t\int_0^1 f(s)/(s - t)ds\right)\overline{f(t)}dt =$$

$$= (2/\pi)\int_0^1\int_0^1 f(s)ds\overline{f(t)}dt = (2/\pi)\left|\int_0^1 f(t)dt\right|^2 \geq 0,$$

for $f \in L^2[0,1]$.

The formal computations above are permitted because the operator S is well defined and bounded on $L^2[0,1]$. This is a classical fact which lies at the heart of the theory of singular integral operators in one dimension. The relevant properties of the operator S, which will be referred to as the *Cauchy singular integral operator*, or the *Hilbert transform*, are quite extensively presented in the Supplement to Chapter VII.

A shlight variation of the above example leads to the following class of hyponormal operators.

Let σ be a compact subset of **R**, and let $a, b \in L^\infty(\sigma)$ with $a = \bar{a}$ a.e. (we continue to work with the usual Lebesgue linear measure dt). Then the operator acting on $L^2(\sigma)$ and given by

$$T = M_t + i(M_a - M_b S M_b^*)$$

is hyponormal with self-commutator of rank equal to one.

The operator T acts explicitly on a function $f \in L^2(\sigma)$ by the formula:

$$(Tf)(t) = (t + ia(t))f(t) - (1/\pi)\int_\sigma [b(t)\bar{b}(s)f(s)/(s - t)]ds.$$

Similarly to the preceding computation one gets

$$\langle Df, f\rangle = (2/\pi)\left|\int_\sigma \bar{b}(t)f(t)dt\right|^2, \quad f \in L^2(\sigma),$$

whence one finds $D \geq 0$ and $\mathbf{rank}\, D = 1$.

We anticipate at this point a result which will be proved later, namely that any pure hyponormal operator with cyclic real part and one dimensional self-commutator is unitarily equivalent to a singular operator as T above. However, different σ, and different functions a and b may produce unitarily equivalent operators, see Exercises 4, 5.

3. Contractions associated to hyponormal operators

In the present section we exhibit other objects of interest associated to hyponormal operators. Since the study of these objects requires a rather specialized framework, the first part of the section deals with some preliminaries.

For reader's convenience we collect below certain facts and conventions on spaces of vector valued functions and distributions. A comprehensive reference is Schwarz [1].

a) Let Ω be an open subset of the complex plane $\mathbf{C}$. As usual one denotes by $D(\Omega)$ the LF space of complex valued, smooth, compactly supported functions on Ω and let $D'(\Omega)$ be its topological dual, that is, the space of scalar distributions. Let $E(\Omega)$ denote the Fréchet space of complex valued smooth functions on Ω. Its dual, denoted by $E'(\Omega)$, is the space of compactly supported scalar distributions.

Given a Hilbert space H, by $D(\Omega,H)$ and $E(\Omega,H)$ we denote the LF space of H-valued smooth functions on Ω, which are compactly supported, respectively, the Fréchet space of H-valued, smooth functions on Ω. Henceforth we shall identify these spaces with the corresponding complete topological tensor products, namely

$$D(\Omega,H) = D(\Omega)\hat{\otimes}H, \quad E(\Omega,H) = E(\Omega)\hat{\otimes}H.$$

Their duals are the *spaces of H-valued distributions* and they will be denoted by $D'(\Omega,H)$ and, respectively, $E'(\Omega,H)$. Since the space $D'(\Omega)$ is nuclear, the identification

$$D'(\Omega,H) = D'(\Omega)\hat{\otimes}H$$

is possible for every complete topological tensor product.

b) In some specific cases we have to deal with the complete injective tensor product, denoted by $\hat{\otimes}_\varepsilon$, or with the complete projective tensor product, denoted by $\hat{\otimes}_\pi$. For instance we shall work with the space $L^\infty(\Omega)\hat{\otimes}_\varepsilon L(H)$, the Banach space of $L(H)$-valued, measurable and essentially bounded functions on Ω, also denoted by $L^\infty(\Omega,L(H))$.

For the integration theory of vector valued functions we refer to the monograph of Dunford and Schwartz [1]. In what follows the notations $L^\infty(\Omega)$, $L^1(\Omega)$ and $L^2(\Omega)$ are constantly used for the Lebesgue spaces corresponding to the usual Lebesgue measure $d\mu$.

c) It is well-known that the predual $L(H)_*$ of the Banach algebra $L(H)$ is the ideal $C_1(H)$ of all trace-class operators on H. The duality pairing is given by the trace functional:

$$(A,T) \to \mathbf{Tr}(AT), \quad A \in C_1(H), \quad T \in L(H).$$

Notice that $L^\infty(\Omega)\hat{\otimes}_\varepsilon L(H)$ is the dual of the Banach space $L^1(\Omega)\hat{\otimes}_\pi C_1(H)$; see for instance Köthe [1]. It follows that $L^{\infty}(\Omega, L(H))$, with the pointwise multiplication, is a von Neumann algebra.

As the dual of $C_1(H)$ the space $L(H)$ carries a weak-$*$ topology. Sometimes we denote the space $L(H)$ endowed with this topology by $L(H)_{\sigma}$. A similar notation, H_{σ}, is used for the space H with the weak-$*$ topology. These weak-$*$ topologies appear when we need the weak continuity of some natural maps. The next two examples are important:

(i) there exists a unique continuous map

$$m_1 : (E(\mathbf{C})\hat{\otimes}L(H)) \times (D'(\mathbf{C})\hat{\otimes}L(H)_{\sigma}) \to D'(\mathbf{C})\hat{\otimes}L(H)_{\sigma},$$

which extends the natural product given by

$$m_1(\phi\otimes T, u\otimes S) = (\phi u)\otimes TS, \quad \phi \in E(\mathbf{C}),\ u \in D'(\mathbf{C}),\ T, S \in L(H).$$

For $\Phi \in E(\mathbf{C})\hat{\otimes}L(H)$ and $U \in D'(\mathbf{C})\hat{\otimes}L(H)$ we shall simply put $m_1(\Phi, U) = \Phi U$;

(ii) there exists also a unique continuous map

$$m_2 : (D'(\Omega)\hat{\otimes}L(H)_{\sigma}) \times (E(\Omega)\hat{\otimes}H) \to D'(\Omega)\hat{\otimes}H_{\sigma},$$

which extends the application

$$m_2(u\otimes T, \phi\otimes\xi) = (\phi u)\otimes(T\xi), \quad u \in D'(\Omega),\ T \in L(H),\ \phi \in E(\Omega),\ \xi \in H.$$

The space $D'(\Omega)\hat{\otimes}L(H)$ will be referred to as the *space of $L(H)$-valued distributions.*

For any $U \in D'(\Omega)\hat{\otimes}L(H)$ and $\phi \in E(\Omega)\hat{\otimes}H$ we shall simply denote $m_2(U,\phi) = U\cdot\phi$. Let us remark that the evaluations $U\cdot\phi$, whith arbitrary ϕ's as above, completely determine the operator valued distribution U.

At this moment let us also point out the natural embedding of the space $L^1_{loc}(\Omega, L(H))$ of all $L(H)$-valued, measurable and locally integrable functions on Ω, into $D'(\Omega)\hat{\otimes}L(H)$.

d) We continue with some remarks concerning the complex derivatives $\partial = \partial/\partial z$ and $\bar{\partial} = \partial/\partial\bar{z}$, in the sense of distributions.

Assume that X is a complex Banach space (for our purpose $X = H$ or $X = L(H)$). Then $\bar{\partial}$ gives rise to a continuous operator

$$\bar{\partial} : L^1_{loc}(\mathbf{C})\hat{\otimes}_{\pi} X \to D'(\mathbf{C})\hat{\otimes}X.$$

The fundamental solution of the differential operator $\bar{\partial}$ is the function

$-\pi^{-1}/z \in L^1_{loc}(\mathbf{C})$, that is,

$$\bar{\partial}(1/z) = -\pi\delta$$

where δ is the Dirac distribution. If one denotes by "$*$" the convolution product then the next identity

(2) $$\bar{\partial}F * (-\pi^{-1}/z) = F$$

holds, whenever $F \in L^1_{loc}(\mathbf{C})\hat{\otimes}_\pi X$, **supp** $\bar{\partial}F$ is compact and $\lim_{|z|\to\infty} \|F(z)\| = 0$. Since $\bar{\partial}F$ has compact support, F must be a smooth function in the neighborhood of the infinity.

We refer to (2) as the (generalized) *Cauchy - Pompeiu formula,* and to the convolution operator with $-\pi^{-1}/z$ as the *Cauchy transform.* The Cauchy transform $v = u * (-\pi^{-1}/z)$ of a compactly supported X-valued distribution u is the unique solution in $D'(\mathbf{C})\hat{\otimes}X$ of the equation $\bar{\partial}v = u$, vanishing at infinity.

We mention also the relation

(3) $$(\bar{\partial}F)\cdot\phi = \bar{\partial}(F\phi) - F(\bar{\partial}\phi),$$

which is considered for $F \in L^\infty(\Omega)\hat{\otimes}L(H)$ and $\phi \in D(\mathbf{C})\hat{\otimes}H$. If in addition $\bar{\partial}F$ is compactly supported, then from (3) one obtains

$$(\bar{\partial}F)\cdot\phi * (-\pi^{-1}/z) = F\phi + F(\bar{\partial}\phi) * (\pi^{-1}/z),$$

a relation in which the right hand term contains only genuine pointwise multiplications between operator valued and vector valued functions.

Similar statements for the operator ∂ can be simply obtained by complex conjugation.

e) Finally let us recall the notation concerning the *regularization of distributions.* Let $\rho \in D(\mathbf{C})$, $\rho \geq 0$, be a fixed function such that $\int_{\mathbf{C}} \rho(z)d\mu(z) = 1$. For any $\varepsilon > 0$ let $\rho_\varepsilon \in D(\mathbf{C})$ be the function $\rho_\varepsilon(z) = \varepsilon^{-2}\rho(z/\varepsilon)$. If $u \in D'(\mathbf{C})\hat{\otimes}X$, one introduces the regularizations u_ε of u by $u_\varepsilon = u * \rho_\varepsilon$, $\varepsilon > 0$. These regularizations are smooth X-valued functions on $\mathbf{C}$ and $\lim_{\varepsilon\downarrow 0} u_\varepsilon = u$, in the weak topology of the space $D'(\mathbf{C})\hat{\otimes}X$.

We are now ready to present the basic part of this section, which, as its title indicates, continues the study of hyponormal operators.

Let $T \in L(H)$ be a hyponormal operator with self-commutator $D = T^*T - TT^*$. Since $D \geq 0$, the following inequalities between positive operators hold:

(4) $$T^*T \geq TT^*,$$

(5) $$T^*T \geq D.$$

Consequently there exist unique contractive operators C and K (i.e. $\|C\|$, $\|K\| \leq 1$) so that

(6) $$T^* = KT, \quad K|\mathbf{Ker}\, T^* = 0,$$

respectively

(7) $$D^{\frac{1}{2}} = T^*C, \quad C^*|\mathbf{Ker}\, T^* = 0.$$

Indeed, from (4) one finds $\|T\xi\| \geq \|T^*\xi\|$ for every vector $\xi \in H$, whence the assignement

$$T\xi \mapsto T^*\xi, \ \xi \in H,$$

extends to a contractive linear operator K defined on $(\mathbf{Ran}\, T)^-$. Then we take K to be the trivial extension by 0 of this operator and the relation (6) becomes obvious.

Analogously, from (5) one obtains a unique contraction C which satisfies the conditions (7).

3.1. Lemma. *Let* T *be a hyponormal operator and denote by* P *the orthogonal projection onto* **Ker** T^*. *Then the contractions* C *and* K *satisfy the identity:*

$$I = K^*K + CC^* + P.$$

Proof. From (6) and (7) one has

$$D = T^*(I - K^*K)T,$$

and respectively,

$$D = T^*CC^*T.$$

Therefore

$$T^*(I - K^*K - CC^*)T = 0.$$

As $KP = 0$ and $C^*P = 0$, from the last identity we infer:

$$T^*(I - P)(I - P - K^*K - CC^*)(I - P)T = 0,$$

and, since the operator T^* is one-to-one on $(I - P)H$, we finally obtain the desired identity.

The factorization (6) has been used by Sz.-Nagy and Foiaş [2], [3] in the construction of a Toeplitz model for hyponormal operators. We shall resume this subject in Chapter VI.

In what follows we exploit mainly the factorization (7). To be more specific, we introduce a family of contractions which will produce complete unitary invariants for pure hyponormal operators.

For the sake of simplicity we adopt the following notations:

$$T_z = T - zI, \quad T_z^* = T^* - \bar{z}I,$$

where $z \in \mathbf{C}$. It is obvious that for every $z \in \mathbf{C}$ the operator T_z is still hyponormal, with

$$[T_z^*, T_z] = D.$$

Accordingly, there is a family of contractions $\{C(z)\,;\ z \in \mathbf{C}\}$ such that

$$T_z^* C(z) = D^{\frac{1}{2}} \quad \text{and} \quad C^*(z) | \mathbf{Ker}\, T_z^* = 0, \quad z \in \mathbf{C}. \tag{8}$$

Note that for $z \notin \sigma(T)$ the operator T_z is invertible, and

$$C(z) = (T_z^*)^{-1} D^{\frac{1}{2}}.$$

Thus $C(\cdot)$ is an operator valued anti-analytic function off $\sigma(T)$.

Next we want some informations on the behaviour of $C(\cdot)$ at points of the spectrum.

3.2. Lemma. *Let* $T \in L(H)$ *be a hyponormal operator and let* $C(z)$, $z \in \mathbf{C}$, *be the family of contractions defined by relations* (8).

Then $C(\cdot) \in L^\infty(\mathbf{C}) \hat{\otimes}_\varepsilon L(H)$.

Proof. Let $\lambda \in \mathbf{C}$ be fixed and denote by E_λ the spectral measure of the self-adjoint operator $T_\lambda^* T_\lambda$. From (8) we deduce the formula:

$$C(\lambda) = \text{so} - \lim_{\varepsilon \downarrow 0} T_\lambda \int_\varepsilon^\infty t^{-1} E_\lambda(dt) D^{\frac{1}{2}}.$$

Indeed, since $\mathbf{Ran}\, C(\lambda)^- = (\mathbf{Ran}\, T_\lambda)^-$ and $D^{\frac{1}{2}} = T_\lambda^* C(\lambda)$, one can use the obvious relation

$$\text{so} - \lim_{\varepsilon \downarrow 0} T_\lambda^* T_\lambda \int_\varepsilon^\infty t^{-1} E_\lambda(dt) D^{\frac{1}{2}} = D^{\frac{1}{2}}.$$

Thus $C(\lambda)$ depends measurably on λ. Because $\| C(\lambda) \| \leq 1$, we find $C \in$ $\in L^\infty(\mathbf{C}) \hat{\otimes}_\varepsilon L(H)$ and the proof is complete.

From now on we shall regard the family of contractions $C(\lambda)$, $\lambda \in \mathbf{C}$, as an element of $L^{\infty}(\mathbf{C},L(H))$, or even of $D'(\mathbf{C})\hat{\otimes}L(H)$, rather than a pointwise parametrized set. Similarly, by T_z and T_z^* we shall often denote the corresponding elements belonging to $E(\mathbf{C})\hat{\otimes}L(H)$. With these conventions the relation (8) above becomes an identity in the space $D'(\mathbf{C})\hat{\otimes}L(H)$, and we can take the derivative $\partial = \partial/\partial z$ in the sense of distributions. One gets the next basic equation:

$$T_z^* \partial C = 0. \tag{9}$$

Hence ∂C is an operator valued eigendistribution of the operator T^*.

Roughly speaking, our next aim is to show that the adjoint of a pure hyponormal operator possesses sufficiently many vector valued eigendistributions. Quite specifically, we shall prove that expresions like $\partial C\cdot\phi \in D'(\mathbf{C},H)$, where $\phi \in D(\mathbf{C},H)$, span the pure subspace $H_p(T)$ of the operator T.

We start with a few technical results. Recall that $u_\varepsilon = u * \rho_\varepsilon$, $\varepsilon > 0$, represent the regularizations of a distribution u, where $\rho \in D(\mathbf{C})$, $\rho \geq 0$, is a fixed function.

3.3. Lemma. *For* $T \in L(H)$ *hyponormal and for any functions* $\phi \in D(\mathbf{C},H)$ *and* $f \in L^2(\mathbf{C},H)$, *one has*

$$\lim_{\varepsilon \downarrow 0} \langle T_z^* \partial C_\varepsilon \phi, f\rangle_{2,\mathbf{C}} = 0.$$

Proof. With the above notations $C_\varepsilon = C * \rho_\varepsilon$. Since

$$T_z^* C(w) = D^{\frac{1}{2}} - (\bar{z} - \bar{w})C(w)$$

for all pairs $(z,w) \in \mathbf{C}^2$, we get

$$T_z^* \partial C_\varepsilon(z) = D^{\frac{1}{2}}\int \partial\rho_\varepsilon(z - w)d\mu(w) - \int(\bar{z} - \bar{w})C(w)\partial\rho_\varepsilon(z - w)d\mu(w).$$

But $\int \partial\rho_\varepsilon(z - w)d\mu(w) = 0$, and, after the linear change of coordinates $\zeta = (z - w)/\varepsilon$, one obtains

$$T_z^* \partial C_\varepsilon(z) = -\int C(z - \varepsilon\zeta)\bar{\zeta}\partial\rho(\zeta)d\mu(\zeta).$$

Thus, for a given function $\phi \in D(\mathbf{C},H)$, the set

$$\{T_z^*(\partial C_\varepsilon)\phi \,;\, \varepsilon > 0\}$$

is bounded in $L^2(\mathbf{C},H)$, hence its closure is weakly compact.

But $\lim_{\varepsilon \downarrow 0} T_z^* \partial C_\varepsilon = 0$ in the weak topology of the space $D'(\mathbf{C})\hat{\otimes}L(H)$. Consequently

$$\lim_{\varepsilon\downarrow 0} T_z^*(\partial C_\varepsilon)\phi = 0$$

in the weak topology of the Hilbert space $L^2(\mathbf{C},H)$, and the proof is complete.

In order to state the next result, for any $z \in \mathbf{C}$ one denotes by $K(z)$ the unique operator which satisfies the conditions

$$T_z^* = K(z)T_z, \quad K(z)|\mathbf{Ker}\, T_z^* = 0.$$

3.4. Lemma. *With the above notations,*

(10) $$T_z = C(z)D^{\frac{1}{2}} + K^*(z)T_z^*, \quad z \in \mathbf{C}.$$

Proof. By Lemma 3.1 the operators $C(z)$ and $K(z)$ are related as follows:

$$I = K^*(z)K(z) + C(z)C^*(z) + P(z),$$

where $P(z)$ is the orthogonal projection onto $\mathbf{Ker}\, T_z^* = (\mathbf{Ran}\, T_z)^{\perp}$. The multiplication on the right with T_z gives (10).

3.5. Lemma. *For any $\phi \in D(\mathbf{C},H)$,*

(11) $$T_z(\partial C)\cdot\phi = \lim_{\varepsilon\downarrow 0} CD^{\frac{1}{2}}\partial C_\varepsilon\cdot\phi$$

in the weak topology of $D'(\mathbf{C})\hat{\otimes}H$.

Proof. Use Lemma 3.3 and formula (10).

In spite of its unpleaseant appearance, the following equation in C turns out to be one of the characteristic properties of an operator valued function $C \in L^\infty(\mathbf{C},L(H))$ arising as the contraction valued function associated to a hyponormal operator.

3.6. Proposition. *Let $w \in \mathbf{C}$ and $\phi \in D(\mathbf{C},H)$ be fixed. The relation*

(12) $$(z - w)\partial(C^*(w)C)\cdot\phi = D^{\frac{1}{2}}\partial C\cdot\phi - \lim_{\varepsilon\downarrow 0} C^*(w)CD^{\frac{1}{2}}\partial C_\varepsilon\cdot\phi$$

holds in the weak topology of the space $D'(\mathbf{C})\hat{\otimes}H$.

Proof. At the operator valued distributions level we have

$$(z - w)\partial(C^*(w)C) = C^*(w)(T_w - T_z)\partial C = D^{\frac{1}{2}}\partial C - C^*(w)T_z\partial C.$$

The proof finishes by substituting (11) in the last expression.

The following result establishes a basic property of the function C and it will enable us to introduce a new complete unitary invariant for pure hyponormal operators.

3.7. Proposition. *Let* $T \in L(H)$ *be a hyponormal operator. Then*

$$H_p(T) = \bigvee \{C(z)\xi\,;\ \xi \in D,\ \ z \in \mathbf{C}\}.$$

where $D = (\mathbf{Ran}\, D)^-$.

Proof. Let $M = \bigvee \{C(z)\xi\,;\ \xi \in D,\ \ z \in \mathbf{C}\}$. Due to the relation $\mathbf{Ran}\, C(z) = \mathbf{Ran}\, C(z)D^{\frac{1}{2}}$ one finds

$$M = \bigvee \{C(z)\xi\,;\ \xi \in H,\ \ z \in \mathbf{C}\}.$$

On the other hand, by definition each of the contractions C(z) takes values in the pure subspace $H_p(T)$, hence $M \subset H_p(T)$. In order to complete the proof it sufficies now to prove the following inclusions:

(i) $D \subset M$; (ii) $TM \subset M$; (iii) $T^*M \subset M$.

Indeed, from (i) - (iii) we derive $H_p(T) \subset M$, because $H_p(T)$ is the smallest closed subspace of H which reduces T and contains the range of D.

The inclusions (i) and (iii) are simple consequences of the relations

$$T_z^* C(z) = D^{\frac{1}{2}} \text{ and } \lim_{|z| \to \infty} zC(z) = -D^{\frac{1}{2}}.$$

The inclusion (ii) follows from Lemma 3.5. More precisely, by Lemma 3.5 we have in $D'(\mathbf{C})\hat{\otimes}H$ the equality

$$T(\partial C)\xi = z\,\partial C \cdot \xi + \lim_{\varepsilon \downarrow 0} CD^{\frac{1}{2}}\,\partial(C_\varepsilon \xi), \quad \xi \in H.$$

Therefore, after a Cauchy transform one finds

$$TC(w)\xi = D^{\frac{1}{2}}\xi + wC(w)\xi + \pi^{-1}(C\xi * (1/\bar{z}))(w) - \pi^{-1}\lim_{\varepsilon \downarrow 0}(CD^{\frac{1}{2}}\,\partial(C_\varepsilon \xi) * (1/\bar{z}))(w),$$

for any $w \in \mathbf{C}$.

This shows that the space M is invariant under T and the proof is complete.

3.8. Proposition. *Let* $T \in L(H)$ *be a pure hyponormal operator. The operator valued function*

$$C^*(\cdot)C(\cdot) : \mathbf{C} \times \mathbf{C} \to L(D)$$

is a complete unitary invariant for T.

Proof. By Proposition 3.7 the vectors $\{C(z)\xi\,;\ \xi \in D,\ z \in \mathbf{C}\}$ span H. Hence the scalar products

$$\langle TC(z)\xi, C(w)\eta\rangle, \quad \xi, \eta \in D, \quad z, w \in \mathbf{C},$$

completely determine the operator T.

Now, if $C^*(w)C(z) \in L(D)$ are known for all $z, w \in \mathbf{C}$, then, by taking residues at infinity, the operator $D \in L(D)$ is known. On the other hand we have

$$\langle TC(z)\xi, C(w)\eta\rangle = \langle C(z)\xi, T^*C(w)\eta\rangle = \langle C(z)\xi, D^{\frac{1}{2}}\eta\rangle + w\langle C(z)\xi, C(w)\eta\rangle =$$

$$= \langle C(z)\xi, D^{\frac{1}{2}}\eta\rangle + w\langle C(w)^*C(z)\xi, \eta\rangle,$$

a remark which ends the proof.

The reader can compare this sketch of proof with a similar argument presented in the proof of Proposition 1.8.

Before ending this section, let us mention that, under the assumption **rank** D = = 1, the function $C \in L^{\infty}(\mathbf{C}, L(H))$ has some additional properties which will be presented in Chapter XI.

4. Unitary invariants

In the first part of this section we refine the unitary invariants listed in Proposition 1.8. By simple combinatorial identities we reduce the sequences of operators depending on four natural numbers to double sequences. At this stage we only establish the properties of these refined invariants, and discuss the corresponding inverse problems. In other words, we characterize the invariants by a minimal set of conditions and then we give a constructive way of recuperating the operator from its invariants.

The second part continues the study of the contraction valued function C. It is devoted to introducing of another unitary invariant derived from C.

Throughout this section $T \in L(H)$ is a pure hyponormal operator with self--commutator D. Let $D_T = (\mathbf{Ran}\, D)^-$.

In what follows we denote by $\iota = (1,0)$, $\kappa = (0,1)$ and $\theta = (0,0)$ the generators, respectively the neutral element of the additive semigroup $\mathbf{N}^2$. We also denote for simplicity:

$$N_T(\alpha,\beta) = D^{\frac{1}{2}}T^m T^{*n} T^p T^{*q} D^{\frac{1}{2}},$$

$$\Delta_T(\alpha) = D^{\frac{1}{2}}T^m T^{*n} D^{\frac{1}{2}},$$

for $\alpha = (n,m)$, $\beta = (p,q) \in \mathbf{N}^2$.

The generating rules of the positive definite kernel N_T can be reduced to the following ones:

4.1. Proposition. *Let* $T \in L(H)$ *be a pure hyponormal operator. The kernel* $N_T : \mathbf{N}^2 \times \mathbf{N}^2 \to L(D_T)$ *is determined by the invariant* $\Delta_T : \mathbf{N}^2 \to L(D_T)$ *by the next rules:*

1) $N_T(\alpha,\beta) = N_T(\beta,\alpha)^*$

2) $N_T(\theta,\alpha) = N_T(m\kappa, n\kappa) = \Delta_T(\alpha)$,

3) $N_T(\alpha + \iota, \beta) - N_T(\alpha, \beta + \kappa) = \sum_{r=0}^{\infty} N_T(\alpha, r\iota)\Delta_T(\beta - (r+1)\iota)$,

for every $\alpha, \beta \in \mathbf{N}^2$, $\alpha = (m,n)$.

By convention one puts $\Delta_T((p,q)) = 0$ if at least one of the entries p, q is negative.

The proof of the Proposition 4.1 is obvious, once Lemma 1.4 is known. The details are left to the reader.

Let us remark also that the equations 1), 2) and 3) above uniquely determine the kernel N_T if the function Δ_T is known. By virtue of Proposition 1.8 the kernel N_T is a complete unitary invariant for the pure operator T, hence Δ_T is also a complete unitary invariant. Similar results are true for all the other unitary invariants listed in Proposition 1.8. We conclude with the next.

4.2. Proposition. *Each of the following double sequences of operators in* $L(D_T)$:

(i) $\{D^{\frac{1}{2}}T^m T^{*n} D^{\frac{1}{2}} ;\ m, n \in \mathbf{N}\}$,

(ii) $\{D^{\frac{1}{2}}T^{*m} T^n D^{\frac{1}{2}} ;\ m, n \in \mathbf{N}\}$,

(iii) $\{D^{\frac{1}{2}}X^m Y^n D^{\frac{1}{2}} ;\ m, n \in \mathbf{N}\}$,

(iv) $\{D^{\frac{1}{2}}Y^m X^n D^{\frac{1}{2}} ;\ m, n \in \mathbf{N}\}$

is a complete unitary invariant for the pure hyponormal operator T, *where* $X = \mathbf{Re}\, T$ *and* $Y = \mathbf{Im}\, T$.

Let C be the contraction valued function which satisfies equations (8). Note that for large values of w and z in $\mathbf{C}$ we have

$$C(w)^*C(z) = D^{\frac{1}{2}}(T - w)^{-1}(T^* - \bar{z})^{-1}D^{\frac{1}{2}} = \sum_{m,n=0}^{\infty} D^{\frac{1}{2}}T^m T^{*n} D^{\frac{1}{2}} w^{(-m-1)}\bar{z}^{(-n-1)} =$$

$$= \sum_{m,n=0}^{\infty} \Delta_T((n,m))w^{(-m-1)}\bar{z}^{(-n-1)}.$$

According to Proposition 4.2 this series determines the unitary orbit of the pure hyponormal operator T, and we obtain a new proof of Proposition 3.8.

In fact we can refine the result stated in Proposition 3.8. Indeed, the uniqueness of the power series expansion in z and $\bar{z}$ of a real analytic function allows us to recuperate all operators $\Delta_T(\alpha)$, $\alpha \in \mathbf{N}^2$, from the expression

$$C(z)^*C(z) = \sum_{m,n=0}^{\infty} \Delta_T((n,m))\, z^{(-m-1)}\bar{z}^{(-n-1)}.$$

Thus we have proved the next result.

4.3. Corollary. *Let* T *be a pure hyponormal operator. Then the stalk at infinity of the function* $C^*(z)C(z)$, $z \in \mathbf{C}$, *is a complete unitary invariant for* T.

At this moment we can introduce another unitary invariant of the operator T, derived also from Proposition 4.2, namely the analytic function of two complex variables

$$E(z,w) = I - 2iD^{\frac{1}{2}}(X - z)^{-1}(Y - w)^{-1}D^{\frac{1}{2}}, \quad z, w \in \mathbf{C} \setminus \mathbf{R}.$$

The properties and the structure of this object, called the *determining function* of the operator T, will be discussed in Chapter X.

Now we state and solve the inverse problem for the invariant Δ_T as follows.

4.4. Proposition. *Let D be a Hilbert space and let* $\Delta : \mathbf{N}^2 \to L(D)$ *be given. Starting with* Δ *one constructs the function* $N : \mathbf{N}^2 \times \mathbf{N}^2 \to L(D)$, *which is related to* Δ *by the equations* 1), 2), 3) *in Proposition* 4.1.

If N *is positive definite and there is a constant* M, *so that*

$$\sum_{\alpha,\beta} \langle N(\alpha + \iota, \beta + \iota)\xi_\beta, \xi_\alpha\rangle \le M \sum_{\alpha,\beta} \langle N(\alpha,\beta)\xi_\beta, \xi_\alpha\rangle, \tag{13}$$

for any finite subset $(\xi_\alpha)_\alpha$ *of D, then there exist a Hilbert space H, a pure hyponormal operator* T *on H and an isometric embedding* $U : D_T \to D$, *so that*

$$\Delta(\alpha) = U\Delta_T(\alpha)U^*, \quad \alpha \in \mathbf{N}^2.$$

Proof. Let us first remark that the kernel N, subject to conditions 1), 2), 3), is uniquely attached to its "boundary data" Δ.

By Kolmogorov's Theorem I.5.1, there is a Hilbert space H and a factorization

$$N(\alpha,\beta) = R(\alpha)^* R(\beta), \quad \alpha, \beta \in \mathbf{N}^2,$$

where $R(\alpha) \in L(D,H)$. Moreover, we may assume

$$H = \bigvee \{R(\beta)\xi \; ; \beta \in \mathbf{N}^2, \xi \in D\}.$$

This shows in particular that the space H is separable whenever D is separable.

Condition (13) allows us to define a unique bounded operator T on H, with the property

$$TR(\beta) = R(\beta + \iota), \quad \beta \in \mathbf{N}^2.$$

Then 3) implies, by a simple computation, that

$$T^* R(\beta) = R(\beta + \kappa) + \sum_{p=0}^{\infty} R(p\iota)\Delta(\beta - (p+1)\iota), \quad \beta \in \mathbf{N}^2.$$

Accordingly, relation 2) yields

$$[T^*, T]R(\beta) = R(\theta)\Delta(\beta) = R(\theta)N(\theta,\beta) = R(\theta)R(\theta)^* R(\beta),$$

for any $\beta \in \mathbf{N}^2$. Consequently

$$[T^*, T] = R(\theta)R(\theta)^*,$$

and T is a hyponormal operator. Moreover, since $\mathbf{Ran}\, D_T = \mathbf{Ran}\, R(\theta)$ and $T^m T^{*n} R(\theta) = R(\alpha)$, where $\alpha = (m,n) \in \mathbf{N}^2$, we infer the equality

$$H = \bigvee \{T^m T^{*n} D_T \xi : \xi \in H, \quad (m,n) \in \mathbf{N}^2\}.$$

Therefore T is a pure hyponormal operator.

Let us consider the polar decomposition $R^*(\theta) = UD_T^{\frac{1}{2}}$. Then the isometry U satisfies :

$$U\Delta_T(\alpha)U^* = UD_T^{\frac{1}{2}} T^m T^{*n} D_T^{\frac{1}{2}} U^* = R^*(\theta)T^m T^{*n} R(\theta) = R^*(m\kappa)R(n\kappa) = N(m\kappa,n\kappa) = \Delta(\alpha).$$

This completes the proof of the proposition.

The inverse problems corresponding to the invariants (ii), (iii) and (iv) listed in

Proposition 4.2 are analogous to Proposition 4.4 above and are left to the reader as simple exercises.

We end this chapter by describing an operator valued distribution supported by the spectrum of a pure hyponormal operator, which is also a complete unitary invariant for this operator. We recall that the function C associated to a hyponormal operator T is anti-analytic off the spectrum $\sigma(T)$ of T.

4.5. Theorem. *The* $L(D)$*-valued distribution* $\Gamma_T = -D^{\frac{1}{2}}\,\partial C$ *is a complete unitary invariant for the pure hyponormal operator* T, *where* $D = (\mathbf{Ran}\, D)^{-}$.

Proof. Let $T \in L(H)$ and $T' \in L(H')$ be pure hyponormal operators. Consider the corresponding subspaces D and D' of H and H', and the operators D, D' and C(z), C'(z), $z \in \mathbf{C}$, defined on them, respectively.

If $U : D \to D'$ is a unitary operator such that $U(D^{\frac{1}{2}}\,\partial C)U^* = D'^{\frac{1}{2}}\,\partial C'$, then, by a Cauchy transform argument one finds

$$UD^{\frac{1}{2}}C(z)U^* = D'^{\frac{1}{2}}C'(z)$$

for all $z \in \mathbf{C}$.

By Lemma 3.5 we know that for every $\phi \in D(\mathbf{C},H)$,

$$T(\partial C)\cdot\phi = z\,\partial C\cdot\phi + \lim_{\varepsilon\downarrow 0} CD^{\frac{1}{2}}\,\partial C_{\varepsilon}\cdot\phi.$$

Therefore

$$D^{\frac{1}{2}}T^{n+1}(\partial C)\cdot\phi = zD^{\frac{1}{2}}T^{n}(\partial C)\cdot\phi + \lim_{\varepsilon\downarrow 0} D^{\frac{1}{2}}T^{n}CD^{\frac{1}{2}}\,\partial C_{\varepsilon}\cdot\phi$$

for every integer $n \geq 0$.

By an obvious induction we obtain

$$UD^{\frac{1}{2}}T^{n}C(z)U^* = D'^{\frac{1}{2}}T'^{n}C'(z)$$

for all $z \in \mathbf{C}$. Consequently, writing the series development of the function C(z) for large $|z|$, one gets

$$UD^{\frac{1}{2}}T^{n}T^{*m}D^{\frac{1}{2}}U^* = D'^{\frac{1}{2}}T'^{n}T'^{*m}D'^{\frac{1}{2}}$$

for all $n, m \in \mathbf{N}$.

The proof ends by applying Proposition 4.2.

Though being a distribution of order -1 on the spectrum $\sigma(T)$, the invariant Γ_T behaves well off the essential spectrum $\sigma_{\mathbf{ess}}(T)$.

4.6. Proposition. *Let* $T \in L(H)$ *be a pure hyponormal operator and denote* $\Omega = \mathbf{C} \setminus \sigma_{ess}(T)$. *Then*

$$\Gamma_T \mid \Omega \in E(\Omega, L(D))$$

and $\Gamma_T(z)^2 = \Gamma_T(z)$ *for every* $z \in \Omega$.

Proof. Since the operator T is pure, $\mathbf{Ker}\, T_z = 0$ for all $z \in \mathbf{C}$. For a point $z \in \Omega$ the operator $T_z^* T_z$ turns out to be invertible, hence

$$C(z) = T_z (T_z^* T_z)^{-1} D^{\frac{1}{2}}.$$

Thus $C(z)$ is a smooth function on Ω, and for this reason Γ_T is a smooth function, too, when restricted to the open set Ω.

By Lemma 3.5 we find the pointwise relation

$$T_z \, \partial C(z) = C(z) D^{\frac{1}{2}} \, \partial C(z), \quad z \in \Omega.$$

By differentiating it, we obtain

$$(14) \qquad - \partial C(z) + T_z \, \partial^2 C(z) = \partial C(z) D^{\frac{1}{2}} \, \partial C(z) + C(z) D^{\frac{1}{2}} \, \partial^2 C(z), \quad z \in \Omega.$$

From $T_z^* \, \partial C(z) = 0$ it follows that $T_z^* \, \partial^2 C(z) = 0$, thus by Lemma 3.4 one finds

$$(15) \qquad T_z \, \partial^2 C(z) = C(z) D^{\frac{1}{2}} \, \partial^2 C(z), \quad z \in \Omega.$$

By (14) and (15),

$$- \partial C(z) = \partial C(z) D^{\frac{1}{2}} \, \partial C(z), \quad z \in \Omega,$$

and finally one obtains

$$\Gamma_T(z) = \Gamma_T(z)^2, \quad z \in \Omega,$$

as desired.

Notice that $\mathbf{rank}\, \Gamma_T(z) = \mathbf{dim\, Ker}\, T_z^*$, whence

$$(16) \qquad \mathbf{Tr}\, \Gamma(z) = - \mathbf{ind}\, T_z ,$$

for all $z \in \mathbf{C} \setminus \sigma_{ess}(T)$.

Notes. The procedure of producing unitary invariants from some generating vectors, like in Propositions 1.8 and 3.8, is not specific to hyponormal operators. For instance, it also appeared to be relevant in the theories of contractive or dissipative operators, cf. Sz.-Nagy – Foiaş [1] and Gohberg – Krein [1]. Theorem 1.11 was well known to experts; for an explicit formulation see Xia [8]. Example 1 is due to Itô – Wong [1]. Example 2 was discovered by Stampfli [3]. The families of contractions C and K, associated to a hyponormal operator, were introduced by Radjabalipour [2], [3], and used by Putnam [10], Clancey - Wadhwa [1], Clancey [6], [8]. The last parts of §3 (Lemma 3.2 - Proposition 3.8) and §4 are reproduced after Martin - Putinar [1]. Previously, Clancey had proved Proposition 3.7 under the additional assumption **rank** $[T^*, T] = 1$, cf. Clancey [8].

EXERCISES

1. Let T be a hyponormal operator. If $H_p(T)$ is finite dimensional then $H_p(T) = \{0\}$.

2. Let $T \in L(H)$ be a hyponormal operator and assume $K \subset H$ is a closed invariant subspace of T. If $T|K$ is normal then K reduces T.

3*. (Putnam). Let T be a hyponormal operator. If $(\mathbf{Re}\, T)\cdot(\mathbf{Im}\, T)$ is normal then T is normal.

4. Let M be a Hilbert space. An operator T acting on $l^2(\mathbf{N})\hat{\otimes} M$ by the rule

$$T(e_n \otimes \xi) = e_{n+1} \otimes A_n \xi,$$

where $A_n \in L(M)$ and $\sup_n \|A_n\| < \infty$ is called an operatorial weighted shift with weights $(A_n)_{n=0}^{\infty}$.

a) Prove that T is hyponormal if and only if $A_{n+1}^* A_{n+1} \geq A_n A_n^*$ for every $n \geq 0$.

b) Find a simple example of hyponormal non-subnormal operatorial weighted shift ($\mathbf{dim}\, M = 2$).

5. Find two distinct singular integral operators of the form $M_t + i(M_a - M_b S M_b^*)$, with $a, b \in L^\infty[0,1]$, $a = \bar{a}$ a.e. on $[0,1]$, which are unitarily equivalent.

Hint. Use the description of the commutant of M_t in $L(L^2[0,1])$, as the set of operators M_f, with $f \in L^\infty[0,1]$.

6. Let σ be a compact subset of $\mathbf{R}$ and $a, b \in L^\infty(\sigma)$, $a = \bar{a}$ a.e. on σ. Consider the hyponormal singular integral operator $T = M_t + i(M_a - M_b S M_b^*)$ acting on $L^2(\sigma)$.

a) Prove that $\mathbf{Ker}\, M_b^*$ reduces T to a normal operator.

b) Prove that the pure space $H_p(T)$ contains $(M_t)^n b$ for every $n \geq 0$, whence $\mathbf{Ran}\, M_b \subset H_p(T)$.

c) The operator T is pure if and only if $b(t) \neq 0$ a.e. on σ.

7. (Putnam). The self-commutator of a hyponormal operator cannot be invertible.

Hint. Use the factorization $D^{\frac{1}{2}} = C^*(z)T_z$ for a point $z \in \sigma_{ap}(T)$.

8. Let T be a weighted shift acting on $l^2(\mathbf{N})$, with weights $(\alpha_n)_{n=0}^{\infty}$.

a) Prove that any weighted shift unitarily equivalent to T has weights $(\beta_n)_{n=0}^{\infty}$ such that $|\beta_n| = |\alpha_n|$, for every $n \geq 0$; and conversely, this condition characterizes the unitary orbit of T.

b) The spectrum of T is circulary symmetric, that is, if $\lambda \in \sigma(T)$, then $e^{i\phi}\lambda \in \sigma(T)$ for $\phi \in [0,2\pi]$.

c) Assume $\alpha_n \geq 0$ for all $n \geq 0$. Find the polar decomposition of T.

9. Let T be a hyponormal weighted shift with weights

$$0 < \alpha_o \leq \alpha_1 \leq \alpha_2 \leq \dots .$$

Prove that:

a) $\sigma_p(T) = \emptyset$ and T is pure.

b) If $|\lambda| < \|T\|$, then $\lambda \in \sigma_p(T^*)$ and $\dim \operatorname{Ker} T^*_\lambda = 1$.

c) $\sigma(T) = \{\lambda \in \mathbf{C} \mid |\lambda| \leq \|A\|\}$, $\sigma_{ess}(T) = \partial\sigma(T)$.

d) T has no square roots.

e) T is subnormal if and only if there exists a probability measure ν on $[0,1]$, with $1 \in \operatorname{supp}(\nu)$, such that

$$(\alpha_o \dots \alpha_n)^2 = \int_0^1 t^{2(n+1)} d\nu(t),$$

for all $n \geq 0$.

f) All powers T^k, $k \geq 0$, are hyponormal operators.

10*. (Stampfli). Let T be a subnormal weighted shift with positive weights $(\alpha_k)_{k=0}^{\infty}$.

If $\alpha_k = \alpha_{k+1}$ for some $k \geq 1$, then $\alpha_1 = \alpha_2 = \dots$.

11*. (Stampfli). If $0 < \alpha_o < \alpha_1 < \alpha_2$ are given, then there exists a subnormal weighted shift with the first three weights α_0, α_1 and α_2.

The similar question for the first four weights has, in general, a negative answer.

12*. (Sun [1]; Cowen, Lory [1]). Let T be a hyponormal weighted shift. Prove that T is unitarily equivalent to a Toeplitz operator with essentially bounded symbol if and only if the weight sequence satisfies

$$|\alpha_n|^2 = c(1 - (1 - \alpha^2)^{n+1}), \quad n \in \mathbf{N},$$

where $c > 0$ and $0 \le \alpha \le 1$.

13. Let U be a non-unitary isometry. Prove that the operator $\alpha U + \beta U^*$ is hyponormal if and only if $|\alpha| \ge |\beta|$, and it is subnormal if and only if $\beta = 0$. Give an example of a hyponormal operator T which is not subnormal, but satisfies $T^* \mathbf{Ran}\, D_T \subset (\mathbf{Ran}\, D_T)^-$. Show that $\mathbf{dim}(\mathbf{Ran}\, D_T) > 1$ for any such operator.

14. (Stochel [1]). Let λ denote the Lebesgue measure on $\mathbf{R}^n$ and let $\rho : \mathbf{R}^n \to (0,\infty)$ be a Borel measurable function. For a given invertible $\mathbf{R}$-linear operator $A : \mathbf{R}^n \to \mathbf{R}^n$, C_A denotes the composition operator

$$(C_A f)(x) = f(Ax); \quad x \in \mathbf{R}^n, f : \mathbf{R}^n \to \mathbf{C}.$$

a) The operator C_A is bounded on $L^2(\rho d\lambda)$ if and only if $\rho/\rho\circ A \in L^\infty(\lambda)$.

b) Suppose C_A is bounded on $L^2(\rho d\lambda)$. Then

$$[C_A^*, C_A]f = |\mathbf{det}\, A|^{-1}(\rho\circ A^{-1}/\rho - \rho/\rho\circ A)f, \quad f \in L^2(\rho d\lambda).$$

c) The operator C_A is (bounded) hyponormal if and only if $\rho^2 \le (\rho\circ A^{-1})(\rho\circ A)$, λ-a.e..

d)* Assume $\rho_0(x) = (2\pi)^n \exp(\|x\|^2/2)$, $x \in \mathbf{R}^n$. Then the operator C_A is bounded on $L^2(\rho_0 d\lambda)$ if and only if $\|A^{-1}\| \le 1$. Moreover, in that case C_A is subnormal if and only if A is a normal transformation of $\mathbf{R}^n$.

15. (Dibrell, Campbell [1]). Let (X,Σ,m) be a σ-finite measure space and let T_1, T_2 be two measurable transformations of X which define L^2-bounded composition operators C_1, C_2, respectively. Denote by $h_i = (dm\circ T_i^{-1})/dm$, $i = 1, 2$, the Radon - Nikodym derivatives.

Prove that, if $h_i \circ T_i \le h_j$ for $i, j = 1, 2$, then the operator $(C_1^m C_2^n)^p$ is hyponormal on $L^2(X,dm)$ for every positive integers m, n and p.

16. Let $T \in L(H)$ be a hyponormal operator and let $K : \Omega \to L(H)$ be the unique function which satisfies the conditions:

$$T_z^* = K(z)T_z\,,\quad K(z)\,|\,\mathbf{Ker}\,T_z^* = 0,\quad z \in \mathbf{C}.$$

Show that $K \in L^{\infty}(\mathbf{C}, L(H))$.

17. Let $T \in L(H)$ be a pure hyponormal operator and let Γ_T be the operator valued distribution introduced in Theorem 4.5.

For any $z \in \mathbf{C} \setminus \sigma_{\mathrm{ess}}(T)$ one defines

$$\Pi_T(z) = \lim_{n\to\infty} (\Gamma_T(z)\Gamma_T(z)^*)^{1/n}.$$

Prove that:

a) $\Pi_T(z) = \Pi_T(z)^* = \Pi_T(z)^2$.

b) $\Gamma_T(z)\Pi_T(z) = \Pi_T(z), \quad \Pi_T(z)\Gamma_T(z) = \Gamma_T(z)$.

c) $\Pi_T \in E(\mathbf{C} \setminus \sigma_{\mathrm{ess}}(T), L(H))$ and

$$\partial \Pi_T = \Pi_T \cdot \partial \Pi_T$$

where $\partial = \partial / \partial z$.

18. Under what conditions does the unitary invariant distribution Γ_T satisfy $\Gamma_T^2 = \Gamma_T$? (Try first to give a meaning to Γ_T^2!).

Chapter III

SPECTRUM, RESOLVENT AND ANALYTIC FUNCTIONAL CALCULUS

From the point of wiev of the general spectral theory of linear operators, the hyponormality condition has several important and rather unexpected consequences. Among these we mention the formula for the spectral radius, the estimates of the resolvent function, as well as other results such as the existence of the scalar extension and Dynkin's analytic functional calculus. All these reveal the richness of the spectral theory of hyponormal operators.

The aim of the present chapter is to develop, away from the more elaborated prospects on this class of operators, these elementary aspects of the analysis of hyponormal operators. Most of the results discussed in this chapter are not specific to hyponormal operators. The reader will easily find, though not always explicitly stated, the natural framework for each of them.

1. The spectrum

Let $T \in L(H)$ be a hyponormal operator, and let us denote by $\sigma(T)$, $\rho(T) = \mathbf{C} \setminus \sigma(T)$, its spectrum, respectively, its resolvent set.

The spectrum $\sigma(T)$ is a compact subset of the complex plane, contained in the ball of radius $\|T\|$ centered at 0. It has the usual decomposition

$$\sigma(T) = \sigma_p(T) \cup \sigma_r(T) \cup \sigma_c(T),$$

where

$$\sigma_p(T) = \{\lambda \in \sigma(T) \,;\, \mathbf{Ker}\, T_\lambda \neq 0\},$$

$$\sigma_r(T) = \{\lambda \in \sigma(T) \,;\, \mathbf{Ker}\, T_\lambda = 0 \text{ and } T_\lambda \text{ has closed range}\},$$

$$\sigma_c(T) = \{\lambda \in \sigma(T) \,;\, \mathbf{Ker}\, T_\lambda = 0 \text{ and } T_\lambda \text{ has not closed range}\}.$$

Above $\sigma_r(T)$ and $\sigma_c(T)$ are the *residual*, respectively the *continuous* spectra of T. The approximate point spectrum, as it was defined in §I.3, is obviously

$$\sigma_{ap}(T) = \sigma_p(T) \cup \sigma_c(T).$$

Recall that $T_\lambda = T - \lambda$, $\lambda \in \mathbf{C}$. Next we adopt the notation

$$\delta^* = \{\bar{\lambda} ; \lambda \in \delta\}$$

for any subset δ of $\mathbf{C}$.

1.1. Proposition. *Let* T *be a hyponormal operator. Then*

$$\sigma_p(T) \subset \sigma_p(T^*)^*,$$

and

$$\sigma_{ap}(T) \subset \sigma_{ap}(T^*)^*.$$

If T *is pure, then* $\sigma_p(T) = \emptyset$.

Proof. For a fixed $\lambda \in \mathbf{C}$, we have $\|T_\lambda^*\xi\| \leq \|T_\lambda\xi\|$, for every vector $\xi \in H$. Then both inclusions are obvious.

If $\lambda \in \sigma_p(T)$, then $\mathbf{Ker}\, T_\lambda$ reduces T and the restriction $T|\mathbf{Ker}\, T_\lambda$ is a normal operator.

1.2. Corollary. *Assume* T *is a hyponormal operator. Then*

$$\sigma(T) = \sigma_{ap}(T^*)^*.$$

Proof. We have only to note the following property of the residual spectrum:

$$\sigma_r(T) \subset \sigma_p(T^*)^*$$

and the proof is complete.

In what follows we denote by $|T|_{sp}$ the *spectral radius* of the operator T, that is

$$|T|_{sp} = \mathbf{sup}\{|\lambda| ; \lambda \in \sigma(T)\}.$$

Recall that it can be computed by the formula:

$$|T|_{sp} = \lim_n \|T^n\|^{1/n}.$$

1.3. Proposition. (Ando, Stampfli). *If* T *is a hyponormal operator, then*

(1) $$\| T^n \| = \| T \|^n$$

for every positive integer n.

Proof. Let $n \in \mathbf{N}$ and $\xi \in H$ be fixed. Then

$$\| T^* T^n \xi \| \leq \| T T^n \xi \| \leq \| T^{n+1} \| \, \| \xi \| ,$$

hence $\| T^* T^n \| \leq \| T^{n+1} \|$.

Assume relation (1) holds for every $p \leq n$, and let us prove it for $n + 1$. First note that

$$\| T^n \|^2 = \| T^{*n} T^n \| = \| T^{*n-1} T^* T^n \| \leq \| T^{*n-1} \| \, \| T^* T^n \| \leq \| T^{*n-1} \| \, \| T^{n+1} \| =$$
$$= \| T^{n-1} \| \, \| T^{n+1} \| .$$

Since $\| T^{n-1} \| = \| T \|^{n-1}$ and $\| T^n \| = \| T \|^n$ we get $\| T \|^{n+1} \leq \| T^{n+1} \|$. The converse inequality being obvious, the proof is complete.

1.4. Corollary. *If* T *is hyponormal, then* $| T |_{sp} = \| T \|$.

1.5. Corollary. *The only quasi-nilpotent hyponormal operator is the zero operator.*

1.6. Corollary. *Every compact hyponormal operator is normal.*

Proof. Assume T is pure hyponormal and compact. Then $\sigma(T) = \sigma_p(T) \cup \{0\} = \{0\}$, whence T is quasi-nilpotent. By Corollary 1.5, T = 0 and the proof is over.

Other deep properties of the spectrum of a hyponormal operator will be described in Chapter VI as consequences of Putnam's inequality.

The reader will easily find examples of pure hyponormal operators T (even subnormal) such that $\sigma_p(T^*)$ is non-empty. Moreover, the set of eigenvectors of T^* tends to span the pure space of T, at least in a generalized sense, see § II.3.

2. Estimates of the resolvent function

The resolvent of a hyponormal operator has a specific behaviour which lies at the origin of several local and global spectral properties. We discuss below two such results.

2.1. Proposition. *Let* λ *be a point in the resolvent set of the hyponormal operator* T. *Then*

(2) $$\|(T - \lambda)^{-1}\| = 1/\mathbf{dist}(\lambda, \sigma(T)).$$

Proof. Let $\alpha \in \sigma(T_\lambda^{-1})$. By the Spectral Mapping Theorem we infer $\beta = \alpha^{-1} + \lambda \in \sigma(T)$, and conversely, every $\beta \in \sigma(T)$ is of the form $\alpha^{-1} + \lambda$, with $\alpha \in \sigma(T_\lambda^{-1})$. Therefore

$$|T_\lambda^{-1}|_{sp} = \mathbf{max}\{|\alpha| ; \alpha \in \sigma(T_\lambda^{-1})\} = (\mathbf{min}\{|\lambda - \beta| ; \beta \in \sigma(T)\})^{-1} = \mathbf{dist}(\lambda, \sigma(T))^{-1}.$$

But T_λ^{-1} is still a hyponormal operator (for instance from the factorization II (6)), so that

$$\|T_\lambda^{-1}\| = |T_\lambda^{-1}|_{sp} = \mathbf{dist}(\lambda, \sigma(T))^{-1},$$

as desired.

Hyponormal operators share the estimate (2) of the growth of the resolvent with other distinguished classes of operators. The next section deals with an application of the growth (2).

Now we turn our attention to an open-local behaviour of the linear function T_z, rather than a pointwise estimate as (2) was.

Let Ω be a bounded open subset of **C**. Recall that $L^2(\Omega, H)$ denotes the Hilbert space of H-valued square summable functions defined on Ω, with respect to the planar Lebesgue measure μ. For any positive integer m we put by definition

$$W_{\bar\partial}^m(\Omega, H) = \{f \in L^2(\Omega, H) ; \bar\partial^k f \in L^2(\Omega, H), \quad k \leq m\}.$$

Here $\bar\partial = \partial/\partial\bar z$ and the derivatives are taken in the sense of distributions. Endowed with the norm

$$\|f\|^2_{W_{\bar\partial}^m} = \sum_{k=0}^{m} \|\bar\partial^k f\|^2_{2,\Omega},$$

the space $W_{\bar\partial}^m(\Omega, H)$ becomes a Hilbert space. The reader will easily compare it with, and distinguish it from, the usual Sobolev space.

When the boundary $\partial\Omega$ is smooth, the functions belonging to $W_{\bar\partial}^m(\Omega, H)$ can be approximated, by standard arguments, with smooth functions defined in neighbourhoods of $\bar\Omega$.

Let us finally denote $P : L^2(\Omega, H) \to A^2(\Omega, H)$ to be the vector-valued analog of the Bergman projection, where

$$A^2(\Omega,H) = \{f \in L^2(\Omega,H) \; ; \; \bar{\partial}f = 0 \text{ on } \Omega\}.$$

2.2. Proposition. *Let* $\Omega \subset \mathbf{C}$ *be a bounded domain with smooth boundary. There exists a constant* C_Ω*, such that for every hyponormal operator* $T \in L(H)$ *and every function* $f \in W^2_{\bar{\partial}}(\Omega,H)$ *one has*

(3) $$\|(I - P)f\|_{2,\Omega} \leq C_\Omega(\|T_z \bar{\partial}f\|_{2,\Omega} + \|T_z \bar{\partial}^2 f\|_{2,\Omega}).$$

Proof. Let $f_n \in E(\bar{\Omega},H)$ be a sequence of smooth functions which approximates f in the norm of the space $W^2_{\bar{\partial}}(\Omega,H)$.

For a fixed $n \in \mathbf{N}$ we have

$$\bar{\partial}(f_n(z) + T_z^* \bar{\partial}f_n(z)) = T_z^* \bar{\partial}^2 f_n(z), \quad z \in \Omega.$$

The Cauchy-Pompeiu formula yields:

$$f_n(z)+T_z^* \bar{\partial}f_n(z)-(1/2\pi i) \int_{\partial\Omega}(f_n(\zeta) + T_\zeta^* \bar{\partial}f_n(\zeta))(d\zeta/(\zeta - z)) = -(1/\pi)\int_\Omega T_\zeta^* \bar{\partial}^2 f_n(\zeta)(d\mu(\zeta)/(\zeta - z)),$$

for every $z \in \Omega$.

Let F_n denote the contour integral above. Then $F_n \in A^2(\Omega,H)$, so that

$$\|F_n - f_n\|_{2,\Omega} \leq \|T_z^* \bar{\partial}f_n\|_{2,\Omega} + \|T_z^* \bar{\partial}^2 f_n\|_{2,\Omega} \cdot \int_{\Omega-\Omega} d\mu(\zeta)/\pi|\zeta|.$$

The second integral was majorized as a convolution with an L^1-function. Take

$$C_\Omega = \mathbf{max}(1 \, , (1/\pi) \int_{\Omega-\Omega} d\mu(\zeta)/|\zeta|).$$

Then

$$\|f - Pf\|_2 \leq \|f - f_n\|_2 + \|f_n - F_n\|_2 \leq$$

$$\leq \|f - f_n\|_2 + C_\Omega(\|T_z^* \bar{\partial}f_n\|_2 + \|T_z^* \bar{\partial}^2 f_n\|_2).$$

By passing to the limit one finds

$$\|f - Pf\|_2 \leq C_\Omega(\|T_z^* \bar{\partial}f\|_2 + \|T_z^* \bar{\partial}^2 f\|_2).$$

Since $\|T_z^* g\|_2 \leq \|T_z g\|_2$ for every function $g \in L^2(\Omega,H)$, the proof is complete.

The last proposition will be the main tool in the construction of a generalized scalar extension of a hyponormal operator, as we shall see in Section 4 below.

Notice finally that Proposition 2.2 still holds for an *M-hyponormal* operator T (i.e. by definition $\|T_z^* \xi\| \leq M\|T_z \xi\|$ for every $z \in \mathbf{C}$), with the constant C_Ω replaced by MC_Ω.

3. A sharpened analytic functional calculus

The growth (2) of the resolvent function of a hyponormal operator leads, via Dynkin's theory of quasi-analytic extensions (cf. the subsequent supplement), to a functional calculus with certain complex-differentiable functions defined only on the spectrum of the operator. Although the method of constructing this calculus is general, we present it only in our specific case. As a corollary we obtain a best estimate, which is analogous to the estimate established for subnormal operators in Corollary I.3.5.

3.1. Lemma. *Let σ be a perfect compact subset of* $\mathbf{C}$. *For every* $1 > \varepsilon > 0$ *there exists a system of piecewise smooth Jordan curves* γ_ε, *surrounding* σ, *and such that*

(4) $$\mathbf{dist}(z,\sigma) < \varepsilon \quad \text{for } z \in \gamma_\varepsilon,$$

(5) $$\mathbf{length}(\gamma_\varepsilon) \leq C \cdot \varepsilon^{-1}$$

where C *is a positive constant depending on* σ.

Proof. Let $\sigma_{\varepsilon/2} = \{z \in \mathbf{C} \,;\, \mathbf{dist}(z,\sigma) \leq \varepsilon/2\}$. We pave the compact set $\sigma_{\varepsilon/2}$ with squares of dimension $\varepsilon/4 \times \varepsilon/4$, and put Ω_ε for the union of those squares which intersect $\sigma_{\varepsilon/2}$. Then $\gamma_\varepsilon = \partial\Omega_\varepsilon$ is the desired curve.

Indeed, $\gamma_\varepsilon \subset \sigma_\varepsilon$ and, if n denotes the number of squares appearing in Ω_ε, one finds

$$n(\varepsilon^2/16) \leq \mu(\sigma_\varepsilon) \leq \mu(\sigma_1),$$

so that

$$\mathbf{length}(\gamma_\varepsilon) \leq n\varepsilon \leq 16\mu(\sigma_1)/\varepsilon.$$

Let us consider a hyponormal operator $T \in L(H)$ and a bounded open set Ω with piecewise smooth boundary, containing $\sigma(T)$. Recall that Riesz – Dunford analytic functional calculus is defined by the formula

$$f(T) = (i/2\pi) \int_{\partial\Omega} f(\zeta)T_\zeta^{-1}d\zeta,$$

where $f \in \mathcal{O}(\overline{\Omega})$ is an analytic function defined in a neighbourhood of $\overline{\Omega}$.

The operator f(T) does not depend on the curve $\partial\Omega$, and the map $f \mapsto f(T)$ is linear, multiplicative and continuous in the following sense:

$$\|f(T)\| \le C_{\Omega,T}\|f\|_{\infty,\Omega}, \quad f \in O(\overline{\Omega}).$$

Notice that the constant $C_{\Omega,T}$ depends on Ω, too.

Assume that the function f is of the form

$$f(z) = \int \phi(\zeta)(d\mu(\zeta)/(\zeta - z)), \tag{6}$$

where ϕ is continuous and compactly supported on $\mathbf{C}$, so that

$$\mathbf{supp}\,(\phi) \cap \overline{\Omega} = \emptyset.$$

By Cauchy's formula at ∞ we find

$$f(T) = (i/2\pi)\int_{\partial\Omega} f(\zeta)T_\zeta^{-1}d\zeta = (i/2\pi)\int_{\partial\Omega}\int_{\mathbf{C}} (\phi(w)/(w - \zeta))d\mu(w)T_\zeta^{-1}d\zeta = -\int\phi(w)T_w^{-1}d\mu(w).$$

This formula suggests the following extension of the standard functional calculus. Let us pick $\alpha > 2$ and consider a continuous function ϕ, compactly supported on $\mathbf{C}$, such that

$$|\phi(\zeta)| \le C_\phi\, \mathbf{dist}\,(\zeta, \sigma(T))^{\alpha-1}, \quad \zeta \in \mathbf{C}, \tag{7}$$

with a suitable constant C_ϕ which depends on ϕ. Then we define

$$f(T) = -\int\phi(w)T_w^{-1}d\mu(w),$$

where the function f is related to ϕ via (6).

The function $\phi \cdot T_w^{-1}$ extends continuously across $\sigma(T)$ because of the estimate (2) of the growth of the resolvent.

Thus with notations introduced in Section 6 below, we can state the next result.

3.2. Theorem. (Dynkin). *Let* $T \in L(H)$ *be a hyponormal operator. For every* $\alpha > 2$ *there exists a continuous, unital morphism of Banach algebras*

$$\Phi : A^\alpha(\sigma(T)) \to L(H),$$

which extends the Riesz – Dunford functional calculus.

Proof. Every function $f \in A^\alpha(\sigma(T))$ can be represented in virtue of Theorem 6.3 like

$$f(z) = \int\phi(\zeta)(d\mu(\zeta)/(\zeta - z)), \quad z \in \sigma(T),$$

where $\phi \in C_o(\mathbf{C})$ and the estimate (7) holds. Put

$$\Phi(f) = -\int \phi(w) T_w^{-1} d\mu(w).$$

Then $\Phi(f)$ does not depend on the choice of ϕ in the representation of f. Indeed, assume that

$$\int \phi(\zeta)(d\mu(\zeta)/(\zeta - z)) = 0$$

for any $z \in \sigma(T)$. Let $\varepsilon > 0$ be small and consider the system γ_ε of Jordan curves as described in Lemma 3.1. Then by Stokes' theorem one finds

$$I_\varepsilon := \int_{\mathbf{C}\setminus\Omega_\varepsilon} \phi(w) T_w^{-1} d\mu(w) = (1/2\pi i) \int_{\partial\Omega_\varepsilon} f(w) T_w^{-1} dw.$$

By Theorem 6.2 we infer

$$|f(\zeta)| \leq C_\phi \, \mathbf{dist}(\zeta, \sigma(T))^\alpha.$$

Then Lemma 3.1 and estimate (2) yield

$$\| I_\varepsilon \| \leq C_\phi \sup_{\zeta \in \partial\Omega_\varepsilon} |\mathbf{dist}(\zeta, \sigma(T))^{\alpha-2}| \leq C_\phi \varepsilon^{\alpha-2}.$$

Accordingly

$$\Phi(f) = \lim_{\varepsilon \downarrow 0} I_\varepsilon = 0,$$

as desired.

The multiplicativity of the map Φ is a standard matter.

Finally, the estimate

$$\| \Phi(f) \| \leq \text{const.} \| f \|_{A^\alpha(\sigma(T))}$$

follows from Theorem 6.3, and the proof is complete.

3.3. Corollary. *Let* T *be a hyponormal operator. For each* $\alpha > 2$ *there exists a constant* C_α*, so that for every function* $f \in O(\sigma(T))$:

$$\| f(T) \| \leq C_\alpha \| f \|_{\mathbf{Lip}(\alpha, \sigma(T))}.$$

Proof. The norm of the space $\mathbf{Lip}(\alpha, \sigma)$ majorizes, on analytic functions defined in neighbourhoods of σ, that of the space $A^\alpha(\sigma)$.

We shall see later that the previous Lipschitz estimate of the functional calculus cannot be replaced by the uniform estimate as in the case of subnormal operators. The next paragraph provides a second explanation for Corollary 3.3.

4. Generalized scalar extensions

The aim of this paragraph is to exploit the second inequality concerning the operator valued function T_z, as it was stated in Proposition 2.2. Similarly to subnormal operators, a hyponormal operator turns out to be the restriction to a closed invariant subspace of an operator behaving in many respects like a normal operator. More specifically, this extension is a so-called C^2-scalar operator, i.e. an operator which possesses a continuous functional calculus with functions of order C^2. This class of operators is well understood, a monograph by Colojoară and Foiaş [1] being devoted to them .

4.1. Definition. An operator $T \in L(H)$ is called *generalized scalar* if there exists a unital, continuous morphism of topological algebras

$$\Phi_T : E(\mathbf{C}) \longrightarrow L(H),$$

such that $\Phi_T(z) = T$.

The map Φ_T is called a *spectral distribution* of T and it is not necessarily unique. The support of Φ_T is exactly the spectrum $\sigma(T)$. Thus, because of the continuity of Φ_T, there exists an integer $r \in \mathbf{N}$ so that

$$\| \Phi_T(f) \| \leq \text{const.} \| f \|_{C^r(\bar{D})}$$

for every function $f \in E(\mathbf{C})$ and for a disk D, with $D \supset \sigma(S)$. Then the distribution Φ extends its domain to the space $C^r(\bar{D})$ and the operator T is called C^r*-scalar*.

Every normal operator is obviously C^o-scalar. The reader will easily find examples of generalized scalar operators among the multipliers on function spaces like the Sobolev or Lipschitz spaces and the L^p-spaces. We confine ourselves to discuss only the next example.

Example 1. *A subnormal generalized scalar operator*

Let Ω be a bounded domain of the complex plane and let m be a fixed natural number. Consider the anisotropic Sobolev space $W^m_{\bar{\partial}}(\Omega)$ introduced in Section 2 and the multiplication operator $M = M_z$ on it.

Since the map

$$W^m_{\bar{\partial}}(\Omega) \longrightarrow \bigoplus_0^m L^2(\Omega),$$

$$\phi \longmapsto (\phi, \bar{\partial}\phi, \dots, \bar{\partial}^m\phi)$$

is an isometrical embedding which intertwines the operators M_z and $\bigoplus_o^m M_z$, the operator M turns out to be subnormal.

On the other hand the map

$$\Phi_M : E(\mathbf{C}) \to L(W^m_{\bar\partial}(\Omega)), \quad \Phi_M(f) = M_f,$$

is a spectral distribution for M, of order m. Thus the operator M is C^m-scalar.

We left to the reader to check that $\sigma(M) = \bar\Omega$ and that M is not normal.

4.2. Definition. The restriction of a generalized scalar operator to a closed invariant subspace is simply called a *subscalar operator.*

4.3. Theorem. *Every hyponormal operator is subscalar.*

Proof. Let Ω be a bounded domain of the complex plane, with smooth boundary and which contains the spectrum of the hyponormal operator T.

The operator T_z naturally acts on the space $W^2_{\bar\partial}(\Omega, H)$. Let us consider the space

$$W_T(\Omega) = W = W^2_{\bar\partial}(\Omega, H) \ominus (\mathbf{Ran}\, T_z)^-,$$

and let Q denote the orthogonal projection of $W^2_{\bar\partial}(\Omega, H)$ onto W.

Let $M = M_z$ be the multiplication operator with z on $W^2_{\bar\partial}(\Omega, H)$. This is a vector valued analogue of Example 1, and the reader will easily check that M is still a C^2-scalar subnormal operator. As above, its spectral distribution is

$$\Phi : E(\mathbf{C}) \to L(W^2_{\bar\partial}(\Omega, H)), \quad \Phi(f) = M_f.$$

Let $S = QM \mid W$ be the compression of the operator M to the space W. Since $(\mathbf{Ran}\, T_z)^-$ is invariant under every operator M_f, $f \in E(\mathbf{C})$, we infer that S is a C^2-scalar operator with spectral distribution $\Phi_S = Q\Phi \mid W$.

We shall prove that S is the desired extension.

Let us consider the natural map

$$V : H \to W_T(\Omega)$$

$V\xi = Q(1 \otimes \xi)$, $\xi \in H$, where $\mathbf{1} \otimes \xi$ stands for the constant function identically equal to ξ. Notice that

(8) $$VT = SV.$$

Indeed, for every $\xi \in H$ one has $VT\xi = Q(\mathbf{1} \otimes T\xi) = Q(z \otimes \xi) = QM(\mathbf{1} \otimes \xi) = QMQ(1 \otimes \xi) = SV\xi$.

Thus it remains to prove that V is a one-to-one operator with closed range. Let $(\xi_n)_n \subset H$ and $(f_n)_n \subset W^2_{\bar\partial}(\Omega, H)$ be two sequences with the property

(9) $$\lim_n \| T_z f_n + 1 \otimes \xi_n \|_{W^2_{\bar\partial}} = 0.$$

The proof will be finished if we prove that $\lim_n \xi_n = 0$.

By assumption (9) we infer

$$\lim_n (\| T_z \bar\partial f_n \|_{2,\Omega} + \| T_z \bar\partial^2 f_n \|_{2,\Omega}) = 0,$$

therefore, in virtue of Proposition 2.2,

$$\lim_n \| (I - P) f_n \|_{2,\Omega} = 0.$$

Recall that P denotes the orthogonal projection onto $A^2(\Omega, H)$.

Then by using (9) again one finds

$$\lim_n \| T_z P f_n + 1 \otimes \xi_n \|_{2,\Omega} = 0.$$

Let Ω' be a relatively compact domain of Ω, containing $\sigma(T)$. Consequently there exists a constant $C > 0$ with the property

$$\| g \|_{\infty,\Omega'} \leq C \| g \|_{2,\Omega}$$

for every function $g \in A^2(\Omega, H)$. Accordingly

$$\lim_n \| T_z P f_n + 1 \otimes \xi_n \|_{\infty,\Omega'} = 0.$$

By the continuity of the map

$$\Psi : O(\Omega) \hat\otimes H \to H, \quad \Psi(f \otimes \xi) = f(T)\xi$$

induced by the Riesz-Dunford functional calculus, we derive the inequality

$$\| \xi_n \| = \| \Psi(T_z P f_n + 1 \otimes \xi_n) \| \leq \text{const.} \| T_z P f_n + 1 \otimes h_n \|_{\infty,\Omega'},$$

and the proof is complete.

By noticing the precise form of the C^2-scalar extension S of the operator T, we can state the following.

4.4. Corollary. *Every hyponormal operator* T *is similar to the compression of the subnormal operator* $M_z \in L(W^2_{\bar\partial}(\Omega, H))$ *to a semi-invariant subspace.*

In other words, T is similar to the marked entry in a suitable block decomposition of the operator M_z:

$$M_z = \begin{bmatrix} * & * & * \\ 0 & \boxed{*} & * \\ 0 & 0 & * \end{bmatrix}$$

Next we study some of the properties of the C^2-scalar extension S of the operator T. First we point out the dependence of the definition of S on the domain Ω. However, it can be proved that only the Hilbert space structure of the space W depends on Ω, not its Fréchet space structure. This pathology is reminiscent from the special Sobolev type space we worked with.

4.5. Proposition. *The generalized scalar extension* $S \in L(W_T(\Omega))$ *of the hyponormal operator* $T \in L(H)$ *has the following properties:*

(i) $\sigma(S) = \sigma(T)$;

(ii) *The linear span of the vectors* $\Phi_S(\phi)\xi$, *where* $\phi \in E(\mathbf{C})$ *and* $\xi \in H$, *is dense in* $W_T(\Omega)$;

(iii) *If* $A \in L(H)$ *and* T_1, T_2 *is a pair of hyponormal operators on H such that* $AT_1 = T_2A$, *then there is an extension* $A' \in L(W_1, W_2)$ *of* A, *with the property* $A'S_1 = S_2A'$, *where* S_1, S_2 *are the scalar extensions of the operators* T_1, T_2, *correspondingly.*

Proof. (i) Assume $\lambda \notin \sigma(T)$. Since we may choose the open set Ω disjoint of λ, one finds the operator S_λ to be invertible. Secondly, it is plain that this algebraic property of S is independent of Ω, whence $\lambda \notin \sigma(S)$.

Since S is an extension of the operator T, $\sigma_{ap}(T) \subset \sigma_{ap}(S)$.

Let us assume $\lambda \in \sigma_r(T) \setminus \sigma(S)$. Since the set $\sigma_r(T)$ is open in $\sigma(T)$ (by the stability of the class of semi-Fredholm operators under small perturbations), there is an open neighbourhood U of λ, such that $U \subset \sigma_r(T) \setminus \sigma(S)$.

Take a vector $\xi \in H \ominus (\mathbf{Ran}\, T_\lambda)^-$, $\xi \neq 0$. Then there exists an analytic function $f \in O(U,W)$, with the property

$$\mathbf{1} \otimes \xi = S_w f(w), \quad w \in U.$$

In other terms

$$\xi - (z - w) f(w,z) \in (\mathbf{Ran}\, T_z)^-,$$

for every $w \in U$.

In virtue of the nuclearity of the space $O(U)$ there exists a sequence of functions $g_n \in O(U, W^2_{\bar\partial}(\Omega, H))$, such that

(10) $$\xi - (z - w)f(w,z) = \lim_n T_z g_n(w,z),$$

in the Fréchet topology of the space $O(U, W^2_{\bar\partial}(\Omega, H))$.

But, for any point $w \in U$, the evaluation at w is a continuous operator on $W^2_{\bar\partial}(\Omega, H)$ (by Sobolev's Embedding Theorem; see Reed and Simon [2]). Therefore, after taking the restriction to the diagonal $z = w$ of $U \times U$ in (10), we obtain

$$\xi = \lim_n T_\lambda g_n(\lambda,\lambda).$$

This contradicts our assumption, and the proof of equality (i) is over.

(ii) This is a consequence of the density of the subspace $E(\bar\Omega)$ in $W^2_{\bar\partial}(\Omega)$.

(iii) The extension A' of the operator A can be easily derived from the definitions of S_1 and S_2.

5. Local spectral properties

A by now classical subject of the general theory of linear opeators is the so called local spectral theory. It consists, roughly speaking, in localizing the spectrum and the functional calculus of an operator $T \in L(H)$ with respect to the vectors of the space H. For instance, if T is the multiplier M_z on $H = W^2_{\bar\partial}(\Omega)$, then, though $\sigma(T) = \bar\Omega$ (see Example 1), the vectors $f \in H$ having their support contained in a given compact subset $F \subset \bar\Omega$, form an invariant subspace $H(F)$ of T so that $\sigma(T|H(F)) \subset F$. Such invariant subspaces which localize T are sufficient for filling up the whole space H when the corresponding F's cover $\bar\Omega$, and so on. For an account on these topics we refer to Dunford and Schwartz [3] and Vasilescu [1].

It is quite obvious that there are operators which may behave much worse than M_z in the above discussed example. A minimal condition required for localizing spectra and functional calculus is Bishop's condition (β). As a consequence of the existence of a scalar extension (Theorem 4.3) we prove in this section that hyponormal operators fulfil this condition.

For the convenience of the reader we recall in the first part of the present section some terminology and known facts from local spectral theory.

5.1. Definition. An operator $T \in L(H)$ is said to satisfy *Bishop's condition* (β) if for any open subset $U \subset \mathbf{C}$, the application

$$T_z : O(U,H) \to O(U,H)$$

is one-to-one with closed range.

In other terms, condition (β) means that, for any open set U, and any sequence of analytic functions $f_n \in O(U,H)$, $\lim_n (f_n) = 0$ whenever $\lim_n (T - z)f_n = 0$. In particular, $T_z g = 0$ if and only if $g = 0$, where $g \in O(U,H)$.

Every vector $\xi \in H$ has a minimal spectrum with respect to the operator T having property (β), described as follows. Let us consider the maximal analytic extension $r_\xi(z)$ of the function $T_z^{-1}\xi$, subject to the condition

$$T_z r_\xi(z) = \xi.$$

It is unique in virtue of assumption (β). The complement of the domain of r_ξ is the *local spectrum* $\sigma_T(\xi)$ of T at ξ.

It is plain that $\sigma_T(\xi) \subset \sigma(T)$ for any ξ. There are simple examples of operators T with property (β) (for instance the unilateral shift U_+) such that $\sigma_T(\xi) = \sigma(T)$ for every vector $\xi \in H$, $\xi \neq 0$. In this case the spectrum of T cannot be localized effectively.

Next we assume only that the operator T possesses property (β). By an easy exercise one checks that the space

$$H_T(F) = \{\xi \in H \ ; \sigma_T(\xi) \subset F\}$$

is a linear invariant subspace of T. It is called the *maximal spectral space* of the operator T, corresponding to the set $F \subset \mathbf{C}$.

5.2. Lemma. *If* $T \in L(H)$ *has property* (β), *then* $H_T(F)$ *is a closed subspace, for every closed set* $F \subset \mathbf{C}$.

Proof. Let $\xi_n \in H_T(F)$ be a convergent sequence in H with the limit ξ. Then

$$\lim_{n,m} T_z(r_{\xi_n} - r_{\xi_m}) = \lim_{n,m} (\xi_n - \xi_m) = 0,$$

so that, by (β), (r_{ξ_n}) is a Cauchy sequence in the Fréchet space $O(\mathbf{C}\setminus F,H)$.

Since the limit $r = \lim_n r_{\xi_n}$ is still analytic on $\mathbf{C} \setminus F$ and satisfies $T_z r = \xi$ for $z \notin F$, we infer $\xi \in H_T(F)$ and the proof is complete.

The restricted operator $T|H_T(F)$ has the spectrum contained in $F \cap \sigma(T)$, hence the analytic functional calculus localizes too. More specifically, expressions like

$$f(T)\xi$$

make sense for any $\xi \in H_T(F)$ and $f \in O(F)$.

If K is an invariant subspace of T, then

$$K \subset H_T(\sigma(T|K)),$$

a fact which justifies the name of maximal spectral space.

5.3. Definition. An operator $T \in L(H)$ is called *decomposable* if for every pair of open sets which cover $\sigma(T)$:

$$\sigma(T) \subset U_1 \cup U_2,$$

there are closed invariant subspaces K_1 and K_2 of T, with the properties

$$H = K_1 + K_2$$

and

$$\sigma(T|K_i) \subset U_i, \quad i = 1, 2.$$

This axiomatic definition markes out the largest class of operators with good spectral decompositions. The reader will easily verify that normal or generalized scalar operators are decomposable. For our purposes we need only the next result.

5.4. Proposition. *Every decomposable operator has property* (β).

Proof. Notice first that condition (β) is local. This means, for instance, that it suffices to be checked only on small discs.

Let D be a bounded disc centered at λ, with radius $r > 0$, and let $f_n \in O(D,H)$ be a sequence with the property $\lim_n T_z f_n = 0$. We have to prove that $\lim_n f_n = 0$. It is sufficient to check that $\lim_n \| f_n \|_{\infty, D_1} = 0$, where D_1 is a disc centered at λ, with radius $r_1 < r$.

Consider a decomposition $H = K + L$ of H into T-invariant subspaces, so that

$$\sigma(T|K) \subset D, \quad \sigma(T|L) \subset \mathbf{C} \setminus D_1.$$

Since the Fréchet space O(D) is nuclear, the sum map

$$O(D,K) \oplus O(D,L) \to O(D,H)$$

is onto. Pick a lifting $f_n = g_n + h_n$, with $g_n \in O(D,K)$, $h_n \in O(D,L)$ for every $n \geq 0$.

Let $\pi : H \to H/L$ denote the projection map. Because $H/L \simeq K/K \cap L$, the spectrum of the quotient operator T^L induced by T on H/L lies into D. But

$$\lim_n T_z^L \pi(g_n) = 0,$$

whence

$$\lim_n \pi(g_n) = 0,$$

uniformly on the compact subset of $D \setminus \sigma(T^L)$. By the maximum principle for analytic functions we deduce $\lim_n \pi(g_n) = 0$ in the topology of the space $O(D,H/L)$.

By invoking again the nuclearity of $O(D)$, we find a sequence $g'_n \in O(D,K \cap L)$, such that $\lim_n (g_n - g'_n) = 0$.

As $\sigma(T^L) \cap D_1 = \emptyset$, from $\lim_n T_z(g'_n + h_n) = 0$ we conclude $\lim_n (g'_n + h_n) = 0$. Finally $\lim_n f_n = \lim_n ((g_n - g'_n) + (g'_n + h_n)) = 0$ and the proof is complete.

5.5. Theorem. *Every hyponormal operator has property* (β).

Proof. Since condition (β) is transmitted from an operator to its restrictions to closed invariant subspaces, we are reduced by Theorem 4.3 to the case of a C^2-scalar operator.

But every C^2-scalar operator is decomposable, and one applies Proposition 5.4.

5.6. Corollary. *Let* $T \in L(H)$ *be a hyponormal operator. For any open set* $U \subset \mathbf{C}$ *and any sequence of smooth functions* $f_n \in E(U,H)$, $\lim_n T_z f_n = 0$ *if and only if* $\lim_n f_n = 0$, *both limits being in the Frechet space* $E(U,H)$.

Proof. One uses Theorem 5.5 and Proposition 2.2.

Thus, hyponormal operators are suitable for localizing the spectrum. However, it may happen as in the case of the shift U_+, that all local spectra are rigid and equal to the whole spectrum.

6. SUPPLEMENT: Pseudo-analityc extensions of smooth functions

In this complement we summarize, after Dynkin [2], the main results of his theory of pseudoanalytic extensions of complex differentiable functions. This theory is the complex analogue of the classical theory of extensions of real smooth functions initiated by Whitney, see Stein [1].

Let K be a perfect compact subset of the complex plane and let α be a positive, non-integer real. Take $k \in \mathbf{N}$ to be the integral part of α, that is $k < \alpha < k + 1$.

The space $A^\alpha(K)$ is defined as the set of those $(k + 1)$-tuples of continuous functions on K:

$$(f^{(0)}, \ldots, f^{(k)}) : K \to \mathbf{C}^{k+1},$$

which are related by the following conditions:

$$f^{(j)}(\zeta) = f^{(j)}(z) + ((\zeta-z)/1!)f^{(j+1)}(z) + \ldots + ((\zeta-z)^{k-j}/(k-j)!)f^{(k)}(z) + R_j(z,\zeta)$$

and

$$|R_j(\zeta,z)| \leq \text{const.}|\zeta - z|^{\alpha-j}, \tag{11}$$

for every $0 \leq j \leq k$ and $z, \zeta \in K$.

Endowed with the norm given by the maximum of the best constants in the estimates (11), plus the uniform norm of $f^{(0)}$ on K, the space $\mathbf{A}^{\alpha}(K)$ becomes a commutative unital Banach algebra.

Since K has no isolated points, the functions $f^{(j)}$ are recurrently determined by the rule

$$f^{(j)}(z) = \lim_{\zeta \to z}(f^{(j-1)}(z) - f^{(j-1)}(\zeta))/(z - \zeta), \quad 0 \leq j \leq k-1,$$

hence by $f^{(0)}$. Notice that $f^{(0)}$ is an analytic function in the interior points of K.

In fact $\mathbf{A}^{\alpha}(K)$ is a closed subalgebra of the Banach algebra $\mathbf{Lip}(\alpha,K)$ of all Lipschitz functions of order α on K, namely that of formal analytic functions. More specifically, an element $f \in \mathbf{Lip}(\alpha,K)$ belongs to $\mathbf{A}^{\alpha}(K)$ if and only if the Taylor expansion of f at every point of K is complex analytic.

Dynkin's idea is to relate the smothness properties of a function $f \in C(K)$ to the growth of $\bar\partial F$ near the outer boundary of K, where F is a privileged extension of f to the whole plane.

Next we state three basic results concerning this relationship.

6.1. Theorem. *Let K be a perfect compact of* $\mathbf{C}$ *and let* $\alpha > 0$ *be a non-integer.*

There exists a constant C, so that every function $f \in \mathbf{A}^{\alpha}(K)$ *extends to a continuous function* $F \in C(\mathbf{C})$, *with the following properties:*

$$|F(z)| \leq C(1 + \mathbf{dist}(z,K)^{\alpha})\|f\|_{\mathbf{A}^{\alpha}(K)},$$

and

$$|\bar\partial F(z)| \leq C\, \mathbf{dist}(z,K)^{\alpha-1}\|f\|_{\mathbf{A}^{\alpha}(K)},$$

for every $z \in \mathbf{C} \setminus K$.

The proof consists in regularizing the remainder $R_k(z,\cdot)$ together with the distance function to K (compare with Stein [1, Chapter VI]).

6.2. Theorem. *Let* K *and* α *be as above. Assume that* F *is a continuously differentiable function on* **C**, *such that* $F|K = 0$ *and*

$$|\bar{\partial}F(z)| \leq \mathbf{dist}(z,K)^{\alpha}.$$

Then there is a universal constant C *such that*

$$|F(z)| \leq C\,\mathbf{dist}(z,K)^{\alpha+1}.$$

The proof relies on a sharp estimate of a Cauchy transform.

6.3. Theorem. *Let* K *and* α *be as above and take* $f \in C(K)$. *The following assertions are equivalent:*

(i) $f \in \boldsymbol{A}^{\alpha}(K)$;

(ii) f *admits an extension* $F \in C^1(\mathbf{C} \setminus K)$, *with the property*

$$|\bar{\partial}F(z)| \leq \text{const.}\mathbf{dist}(z,K)^{\alpha-1}, \quad z \notin K;$$

(iii) *There exists* $\phi \in C_0(\mathbf{C})$, *such that*

$$f(z) = \int(\phi(\zeta)/(\zeta - z))d\mu(\zeta),$$

and

$$|\phi(\zeta)| \leq \text{const.}\,\mathbf{dist}(\zeta,K)^{\alpha-1}, \quad \zeta \notin K.$$

The proof of the theorem reveals that the function ϕ above may be chosen so that

$$|\phi(\zeta)| \leq C\,\mathbf{dist}(\zeta,K)^{\alpha-1}\,\|f\|_{\boldsymbol{A}^{\alpha}(K)},$$

where C is a universal constant.

The above theorems are proved in the paper of Dynkin for a large variety of complex differentiable functions, like, for instance, for regularly weighted Hölder and Carleman classes.

Notes. Two pioneering works on the spectral theory of hyponormal operators are Ando [1] and Stampfli [1]. These include independent proofs of Proposition 1.3 and a few of its corollaries; see also Berberian [1].

Proposition 2.2 is taken from Putinar [1]. Section 3 is entirely due to Dynkin [1], [2]. The theory of generalized spectral operators is treated in Colojoară – Foiaş [1] and

Vasilescu [1]. Theorem 4.3 has been proved in Putinar [1]; see also Putinar [2], [3]. For the local spectral theory consult Bishop [1], Dunford-Schwartz [3], Stampfli [4]-[8] and Vasilescu [1]. Decomposable operators are treated in Foiaş [1], Radjabalipour [4], Vasilescu [1], Albrecht [1]. A weaker form of Theorem 5.5 has been obtained by Stampfli [8], see also Clancey [5].

EXERCISES

1. If T is a hyponormal operator and the set $\sigma(T)$ has finite perimeter, then Theorem 3.2 is valid for every real number $\alpha > 1$.

2. If T is hyponormal and $\sigma(T) \subset \Omega$, where $\Omega \subset \mathbf{C}$ is a bounded open set with smooth boundary, then for every $\alpha > 1$ there exists a constant $C > 0$ so that

$$\|f(T)\| \leq C\|f\|_{W^{\alpha},\Omega}.$$

Here $\|f\|_{W^{\alpha},\Omega}$ is the usual Hilbertian Sobolev norm.

3. Generalize Theorem 3.2 to operators $T \in L(H)$ whose resolvent satisfies

$$\|T_z^{-1}\| = O(\mathbf{dist}(z,\sigma(T))^{-m}), \quad z \in \mathbf{C} \setminus \sigma(T)$$

for a given $m > 1$.

4. Let S denote the generalized scalar extension of a hyponormal operator T, as constructed in Theorem 4.3. Prove that:

a) The operator S is normal whenever T is normal.

b) If T is subnormal, the extension S is not similar, in general, to the minimal normal extension of T.

c) The operator T_z has not, in general, closed range on the space $W^2_{\bar\partial}(\Omega,H)$, $\Omega \supset \sigma(T)$. (**Hint.** Consider the case $\mathbf{dim}\, H = 1$).

5. Let $N \in L(H)$ be a normal operator with spectral measure E. Then the maximal spectral spaces of N are:

$$H_N(\delta) = E(\delta)H, \quad \delta = \bar{\delta} \subset \mathbf{C}.$$

In particular $H_N(\delta) \perp H_N(\delta')$ whenever $\delta \cap \delta' = \emptyset$.

6. a) Prove that the maximal spectral spaces of the multiplication operator M_z on $W^m_{\bar\partial}(\Omega)$, corresponding to two disjoint closed subsets of $\mathbf{C}$, are orthogonal.

b) Let M denote the operator M_z above. Prove that

$$\sigma(M) = \sigma_{ess}(M) = \sigma_{ap}(M) = \overline{\Omega},$$

$$\sigma_r(M) = \Omega, \quad \sigma_c(M) = \partial\Omega.$$

c) Compute the adjoint of the operator M and show that its maximal spectral spaces are not orthogonal for disjoint sets.

7. Let $T \in L(H)$ be a hyponormal operator and let $f \in O(\sigma(T))$ be an analytic function. Prove that the operator f(T) is subscalar.

8. Let $N \in L(H)$ be a normal operator and take a closed set $F \subset \mathbf{C}$. Then

$$H_N(F) = \bigcap_{z \notin F} N_z H.$$

Hint. One may assume $\sigma_p(N) = \emptyset$. Take $\xi \in \bigcap_{z \notin F} N_z H$ and consider the unique solution f(z) of the equation $N_z f(z) = \xi$. Let E denote the spectral measure of N. For $\delta \subset \mathbf{C} \setminus F$ a closed set and $z \in \delta$, prove that $\langle E(\delta)\xi, \xi\rangle = \langle E(\delta)N_z f(z), N_z f(z)\rangle \leq \leq \mathbf{diam}(\delta)^2 \langle E(\delta)f(z), f(z)\rangle$ and conclude that $\operatorname{supp}\langle dE\xi, \xi\rangle \subset F$.

9*. (Radjabalipour [3]). Let $T \in L(H)$ be a pure hyponormal operator. For any closed subset $F \subset \mathbf{C}$, one has

$$H_T(F) = \bigcap_{z \notin F} T_z H.$$

10*. (Hartman [1]). The spectrum of the hyponormal weighted shift

$$Te_n = \begin{cases} (1/2)e_{n+1}, & n < 0 \\ e_{n+1}, & n \geq 0, \end{cases}$$

is not a spectral set (σ is called a *spectral set* for the operator T if $\|r(T)\| \leq \|r\|_{\infty,\sigma}$ for $r \in \mathbf{Rat}(\sigma)$).

11*. (Albrecht-Eschmeier [1], Eschmeier-Putinar [1]). Let $T \in L(X)$ be a Banach space operator and $T' \in L(X')$ denote its topological dual. Then

a) T is decomposable iff T and T' have Bishop's property (β).

b) T is subdecomposable iff T has (β).

c) T is subscalar iff the map

$$T_z : E(\mathbf{C},X) \to E(\mathbf{C},X)$$

is one-to-one with closed range.

12. Let T be a hyponormal operator and denote by W(A) the numerical range of an operator A, see §IX.5.

a) Prove that $0 \in W([T,A])^-$ for every operator A.

b) The set $W(T)^-$ is the closed convex hull of the spectrum of T.

13*. (Clary [1]). Two operators T and S are said to be *quasi-similar* if there exist operators Q and R, one-to-one and with closed range, such that TQ = QS, and RT = SR.

Prove that two quasi-similar hyponormal operators have equal spectra.

Chapter IV

SOME INVARIANT SUBSPACES FOR HYPONORMAL OPERATORS

This chapter is entirely devoted to some recent progresses achieved in the invariant subspace problem by means of the new method originating in Scott Brown's paper [1]. This method turned out to be extremely generous, a large area of its applications being already registred, see for instance the book by Bercovici-Foiaş-Pearcy [1].

In the sequel we shall be concerned only with those invariant subspace results which refer to hyponormal operators. The main theorem is also due to Scott Brown [2], its proof consisting of an adaptation of his original technique to the class of subdecomposable operators. The application to hyponormal operators was possible only after Theorem III.4.3 had been proved.

A second result concerning the richness of the invariant subspace lattice of a hyponormal operator will be presented after Albrecht and Chevreau [1].

A remarkably simpler proof of the existence of (hyper)invariant subspaces in the case of subnormal operators is presented in Section 3 after J. E. Thomson [1]. This approach is also reminiscent to S. Brown [1].

The present chapter is not related to the rest of the book, so it may be omitted without loss of continuity.

1. Preliminaries

In this section we collect some technical facts on the Banach algebra of uniformly bounded analytic functions and on decomposable operators.

Let U be a bounded open subset of $\mathbf{C}$. As usually one denotes by $H^{\infty}(U)$ the Banach algebra of those analytic functions $f \in O(U)$, with the property

$$\|f\|_{\infty,U} = \sup_{z\in U} |f(z)| < \infty.$$

Since $L^1(U,\mu)^* \simeq L^{\infty}(U,\mu)$ and $H^{\infty}(U)$ embeds isometrically into $L^{\infty}(U,\mu)$, $H^{\infty}(U)$ is a dual Banach space. We shall adopt the following ad-hoc notation for its dual:

$$Q(U) = L^1(U,\mu)/\perp H^\infty(U), \quad Q(U)^* \simeq H^\infty(U).$$

Again μ above denotes the planar Lebesgue measure.

As a dual Banach space, $H^\infty(U)$ carries a weak-$*$ topology (given by the seminorms $|\langle\gamma,f\rangle|$, where $f \in H^\infty(U)$ and $\gamma \in Q(U)$). Because the space $L^1(U,\mu)$, and a fortiori $Q(U)$, is separable, the unit ball of $H^\infty(U)$ is compact and metrizable in the weak-$*$ topology.

Let $\lambda \in U$. Then the Dirac measure

$$\delta_\lambda(f) = f(\lambda), \quad f \in H^\infty(U)$$

is w-$*$ continuous on $H^\infty(U)$. Indeed

$$f(\lambda) = (1/\mu(D)) \int_D f d\mu,$$

where D is a disk centered at λ and contained in U. Accordingly, we shall denote by ε_λ the class of the functional δ_λ in $Q(U)$.

A subset $\Delta \subset U$ is called *dominating* for U, if

$$\|f\|_{\infty,U} = \sup_{\lambda\in\Delta} |f(\lambda)|, \quad f \in H^\infty(U).$$

1.1. Lemma. *Let $\Delta \subset U$ be a dominating subset of the bounded open set U. Then the absolutely convex hull of the functionals ε_λ, $\lambda \in \Delta$, denoted by*

$$\mathbf{aco}\{\varepsilon_\lambda \,;\, \lambda \in \Delta\}$$

is dense in the unit ball of $Q(U)$.

Proof. Assume γ is a point in the unit ball of $Q(U)$, which cannot be approximated by elements of the set $\mathrm{aco}\{\varepsilon_\lambda \,;\, \lambda \in \Delta\}$. By the Hahn-Banach theorem, there exists a function $f \in H^\infty(U)$, such that

$$\mathbf{Re}\langle\gamma, f\rangle > 1 \geq \mathbf{Re}\langle\sigma, f\rangle,$$

for every $\sigma \in \mathrm{aco}\{\varepsilon_\lambda \,;\, \lambda \in \Delta\}$. In particular $1 \geq \mathrm{Re}(e^{i\theta}f(\lambda))$ for every $\lambda \in \Delta$ and $\theta \in [0\, ,\, 2\pi]$, whence $\|f\|_{\infty,U} \leq 1$. On the other hand

$$\|f\|_{\infty,U} \geq |\langle\gamma,f\rangle| \geq \mathrm{Re}\langle\gamma, f\rangle > 1,$$

a contradiction!

In conclusion, the set $\mathrm{aco}\{\varepsilon_\lambda ; \lambda \in \Delta\}$ is dense in the unit ball of Q(U).

An immediate application of Cauchy's inequalities yields the next result.

1.2. Lemma. *Let* $\lambda \in U$ *be fixed. Then any function* $h \in H^\infty(U)$ *can be decomposed into*

$$h(z) = h(\lambda) + (z - \lambda)g(z),$$

with $g \in H^\infty(U)$, *and* $\|g\|_{\infty,U} \leq 2\|h\|_{\infty,U}/\mathbf{dist}(\lambda, \partial U)$.

A result with a more involved proof, which will provide a restatement of one of the main results below is the following.

1.3. Proposition. *Let* σ *be a compact subset of* $\mathbf{C}$. *If* $\sigma \cap U$ *is not dominating for any nonempty open set* $U \subset \mathbf{C}$, *then* $R(\sigma) = C(\sigma)$.

Recall that $R(\sigma)$ is the uniform closure (in $C(\sigma)$) of the algebra $\mathbf{Rat}(\sigma)$ of all rational functions with poles off σ.

For a proof of Proposition 1.3, see Apostol [1] or S. Brown [2].

Next we discuss the relationship between decomposability and duality for Hilbert space operators.

Let H be a complex Hilbert space and $T \in L(H)$ be a linear bounded operator on H. The question of how the spectral decomposition properties of T affect those of its adjoint T^* has been treated by several authors, beginning with Bishop [1]. For further references, see Vasilescu [1] and Eschmeier [1]. From these topics we need only the following basic fact.

1.4. Proposition. *The Hilbert space operator* T *is decomposable if and only if* T^* *is decomposable.*

Proof. Due to the fact that $T^{**} = T$, we only have to prove T decomposable implies T^* decomposable.

Assume the operator $T \in L(H)$ to be decomposable, see §III.5 for the definition. Recall that every vector $\xi \in H$ is carried on a local spectrum $\sigma_T(\xi)$, and the linear subspace

$$H_T(F) = \{\xi \in H \; ; \; \sigma_T(\xi) \subset F\}$$

is a closed T-invariant subspace, for every closed set $F \subset \mathbf{C}$. Moreover

$$\sigma(T \mid H_T(F)) \subset F,$$

cf. §III.5.

Fix two closed sets $F, G \subset \mathbf{C}$, such that $\mathbf{C} \setminus G \subset F$. Let P denote the orthogonal projection onto $H_T(F)^{\perp}$. Because $H_T(F)$ is an invariant subspace for T, its orthogonal is invariant under T^*, hence

$$PT^*P = T^*P.$$

Our first aim is to prove the inclusion

$$\sigma(T^* \mid H_T(F)^{\perp}) \subset G^* := \{\bar{z}\,;\, z \in G\}. \tag{1}$$

But $\sigma(T^* \mid H_T(F)^{\perp}) = \sigma(T^*P, PH) = \sigma(PT^*P, PH) = \sigma(PTP, PH)^*$, and the compressed operator PTP is unitarily equivalent to the operator induced by T onto the quotient $H/H_T(F)$. By the decomposability assumption we infer

$$H = H_T(F) + H_T(G),$$

therefore, the second isomorphism theorem yields

$$H/H_T(F) = (H_T(F) + H_T(G))/H_T(F) \simeq H_T(G)/H_T(F \cap G).$$

Since this isomorphism is compatible with the action of T, one gets

$$\sigma(PTP, PH) = \sigma(T, H/H_T(F)) = \sigma(T, H_T(G)/H_T(F \cap G)). \tag{2}$$

As $\sigma(T \mid H_T(G)) \subset G$ and $\sigma(T \mid H_T(F \cap G)) \subset G$, it is immediate that the spectrum of the quotient operator is still contained in G. Thus inclusion (1) follows from (2).

In order to prove that the operator T^* is decomposable, take two closed sets $G, G' \subset \mathbf{C}$, with the property $G \cup G' = \mathbf{C}$, and such that there exist other two closed sets $F, F' \subset \mathbf{C}$, so that

$$\mathbf{C} \setminus G \subset F, \quad \mathbf{C} \setminus G' \subset F', \quad F \cap F' = \emptyset.$$

From $H_T(F) \cap H_T(F') = \emptyset$, it follows that

$$H = H_T(F)^{\perp} + H_T(F')^{\perp}.$$

The inclusion (1) shows that this is a decomposition of the space H into T^*-invariant subspaces, subordinated to the covering $\{G, G'\}$. This completes the proof of the proposition.

Proposition 1.4 remains valid on arbitrary Banach spaces, but in that case the proof is much more complicated, cf. Eschmeier [1].

2. Scott Brown's theorem

Throughout this section $T \in L(H)$ is a *subdecomposable operator*. This means that there is a larger Hilbert space K, containing H as a closed subspace, and a decomposable operator $S \in L(K)$, which leaves invariant the subspace H, so that $S|H = T$. In virtue of Theorem III.4.3, any hyponormal operator is similar to a subdecomposable operator.

We shall prove a series of invariant subspace theorems, which will culminate with the next result.

2.1. Theorem. (Scott Brown). *Let* $T \in L(H)$ *be a hyponormal operator. If there exists a nonempty open set* $U \subset \mathbf{C}$, *such that the set* $\sigma(T) \cap U$ *is dominating for* U, *then* T *has a nontrivial invariant subspace.*

According to the function theoretic Proposition 1.3, the above statement can be reformulated as follows.

2.1'. Theorem. *Any hyponormal operator* T *with the property* $R(\sigma(T)) \neq C(\sigma(T))$ *has a nontrivial invariant subspace.*

The main body of the proof refers to any subdecomposable operator, the more restrictive hyponormality assumption being used only in the final part of it.

Before going on, let us notice a remarkable corollary of Theorem 2.1'.

2.2. Corollary. *Any subnormal operator has a nontrivial invariant subspace.*

Proof. Because of Theorem 2.1' it suffices to consider a subnormal operator T with $R(\sigma(T)) = C(\sigma(T))$. By taking into account the estimate asserted by Corollary I.3.5, one obtains that the operator T admits a continuous functional calculus with continuous functions. Then obviously T has nontrivial invariant subspaces, and the proof of Corollary 2.2 is over.

The next section contains a simpler proof of a stronger form of Corollary 2.2.

An intermediate result toward S. Brown's theorem is the following.

2.3. Theorem. *Any subdecomposable operator with interior points in its spectrum has a rationally invariant subspace.*

In the proof of this theorem we will follow closely Albrecht and Chevreau [1].

First, a few notations. Let $S \in L(K)$ be a decomposable operator with an invariant subspace H. Assume $U \subset \sigma(T)$ is a nonempty open set contained in the spectrum of the subdecomposable operator $T = S|H$.

The idea of the proof of Theorem 2.3 is the following. Pick a point $\lambda \in U$. We shall prove that there are vectors $\xi, \eta \in H$, such that

(3) $$r(\lambda) = \langle r(T)\xi, \eta\rangle$$

for every rational function $r \in \mathbf{Rat}(\sigma(T))$. Consider the following subspace of H:

$$M = \bigvee\{r(T)\xi \; ; \; r \in \mathbf{Rat}(\sigma(T)), \; r(\lambda) = 0\}.$$

It is plain that M is rationally T-invariant. Since $1 = \langle \xi, \eta\rangle$ it follows $\eta \neq 0$. On the other hand $M \perp \eta$, again by relation (3). Therefore $M \neq H$. Then either $M \neq 0$ or $(T - \lambda)\xi = 0$. But in any case M or $\mathbf{C}\cdot\xi$ is a nontrivial rationally invariant subspace for T.

In order to find the vectors ξ, η which satisfy (3) we shall use an approximation device in the predual $Q(U)$ of $H^\infty(U)$.

Let $V \subset U$ be a fixed open subset. For any couple of vectors $\xi, \eta \in K$, with the property $\sigma_{S^*}(\eta) \subset V$, a linear functional on $H^\infty(V)$ is defined by

(4) $$\langle \xi \otimes \eta, f\rangle = \langle f(S^*)\eta, \xi\rangle, \quad f \in H^\infty(V).$$

The expression $f(S^*)\eta$ has a good meaning because the operator S^* is decomposable by Proposition 1.4 and the vector η belongs to the space $K_{S^*}(\sigma_{S^*}(\eta))$, and

$$\sigma(S^* | K_{S^*}(\sigma_{S^*}(\eta)) \subset V.$$

Our next aim is to establish some properties of the functionals defined by (4).

2.4. Lemma. *For any $\xi, \eta \in K$ and V an open subset such that $\sigma_{S^*}(\eta) \subset V$, the functional $\xi \otimes \eta$ is weak-$*$ continuous on $H^\infty(V)$.*

Proof. Take an open set W, such that

$$\sigma_{S^*}(\eta) \subset W \subset \overline{W} \subset V.$$

By the metrizability of the unit ball of $H^\infty(V)$ in the weak-$*$ topology, it suffices to prove that $\lim_{n\to\infty} \langle \xi \otimes \eta, f_n\rangle = 0$, whenever w-$*$-$\lim_{n\to\infty} f_n = 0$.

If $f_n \in H^\infty(V)$ and w-$*$-$\lim_{n\to\infty} f_n = 0$, then the compactness of the restriction map $H^\infty(V) \to H^\infty(W)$ shows that $\lim_{n\to\infty} \| f_n \|_{\infty, W} = 0$. Further the continuity of the functional calculus map $f_n \to f_n(S^*)\eta$ ends the proof of the lemma.

The class of $\xi\otimes\eta \in H^\infty(V)^*$ in $Q(V)$ will be denoted by the same symbol $\xi\otimes\eta$. The naturality of the analytic functional calculus yields the compatibility between the elements $\xi\otimes\eta \in Q(V)$ and $\xi\otimes\eta \in Q(V')$, via the corestriction map $Q(V') \to Q(V)$, whenever $\sigma_{S^*}(\eta) \subset V' \subset V$.

2.5. Lemma. *Let* $F \subset V$ *be a closed subset. If* $\xi_n \in K$ *and* $\eta_n \in K_{S^*}(F)$ *are two sequences weakly convergent to zero, then*

$$\lim_{n\to\infty} \| \xi_n \otimes \eta_m \|_{Q(V)} = \lim_{n\to\infty} \| \xi_m \otimes \eta_n \|_{Q(V)} = 0$$

for every fixed $m \in \mathbf{N}$.

Proof. An argument in the proof of Lemma 2.4 shows that the map

$$K \times K_{S^*}(F) \to Q(V), \quad (\xi,\eta) \mapsto \xi\otimes\eta,$$

is compact. Therefore it maps weakly convergent sequences into norm convergent sequences.

Next we introduce a notation. Let V, V' be open sets such that $\overline{V} \subset V'$. We denote by $A_H(V,V')$ the set of all those functionals $\gamma \in Q(V')$, for which there are sequences $\xi_n \in H$, $\eta_n \in K_{S^*}(\overline{V})$ weakly convergent to zero, such that $\|\xi_n\| \le 1$, $\|\eta_n\| \le 1$ and

$$\lim_{n\to\infty} \| \gamma - \xi_n\otimes\eta_n \|_{Q(V')} = 0.$$

2.6. Lemma. *For any open sets* $V \subset \overline{V} \subset V'$, *the set* $A_H(V,V')$ *is closed and absolutely convex in* $Q(V')$.

Proof. A diagonal selection into double sequences shows that the set $A_H(V,V')$ is closed.

It is obvious that $e^{i\theta}\gamma \in A_H(V,V')$ whenever $\gamma \in A_H(V,V')$ and $\theta \in [0,2\pi]$. Thus it suffices to prove that

$$t\gamma + (1-t)\gamma' \in A_H(V, V')$$

for any $\gamma, \gamma' \in A_H(V,V')$, $t \in [0,1]$.

Let $\xi_n\otimes\eta_n$ and $\xi'_n\otimes\eta'_n$ be sequences which approximate by definition γ and γ', respectively. We may assume that

$$\max\{\|\xi_n\|, \|\eta_n\|, \|\xi'_n\|, \|\eta'_n\|\} < 1$$

for every $n \in \mathbf{N}$.

Let us denote for n, m $\in$ **N**:

$$\xi_{m,n} = \sqrt{t}\xi_m + \sqrt{1-t}\xi'_n,$$

$$\eta_{m,n} = \sqrt{t}\eta_m + \sqrt{1-t}\eta'_n.$$

By using Lemma 2.5 one finds an integer $m_1 > 0$, with the property

$$\|\gamma - \xi_{m_1} \otimes \eta_{m_1}\| < 1/3.$$

Next choose a sufficiently large $n_1 > 0$, so that

$$\|\xi_{m_1,n}\| < 1, \quad \|\eta_{m_1,n}\| < 1, \quad n \geq n_1,$$

$$\|\gamma' - \xi'_{n_1} \otimes \eta'_{n_1}\| < 1/3$$

$$\|\xi_{m_1} \otimes \eta'_{n_1}\| + \|\xi'_{m_1} \otimes \eta_{n_1}\| < 1/3.$$

By putting together these inequalities one easily deduces

$$\|t\gamma + (1-t)\gamma' - \xi_{m_1,n_1} \otimes \eta_{m_1,n_1}\| < 1.$$

Analogously one finds increasing sequences $m_j \uparrow \infty$ and $n_j \uparrow \infty$, so that

$$\|\xi_{m_j,n}\| < 1, \quad \|\eta_{m_j,n}\| < 1, \quad n \geq n_j$$

$$\|\gamma - \xi_{m_j} \otimes \eta_{m_j}\| < 1/3j,$$

$$\|\gamma' - \xi'_{n_j} \otimes \eta'_{n_j}\| < 1/3j$$

and

$$\|t\gamma + (1-t)\gamma' - \xi_{m_j,n_j} \otimes \eta_{m_j,n_j}\| < 1/j.$$

Since $\text{w-}\lim_{j\to\infty} \xi_{m_j,n_j} = \text{w-}\lim_{j\to\infty} \eta_{m_j,n_j} = 0,$

the proof of the lemma is over.

At this point we should remember that one seeks invariant subspaces for the subdecomposable operator T, with the open set U contained in its spectrum. Accordingly, we may assume that

$$\mathbf{Ker}(T - \lambda) = \mathbf{Ker}(T^* - \bar{\lambda}) = 0$$

for every $\lambda \in \sigma(T)$. Otherwise $\mathbf{Ker}(T - \lambda)$ or $\mathbf{Ran}(T - \lambda)^-$ would be (non-trivial) rationally invariant subspaces of T.

In particular we may assume without loss of generality that the operator T has only continuous spectrum, $\sigma(T) = \sigma_c(T)$, see §.III.1 for details.

2.7. Proposition. *Let* $T \in L(H)$ *be subdecomposable and* W, V, U *be open sets, such that* $\overline{W} \subset V \subset \overline{V} \subset U$. *Then there is a positive constant* δ, *depending on* W, V, U *and* T, *with the property*

$$A_H(V^*,U^*) \supset \delta\cdot\mathbf{aco}\{\varepsilon_{\overline{\lambda}}\,;\,\lambda \in \overline{W} \cap \sigma_c(T)\}.$$

Proof. Recall that $\varepsilon_\lambda \in Q(U)$ denotes the evaluation functional at the point λ.

By Lemma 2.6 it suffices to prove that $\varepsilon_\lambda \in A_H(V,U)$ for a fixed point $\lambda \in \overline{W} \cap \sigma_c(T)$.

Since $\lambda \in \sigma_c(T)$, there is a sequence of vectors $\xi_n \in H$, weakly convergent to zero, and with the properties

$$\lim_{n\to\infty}(T - \lambda)\xi_n = 0,$$

$$\|\xi_n\| = 1, \quad n \geq 1.$$

The idea of the proof is to embed H into the space K of the decomposable extension S of T, and then to approximate $\varepsilon_{\overline{\lambda}}$ in $Q(U^*)$ by functionals like $\xi_n \otimes \eta'_n$, where η'_n is the localization of the vector η_n at $\overline{V}^*$, with respect to S^*.

Let W' be a relatively compact neighbourhood of $\overline{W}$ in V. By the decomposability of the operator S^* on K we infer

$$K = K_{S^*}(\overline{V}^*) + K_{S^*}(\mathbf{C} \setminus W'^*).$$

In virtue of the open map principle, there is a constant $C > 0$ and decompositions

$$\xi_n = \eta'_n + \eta''_n; \quad \sigma_{S^*}(\eta'_n) \subset V^*, \sigma_{S^*}(\eta''_n) \subset \mathbf{C} \setminus W'^*,$$

such that

$$\|\eta'_n\| \leq C, \quad \|\eta''_n\| \leq C,$$

for every $n \in \mathbf{N}$.

Moreover, by passing to subsequences we may assume that the sequences η'_n and η''_n are weakly convergent. Say $\eta' = w\text{-}\lim_{n\to\infty} \eta'_n$. By replacing η'_n with $\eta'_n - \eta'$ we can assume that

$$\text{w-}\lim_{n\to\infty} \eta'_n = \text{w-}\lim_{n\to\infty} \eta''_n = 0$$

and

$$\|\eta'_n\| \leq 2C, \quad \|\eta''_n\| \leq 2C, \quad n \in \mathbf{N}.$$

Due to the fact that $\lim_{n\to\infty} (T - \lambda)\xi_n = 0$, one gets

$$\lim_{n\to\infty} \langle \xi_n, \eta''_n \rangle = \lim_{n\to\infty} \langle (T-\lambda)\xi_n, (S^* | K_{S^*}(\mathbf{C} \setminus W') - \overline{\lambda})^{-1}\eta''_n \rangle = 0,$$

hence

(5) $$\lim_{n\to\infty} \langle \xi_n, \eta'_n \rangle = 1.$$

Now consider a function $f \in H^\infty(U^*)$, and the decomposition

$$f(z) = f(\overline{\lambda}) + (z - \overline{\lambda})g(z), \quad \|g\|_{\infty,U} < \textbf{const.} \|f\|_{\infty,U},$$

asserted in Lemma 1.2. Since $\eta'_n \in K_{S^*}(\overline{V}^*)$, one finds

$$f(S^*)\eta'_n = f(\overline{\lambda})\eta'_n + (S^* - \lambda)g(S^*)\eta'_n, \quad n \in \mathbf{N}.$$

Accordingly

$$|\langle \xi_n \otimes \eta'_n - \langle \eta'_n, \xi_n \rangle \varepsilon_{\overline{\lambda}}, f \rangle| = |\langle f(S^*)\eta'_n, \xi_n \rangle - \langle \eta'_n, \xi_n \rangle f(\overline{\lambda})| =$$

$$= |\langle (S^* - \overline{\lambda})g(S^*)\eta'_n, \xi_n \rangle| \leq \textbf{const.} \|(T - \lambda)\xi_n\|,$$

with a suitable constant depending also on f.

By relation (5) one finally obtains

$$\lim_{n\to\infty} \|\xi_n \otimes \eta'_n - \delta\varepsilon_{\overline{\lambda}}\|_{Q(U^*)} = 0$$

for a suitable $\delta > 0$. This ends the proof of the proposition.

2.8. Corollary. *With the notations of Proposition* 2.7, *assume* $U \subset \sigma_c(T)$. *Then for every* $\alpha, \beta \in K$, $\gamma \in Q(W^*)$ *and* $\varepsilon > 0$, *there are* $\xi \in H$ *and* $\eta \in K_{S^*}(V^*)$, *so that*

$$\|\gamma - \xi \otimes \eta\|_{Q(U^*)} < \varepsilon,$$

$$\|(\alpha + \xi) \otimes (\beta + \eta) - \alpha \otimes \beta - \xi \otimes \eta\| < \varepsilon,$$

and

$$\max(\|\xi\|, \|\eta\|) < \sqrt{\|\gamma\|_{Q(W^*)}/\delta}.$$

Proof. Since the set $W \cap \sigma_c(T)$ is dominating for W, Lemma 1.1 shows that $\gamma/\|\gamma\|$ belongs to the closure of the set $\mathbf{aco}\{\varepsilon_{\bar{\lambda}}\,;\,\lambda \in W\}$. Then one applies Proposition 2.7 and Lemma 2.5.

The above corollary constitutes the basic tool for the rest of this section.

Proof of Theorem 2.3. As in other approximation proofs, one starts with an exhaustion of the open set $U \subset \sigma(T)$ with relatively compact open subset U_n, which satisfy

$$U_n \subset \bar{U}_n \subset U_{n+1}, \quad n \in \mathbf{N}.$$

For every natural n, choose an open subset V_n, with the properties

$$\bar{U}_n \subset V_n \subset \bar{V}_n \subset U_{n+1}, \quad n \in \mathbf{N}.$$

For every triple $\bar{U}_n \subset V_n \subset U_{n+1}$, denote by δ_n the positive constant whose existence and property are asserted by Proposition 2.7. We shall denote for short

$$\|\gamma\|_n = \|\gamma\|_{Q(U_n^*)}, \quad \gamma \in Q(U_n^*).$$

Fix a point $\lambda \in U$. We may assume without loss of generality that $\lambda \in U_0$. We shall recurrently construct two Cauchy sequences

$$\xi_n \in H, \ \eta_n \in K_{S^*}(\bar{V}_{n-1}^*), \quad n \geq 1$$

with the properties

(6)
$$\lim_{n\to\infty} \|\varepsilon_{\bar{\lambda}} - \xi_n \otimes \eta_n\|_n = 0,$$

and

$$\mathbf{max}(\|\xi_n\|\,,\,\|\eta_n\|) < 2/\sqrt{\delta_0}.$$

Let us consider a decreasing sequence of positive numbers $\varepsilon_j > 0$, with the property

$$\sum_{j=0}^{\infty} \sqrt{\varepsilon_j/\delta_j} < 1/\sqrt{\delta_0}.$$

We put by convention $\xi_0 = \eta_0 = 0$ and $\varepsilon_0 = 0$.
By Corollary 2.8 there are $\xi_1 \in H$ and $\eta_1 \in K_{S^*}(\bar{V}_0^*)$, such that

$$\|\varepsilon_{\bar{\lambda}} - \xi_1 \otimes \eta_1\|_1 < \varepsilon_1,$$

$$\mathbf{max}(\|\xi_1\|\,,\,\|\eta_1\|) < 1/\sqrt{\delta_0}.$$

Suppose that $\xi_1,\dots,\xi_m \in H$ and $\eta_j \in K_{S^*}(\bar{V}^*_{j-1})$, $j = 1,2,\dots,m$ have already been found, so that

$$(7)\qquad \begin{cases} \| e_{\bar\lambda} - \xi_j \otimes \eta_j \|_j < \varepsilon_j \\ \max(\|\xi_j - \xi_{j-1}\|, \|\eta_j - \eta_{j-1}\|) < \sqrt{\varepsilon_{j-1}/\delta_{j-1}},\ 1 \le j \le m. \end{cases}$$

By applying again Corollary 2.8, we find vectors $\xi_{m+1} \in H$ and $\eta_{m+1} \in K_{S^*}(\bar{V}^*_m)$, so that (7) holds up to $j = m + 1$.

Thus relation (6) is proved. Moreover

$$\|\xi_m\| \le \sum_{j=1}^{m} \|\xi_j - \xi_{j-1}\| \le \sum_{j=1}^{m} \sqrt{\varepsilon_{j-1}/\delta_{j-1}} < 2\sqrt{\delta_0},$$

and analogously for $\|\eta_m\|$.

If $\xi = \lim\limits_{j\to\infty} \xi_j$ and $\eta = \lim\limits_{j\to\infty} \eta_j$, then (6) yields $e_{\bar\lambda} = \xi \otimes \eta$ as elements of $Q(U^*)$. In particular one gets the identity

$$f(\bar\lambda) = \langle f(S^*)\eta, \xi\rangle, \quad f \in H^\infty(U^*).$$

By taking $f(z) = \overline{r(\bar z)}$, where $r \in \mathbf{Rat}(\sigma(T))$, one obtains

$$\overline{r(z)} = \langle r(S)^*\eta, \xi\rangle = \overline{\langle r(S)\xi, \eta\rangle}.$$

Consequently relation (3) follows and the proof of Theorem 2.3 is finished.

Proof of Theorem 2.1. The proof goes along the same lines as the proof of Theorem 2.3.

As above, we may assume T is a hyponormal operator with $\sigma(T) = \sigma_c(T)$. Suppose U is a bounded domain of $\mathbf{C}$, with the property that $\sigma_c(T) \cap U$ is dominating for U. According to Theorem 2.3 we may also assume $\operatorname{int}\sigma(T) = \emptyset$.

First we prove that in this case the constant δ appearing in Proposition 2.7 can be chosen to be $\delta = 1$. Quite specifically we state the next lemma.

2.9. Lemma. *With the above notations, there is a dense subset $\Delta \subset \sigma_c(T) \cap U$, such that for any $\lambda \in \Delta$ and every disk $D \subset U$ centered at λ, there are sequences $\xi_n \in H$, $\eta_n \in K_{S^*}(D^*)$, with the properties*

$$(8)\qquad \begin{cases} \text{w-}\lim_{n\to\infty} \xi_n = \text{w-}\lim_{n\to\infty} \eta_n = 0, \\ \lim_{n\to\infty} \| \varepsilon_{\overline{\lambda}} - \xi_n \otimes \eta_n \|_{Q(U^*)} = 0, \\ \|\xi_n\| \le 1,\ \|\eta_n\| \le 1, \textit{ and } \lim_{n\to\infty} \langle \eta_n, \xi_n \rangle = 1. \end{cases}$$

Proof. Fix a point $\lambda \in \sigma_c(T) \cap U$, and a disk D centered at λ and relatively compact in U. Of course there exists a sequence of unit vectors $\xi_n \in H$, so that $\lim_{n\to\infty} (T - \lambda)\xi_n = 0$ and $\text{w-}\lim_{n\to\infty} \xi_n = 0$.

Let P denote the orthogonal projection of K, the space of the scalar extension S of the operator T, onto $K_{S^*}(D^*)$. By repeating the argument given in the proof of Proposition 2.7, it suffices to prove that $\lim_{n\to\infty} \| (I - P)\xi_n \| = 0$, in order to achieve (8) for the sequences ξ_n and $\eta_n = P\xi_n$.

So, what remains to be proved is that for every point $\lambda_o \in \sigma_c(T) \cap U$ and $\varepsilon > 0$, there is a $\lambda \in \sigma_c(T) \cap U$, so that for any disk D as above:

$$(9)\qquad \begin{cases} |\lambda - \lambda_o| < \varepsilon, \text{ and} \\ \lim_{n\to\infty} \| (I - P)\xi_n \| = 0, \text{ whenever } \lim_{n\to\infty} \| (T - \lambda)\xi_n \| = 0,\ \|\xi_n\| \le 1. \end{cases}$$

At this point we use the hyponormality assumption again. Namely recall from § III.2 that

$$\| (T - z)^{-1} \| = 1/\mathbf{dist}(z,\sigma(T)), \quad z \notin \sigma(T).$$

Also, if $S \in L(K)$ denotes the canonical scalar extension of the operator T, cf. § III.4, then Corollary III.4.4 yields $\sigma(S) \subset \sigma(T)$, and

$$(10)\qquad \| (S - z)^{-1} \| \le 1/\mathbf{dist}(z,\sigma(T)), \quad z \notin \sigma(T).$$

Now choose for the point $\lambda_o \in \sigma_c(T) \cap U$, a $\lambda \in \partial\sigma(T) \cap U$ such that λ is the closest point in $\sigma(T)$ to some $\zeta \in \mathbf{C} \setminus \sigma(T)$ and $|\lambda_o - \lambda| < \varepsilon$. Assume the disk $D \ni \lambda$ fixed and consider a sequence of unit vectors $\xi_n \in H$, with the property $\lim_{n\to\infty} \| (T - \lambda)\xi_n \| = 0$. Then condition (9) must be fulfilled. This means that for a fixed δ, there is an $N > 0$, such that $\|\xi_n\| > 1 - \delta$ for $n > N$.

By Proposition 1.4 we learn that $(I - P)K$ is an invariant subspace for the operator S, and

$$\sigma(S|(I - P)K) = \sigma(S(I - P),(I - P)K) \subset \mathbf{C} \setminus D.$$

Thus there exists a point z on the line segment joining ζ to λ, so that

$$\|(\lambda - z)(S(I - P) - z)^{-1}\| < \delta/2.$$

Also (10) implies the estimate

$$\|(\lambda - z)(S - z)^{-1}\| \leq 1.$$

Therefore

$$\|(\lambda - z)(S - z)^{-1}\xi_n\| \leq \|(\lambda - z)(S - z)^{-1}P\xi_n\| +$$

$$+ \|(\lambda - z)(S - z)^{-1}(I - P)\xi_n\| \leq \|P\xi_n\| + \delta/2.$$

As the left term converges to 1, there exists an $N > 0$ with the property

$$1 - \delta/2 < \|P\xi_n\| + \delta/2$$

for $n > N$.

This completes the proof of lemma.

Now, a variant of Corollary 2.8 holds with $\delta = 1$. Finally one approximates $\varepsilon_{\overline{\lambda}}$, $\lambda \in \Delta$, with $\xi \otimes \eta$, $\xi \in H$, $\eta \in K$, as in the proof of Theorem 2.3 by working only on the set U instead of the sequence U_n.

Thus the proof of Theorem 2.1 is finished.

We shall see later in Chapter XI, that there are pure hyponormal operators with no dominating spectrum in any open subset of the complex plane.

3. Hyperinvariant subspaces for subnormal operators

The functional model of a cyclic subnormal operator (Proposition I.3.8) has oriented for quite a long while the search for invariant subspaces of subnormal operators to the function theory of two real variables methods, cf. J.E. Brennan [1], [2]. The solution to this (partial) invariant subspace problem came through a rather involved technique (S. Brown [1]), originating in D. Sarason's work.

Recently, J.E. Thomson [1] discovered a new and very simple proof for the existence of hyperinvariant subspaces for any rationally cyclic subnormal operator. By its simplicity, Thomson's proof is definitive. It is the aim of this section to present it.

Recall that a rationally cyclic subnormal operator is unitarily equivalent to the multiplication operator $M = R_{\nu,\sigma}$, acting on the space $R^2_\sigma(\nu) \subset L^2(\nu)$ (Proposition I.3.8). Here ν is a positive Borel measure, compactly supported in $\mathbf{C}$, and $R^2_\sigma(\nu)$ is the

$L^2(\nu)$-closure of the algebra of all rational functions with poles off $\sigma = \sigma(M) \supset \mathbf{supp}(\nu)$.

The following result, which characterizes the commutant of the operator M in $L(R^2_\sigma(\nu))$, is well known, see Bram [1] and Conway [2].

3.1. Lemma. *Assume that the operator* $A \in L(R^2_\sigma(\nu))$ *commutes with* M. *Then there is a function* $f \in L^\infty(\nu) \cap R^2_\sigma(\nu)$, *such that* $Ag = fg$, $g \in R^2_\sigma(\nu)$.

Proof. Since $[A , M] = 0$, it follows that $[A , h(M)] = 0$ for any rational function h with poles off σ.

Denote $f = A(\mathbf{1}) \in R^2_\sigma(\nu)$. Then

$$Ah = Ah(M)\mathbf{1} = h(M)A\mathbf{1} = h(M)f = f\cdot h$$

for any $h \in \mathbf{Rat}(\sigma(M))$.

It remains to prove that $f \in L^\infty(\nu)$. In order to do this it is more convenient to work abstractly with the minimal normal extension N (acting on $L^2(\nu)$) of the operator M. For any $g \in R^2_\sigma(\nu)$ and $k \in \mathbb{N}$, one has

$$\| N^{*k}Ag \|^2 = \langle N^{*k}Ag , N^{*k}Ag\rangle = \langle N^kAg , N^kAg\rangle =$$

$$= \langle N^{*k}A^*AN^kg , g\rangle \le \| A\|^2 \| N^{*k}g\| .$$

Thus the transformation

$$B(N^{*k}g) = N^{*k}Ag$$

extends continuously to $L^2(\nu) = \bigvee_{k=0}^{\infty} N^{*k}R^2_\sigma(\nu)$, and yields an operator $B \in L(L^2(\nu))$ which satisfies $B| R^2_\sigma(\nu) = A$ and $[B , N] = 0$. But the commutant of the $*$-cyclic normal operator $N = M_z$ on $L^2(\nu)$ is well-known to be equal to $L^\infty(\nu)$, cf. for instance Halmos [2].

Returning to the operator $A \in L(R^2_\sigma(\nu))$, we have proved that $f = A\cdot\mathbf{1} = B\cdot\mathbf{1} \in L^\infty(\nu)$. Hence $A = M_f$ and the proof is complete.

To finish the technical preliminaries we state the following.

3.2. Lemma. *Let, with the above notation,* $g \in L^q(\nu)$, $g \ne 0$, $1 < q < 2$. *Then there exists a point* $w \in \mathbf{C}$, *such that the function* $g(z)/(z - w)$ *still belongs to* $L^q(\nu)$ *and* $\int g(z)/(z - w)\, d\nu(z) \ne 0$.

Proof. Since the function $1/|z|^q$ is locally integrable with respect to the Lebesgue measure $d\mu(z)$, Fubini's theorem shows that

$$\int_{\mathbf{C}} (|g(z)|^q/|z - w|^q)d\nu(z) < \infty$$

for μ-almost all $w \in \mathbf{C}$, and the function

$$h(w) = \int (g(z)/(z - w))d\nu(z)$$

is μ-integrable.

By taking the $\bar{\partial}$ derivative in the sense of distributions one gets

$$\bar{\partial}h = \bar{\partial}((1/z)*g) = -\pi ig.$$

This shows that h cannot be identically equal to zero, and the proof is finished.

The phenomenon discovered by J.E. Thomson does not refer explicitly to subnormal operators. We present it below in the original form, which may be of an independent interest.

3.3. Theorem. (J.E. Thomson). *Let* ν *be a Borel positive measure, compactly supported on* $\mathbf{C}$. *Let* H *be a closed subspace of* $L^2(\nu)$, *which contains* **1** *and let* A *be a subalgebra of* $L^\infty(\nu)$, *such that* $z \in A$ *and* $AH \subset H$.

Then there exists a nontrivial closed subspace $K \subset H$, *with the property* $AK \subset K$.

Proof. The cases $H = L^2(\nu)$ or supp(ν) reduced to a point are trivial.

Assume $H \neq L^2(\nu)$. Then there is a function $g \in L^2(\nu)$, $g \neq 0$ and $\bar{g} \perp H$.

Throughout the proof we fix the conjugate numbers $q = 3/2$ and $p = 3$.

Obviously $g \in L^q(\nu)$, and by Lemma 3.2 there is a point $w \in \mathbf{C}$, such that $g(z)/(z - w) \in L^q(\nu)$ and

$$\int (g(z)/(z - w))d\nu(z) \neq 0. \tag{11}$$

Since the inclusion $L^p(\nu) \subset L^2(\nu)$ is continuous and $\mathbf{1} \in H$, the $L^p(\nu)$-closure of A, denoted by $\tilde{A}$, is contained in H. Consider the bounded linear functional L on $\tilde{A}$ given by

$$L(f) = \int (f(z)g(z)/(z - w))d\nu(z), \quad f \in \tilde{A} \subset H.$$

By the Hahn-Banach theorem there exists a functional with the same norm which extends L to $L^p(\nu)$. This extension is represented by a function $h \in L^q(\nu)$:

$$L(f) = \int fh d\nu, \quad f \in \tilde{A}.$$

By duality arguments (the Banach space $\tilde{A}$ is reflexive) there exists a function $r \in \tilde{A}$

such that $\|r\|_p = 1$ and $L(r) = \|L\|$. Then

$$\|h\|_q = \|L\| = L(r) = \int rh d\nu \leq \|r\|_p \|h\|_q = \|h\|_q.$$

Consequently there is a constant $\lambda > 0$ so that:

$$|r|^p = \lambda^{2/3} |h|^q$$

or

(12) $$|r|^2 = \lambda |h|,$$

because the above Hölder inequality is an equality.

Let us define the function $u = h/r$ on the set where $r \neq 0$ and zero elsewhere. By (12) we infer $u \in L^2(\nu)$.

Let K be the closure in H of the linear space $(z - w)Ar$. Our next aim is to prove that $\bar{u} \in K^{\perp}$, but $\bar{u} \notin H^{\perp}$. Indeed,

$$\int (z - w) f r u d\nu = \int (z - w) f h d\nu = \int (z - w) f (g(z)/(z - w)) d\nu = \int f g d\nu = 0$$

for every $f \in A$. On the other hand

$$\int r u d\nu = \int h d\nu = \int (g(z)/(z - w)) d\nu \neq 0.$$

In conclusion $K \neq H$ and the proof is complete.

3.4. Corollary. *Let* $S \in L(H)$ *be a rationally cyclic subnormal operator. Then there is a proper closed subspace of H, invariant to any operator in the commutator of* S.

When it exists, such a subspace is called a *hyperinvariant* subspace of S.

Proof. Apply Theorem 3.3 to the commutant of S, described in Lemma 3.1.

Of course, this result settles the invariant subspace problem for arbitrary subnormal operators, too.

4. The lattice of invariant subspaces

A variation of the method used in the proof of Scott Brown's theorem gives some interesting information on the richness of the invariant subspace lattice of a hyponormal operator with thick continuous spectrum. The rather long computations involved in the proof of the main result below are originating in the work of

Bercovici, Foiaş and Pearcy, see their book [1] for further details. In our exposition we follow Albrecht and Chevreau [1]. Instead of giving full details of this proof we shall confine ourselves to sketching its main steps, in comparison with the technical lemmas of §2.

For any operator $T \in L(H)$, one denotes by **Lat**(T) the set of all closed invariant subspaces of T. Endowed with the inclusion as order relation, **Lat**(T) is obviously a lattice.

Scott Brown's theorem shows that **Lat**(T) contains chains of infinite length, whenever T is a pure hyponormal operator with $R(\sigma(T)) \neq C(\sigma(T))$. If one imposes the same condition on the continuous spectrum of T, then **Lat**(T) becomes a huge lattice, as follows.

4.1. Theorem. *Assume* T *is a hyponormal operator whose continuous spectrum is dominating in a bounded nonempty open set.*

Then **Lat**(T) *contains a sublattice order isomorphic to the lattice of all closed subspaces of an infinite dimensional Hilbert space.*

The idea which lies at the heart of the proof of this theorem is as follows. Instead of expressing the evaluation functional $\varepsilon_\lambda = \xi \otimes \eta$ with the aid of only two vectors ξ, η, one performs a countable set of such representations corresponding to pairwise orthogonal couples of vectors. More specifically, assume that there are sequences $\xi_n, \eta_n \in H$, so that

$$p(\lambda)\delta^n_m = \langle p(T)\xi_m, \eta_n \rangle, \quad n, m \geq 0, \ p \in \mathbf{C}[z], \tag{13}$$

where δ^n_m denotes the Kronecker symbol. Now define the subspaces

$$M = \bigvee \{T^k \xi_m \ ; \ k, m \geq 0\},$$

$$N = \{\theta \in M \ ; \ \langle T^k \theta, \eta_n \rangle = 0, \quad k, n \geq 0\}.$$

Of course $N \subset M$ and $\eta_n \perp N$ for any $n \geq 0$. Because of (13) the vectors (η_n) are linearly independent, whence $\dim M/N = \infty$.

It is plain that $M, N \in$ **Lat**(T).

Let $\pi : M \longrightarrow M/N$ denote the canonical projection, and choose a closed subspace $L \subset M/N$. Since

$$\langle (T - \lambda)\xi_m, \eta_n \rangle = 0$$

for any $m, n \geq 0$, we infer

$$(T - \lambda)M \subset N.$$

In particular

$$T(\pi^{-1}L) \subset \lambda(\pi^{-1}L) + N = \pi^{-1}L.$$

This inclusion shows that $\pi^{-1}L \in \mathbf{Lat}(T)$. In conclusion the sublattice of $\mathbf{Lat}(T)$:

$$\{K \in \mathbf{Lat}(T) \; ; \; N \subset K \subset M\}$$

is order isomorphic to the lattice of all closed subspaces of M/N.

In order to get the representation (3) one needs to enlarge the framework developed in the preceding section from one functional on $H^{\infty}(U)$, to matrices of functionals on $H^{\infty}(U)$. First, a few notations.

Let $N \in \mathbf{N} \cup \{\infty\}$. The Hilbert space of square summable sequences with entries in H is denoted by $l^2(H)$. We put

$$l^2_N(H) = \{\xi \in l^2(H) \mid \xi = (\xi_j), \;\; \xi_j = 0 \text{ for } j > N\}.$$

It is clear that the $l^2_N(H)$ form an increasing sequence of closed subspaces of $l^2_{\infty}(H) = l^2(H)$.

For a given bounded open set $U \subset \mathbf{C}$, one denotes by $H^{\infty}(U,N)$ the Banach space of all matrices $f = (f_{ij})^N_{i,j=1}$, such that $f_{ij} \in H^{\infty}(U)$ and

$$\|f\|_{H^{\infty}(U,N)} = \sum_{i,j=1}^{N} \|f_{ij}\|_{\infty,U} < \infty.$$

Similarly, $Q(U,N)$ stands for the space of all matrices $\gamma = (\gamma_{ij})^N_{i,j=1}$, such that $\gamma_{ij} \in Q(U)$ and

$$\|\gamma\|_{Q(U,N)} = \sup_{i,j=1}^{N} \|\gamma_{ij}\| < \infty.$$

Thus the unit ball in $Q(U,N)$ is just the product of the unit balls of the factors. For $N < \infty$, it is straightforward to check that $Q(U,N)^* \simeq H^{\infty}(G,N)$, via the bilinear pairing

$$\langle f , \gamma \rangle = \sum_{i,j=1}^{N} \langle f_{ij} , \gamma_{ij} \rangle.$$

Let S be the scalar extension of the hyponormal operator T, acting on the supraspace $K \supset H$. For every pair of bounded open sets V, U satisfying $\overline{V} \subset U$ one introduces the map

$$l^2_N(H) \times l^2_N(K_{S^*}(\overline{V})) \rightarrow Q(U,N)$$

by the formula

$$\langle \xi\otimes\eta \, , f\rangle = \sum_{i,j=1}^{N} \langle f_{ij}(S^*)\eta_j \, , \xi_i\rangle,$$

where $\xi = (\xi_i)$, $\eta = (\eta_j)$, $f = (f_{ij})$, $1 \leq i, j \leq N$.

It is immediate to verify that the analogues of Lemmas 2.4 and 2.5 still hold for the functionals $\xi\otimes\eta$, for every finite N.

Further one denotes by $A_H^N(V,U)$ the set of those elements $\gamma \in Q(U,N)$ such that there exist sequences $\xi_n \in l_N^2(H)$ and $\eta_n \in l_N^2(K_{S^*}(\bar{V}))$, with the properties

$$\max(\|\eta_n\| , \|\xi_n\|) \leq 1, \quad n \in \mathbf{N},$$

and

$$\lim_{n\to\infty} \|\gamma - \xi_n\otimes\eta_n\|_{Q(U,N)} = 0.$$

An inspection of the proof of Lemma 2.6 will show again that the set $A_H^N(V,U)$ is absolutely convex and closed in $Q(U,N)$.

4.2. Lemma. *With the above notations, if* $\gamma \in Q(U,N)$ *satisfies* $\gamma_{ij} \in A_H(V,U)$, $1 \leq i, j \leq N$, *then* $N^{-2}\gamma \in A_H^N(V,U)$.

Proof. If only one of the entries of γ, say $\gamma_{k\ell}$, is different from zero, then the preceding definition shows that $\gamma \in A_H^N(V,U)$.

In the general case decompose $N^{-2}\gamma$ into the convex combination

$$N^{-2}\gamma = \sum_{k,\ell=1}^{N} \gamma^{k\ell},$$

where

$$\gamma_{ij}^{k\ell} = N^{-2}\delta_i^k\delta_j^\ell\gamma_{ij}.$$

4.3. Lemma. *If* W, V, U *are bounded open sets with* $\bar{W} \subset V \subset \bar{V} \subset U$, *then there is a constant* $\delta > 0$, *depending on* W, V, U *and* T, *with the property*

$$A_H^N(V^*,U^*) \supset \delta\cdot N^{-2}\cdot \mathbf{aco}\{\varepsilon_{\bar{\lambda}} \, ; \lambda \in \bar{W} \cap \sigma_c(T)\},$$

for all $N \in \mathbf{N}$.

This lemma is obtained directly from Proposition 2.7 and Lemma 3.2.

As in the preceding section, we can now derive the basic approximation result, namely:

4.4. Corollary. *Let* W, V, U, *be as in Lemma* 3.3. *If the set* $W \cap \sigma_c(T)$ *is dominating for* W, *then for any* $N \in \mathbf{N}$, $\gamma \in Q(W^*)$, $\alpha \in l^2_N(H)$, $\beta \in l^2_N(K_{S^*}(\bar{V}^*))$ *and* $\varepsilon > 0$, *there are* $\xi \in l^2_N(H)$, $\eta \in l^2_N(K_{S^*}(\bar{V}^*))$, *such that:*

$$(14)\qquad \begin{cases} \|\gamma - \xi\otimes\eta\|_{Q(U^*,N)} < \varepsilon, \\ \|(\alpha+\xi)\otimes(\beta+\eta) - \alpha\otimes\beta - \xi\otimes\eta\|_{Q(U^*,N)} < \varepsilon, \\ \max(\|\xi\|, \|\eta\|) < N\delta^{-\frac{1}{2}}\|\gamma\|^{\frac{1}{2}}_{Q(W^*,N)}. \end{cases}$$

The proof is similar to that of Corollary 2.8 and we omit it.

With these tools at hand we can proceed to the

Proof of Theorem 4.1. Assume that U is a nonempty open subset of $\sigma_c(T)$. Choose the open exhaustions $U_n \uparrow U$, $V_n \uparrow U$, so that

$$\bar{U}_n \subset V_n \subset \bar{V}^n \subset U_{n+1}, \quad n \in \mathbf{N}.$$

We shall prove that any infinite matrix $\gamma \in Q(U_1,\infty)$ can be represented as

$$\gamma_{ij} = \xi_i \otimes \eta_j, \quad i, j \in \mathbf{N},$$

where $\xi_i \in H$, $\eta_j \in K$.

This of course will imply (13) and the proof of this case will be finished.

In order to simplify notation we denote by δ_n the constant attached by Lemma 3.3 to the triple $U_n \subset V_n \subset U_{n+1}$, and let us put

$$\|\gamma\|_{n,N} = \|\gamma\|_{Q(U_n,N)}, \quad \gamma \in Q(U,N).$$

The extension by zero of a sequence $\xi \in l^2_N(H)$ will be denoted

$$\hat{\xi} = (\xi_1, \ldots, \xi_n, 0) \in l^2_{N'}(H), \quad N' > N.$$

Thus, let $\gamma \in Q(U,\infty)$ be fixed. After a normalization like $\gamma_{ij} \mapsto ((1/a_i b_j)\gamma_{ij})$, we may assume

$$\sum_{n=1}^{\infty} n\sqrt{d_n/\delta_n} < \infty,$$

where $d_n = \sum_{k=1}^{n} \|\gamma_{nk}\|_1 + \sum_{j=1}^{n-1} \|\gamma_{jn}\|_1$.

Choose a decreasing sequence $\varepsilon_n \downarrow 0$, so that

$$\sum_{n=1}^{\infty} n\sqrt{\varepsilon_{n-1}/\delta_n} < \infty .$$

We shall construct successively sequences (ξ_n), (η_n) with the properties

$$(15)\qquad \begin{cases} \xi_n \in l^2_n(H),\ \eta_n \in l^2_n(K_{S^*}(\bar{V}^*_n)), \\ \|\gamma - \xi_n \otimes \eta_n\|_{n+1,n} < \varepsilon_n \\ \|\hat{\xi}_n - \hat{\xi}_{n-1}\|_{l^2(H)},\ \|\hat{\eta}_n - \hat{\eta}_{n-1}\|_{l^2(K)} < n\sqrt{(d_n + \varepsilon_{n-1})/\delta_n}. \end{cases}$$

We put by convention $\xi_0 = \eta_0 = 0$, $\varepsilon_0 = 1$.

For $n = 1$, just apply Corollary 2.8.

Suppose (14) holds for n. Next one applies Corollary 3.4 to $N = n + 1$, $\gamma = \hat{\xi}_n \otimes \hat{\eta}_n$, $\alpha = \hat{\xi}_n$, $\beta = \hat{\xi}_n$, and one gets some elements $\xi \in l^2_{n+1}(H)$ and $\eta \in l^2_{n+1}(K_{S^*}(\bar{U}_n))$, such that (12) is fulfilled for $\varepsilon = \frac{1}{2}\varepsilon_{n+1}$. Then we take

$$\xi_{n+1} = \xi_n + \xi,\ \eta_{n+1} = \eta_n + \eta.$$

The only quantity to be computed in order to derive (15) from (14) is:

$$\|\gamma - \hat{\xi}_n \otimes \hat{\eta}_n\|_{n+1,n+1} = \|\gamma - \gamma_n\|_{n+1,n+1} + \|\gamma_n - \hat{\xi}_n \otimes \hat{\eta}_n\|_{n+1,n+1} \le d_{n+1} + \varepsilon_{n+1},$$

where γ_n is the projection of the double sequence γ onto the first $n \times n$ entries.

Thus the induction step in our construction is verified. In conclusion we have obtained two Cauchy sequences ξ_n, η_n converging to some elements $\xi \in l^2(H)$, $\eta \in l^2(K)$ which fulfil

$$\gamma = \xi \otimes \eta$$

in $Q(U,\infty)$.

This proves (11) and Theorem 3.1 in the case when $\operatorname{int} \sigma_c(T) \neq \emptyset$.

If $\mathbf{int}\, \sigma_c(T) = \emptyset$, then one argues as in Lemma 2.9 and one derives the decomposition (13).

This finishes the proof of Theorem 3.1.

The first part of the above proof gives a counterpart of Theorem 3.1 for general subdecomposable operators with interior points in their continuous spectrum. We leave to the reader the details, see Albrecht and Chevreau [1] (even for the Banach space case).

Originally, Theorem 3.1 was established for the *left essential spectrum* $\sigma_{le}(T)$:

$$\sigma_{le}(T) = \{z \in \sigma(T);\ \mathbf{dim\,Ker}\, T_z = \infty \text{ or } \mathbf{Ran}\, T_z \text{ is not closed}\}.$$

Actually $\sigma_{le}(T) = \sigma_c(T)$ for pure hyponormal operators, as follows from Proposition III.1.1.

Notes. There is an extensive literature (more than 150 titles) devoted to Scott Brown's technique of producing invariant subspaces, whose main contributors are: C. Apostol, H. Bercovici, S. Brown, B. Chevreau, C. Foiaş, C. Pearcy and many others. Their works led to important progresses in the structure theory of dual (non self-adjoint) algebras of operators, cf. Bercovici-Foiaş-Pearcy [1].

EXERCISES

1. Let T be a Hilbert space decomposable operator. Prove that for every closed set $F \subset \mathbf{C}$,

$$H_{T^*}(F^*) = H_T(\mathbf{C} \setminus F)^{\perp}.$$

2. Let $S \in L(K)$ be a decomposable operator with an invariant subspace K'. Denote by S' the quotient operator induced by S on K/K'.

a) Prove that for every open covering $(U_j)_{j=0}^{n}$ of $\mathbf{C}$, such that $\sigma(S|K') \subset U_0$, there are S'-invariant subspaces $L_j \subset K/K'$, with the properties

$$\sigma(S'|L_j) \subset U_j, \quad 0 \le j \le n,$$

$$\sum_{j=0}^{n} L_j = K/K'.$$

b) Assume $H \subset L/L'$ is an invariant subspace for S', and denote $T = S'/H$.

If $\mathbf{int}(\sigma(T) \setminus \sigma(S|K')) \neq \emptyset$, then the operator T possesses a rationally invariant subspace.

3. Let T be a subdecomposable operator with the property that there are open sets $(U_n)_{n+1}^{\infty}$ such that:

(i) $\bar{U}_n \subset U_{n+1}$, $n \ge 1$,

(ii) $\sigma(T) \cap U_n$ is dominating for U_n.

Then the operator T has non-trivial invariant subspaces.

4. a) Prove that Thomson's theorem remains valid if $L^2(\nu)$ is replaced by any space $L^p(\nu)$, $1 < p < \infty$.

b*) (Brennan [2]). For $p > 2$, there are continuous point evaluations on $R^p_\sigma(\nu)$, i.e. there exist points $w \in \mathbf{C}$, so that

$$|f(w)| \leq \mathbf{const} \|f\|_p, \quad f \in \mathbf{Rat}(\sigma).$$

c) If S is a rationally cyclic subnormal operator then there is a point $w \in \mathbf{C}$ and vectors $\xi, \eta \in H$ with the property

$$f(w) = \langle f(S)\xi, \eta \rangle, \quad f \in \mathbf{Rat}(\sigma(S)).$$

Hint. For b) use Lemma 3.2 and for a) and c) repeat the proof of Theorem 3.3.

Chapter V

OPERATIONS WITH HYPONORMAL OPERATORS

It is the aim of the present chapter to collect some of the operations which preserve the class of hyponormal operators. We have already seen in Example II.2.1 that the square of a hyponormal operator may not be hyponormal. Thus the analytic functional calculus is excluded from these operations. Instead, the main tools in transforming hyponormal operators are the rectangular (and polar) cuttings, which originate in the work of Putnam. The main part of this chapter is devoted to them. We shall also be interested in spectral mappings results, i.e. in how the spectrum transforms under these operations. For an excellent and comprehensive report on these topics we refer to the book of Xia [3].

1. Operations

Let $T \in L(H)$ be a hyponormal operator with Cartesian decomposition $T = X + iY$, $X = \mathbf{Re}(T)$, $Y = \mathbf{Im}(T)$.

Below we list some operations which transform T into a hyponormal operator.

1) *If* $\alpha, \beta \in \mathbf{C}$, *then the operator* $\alpha T + \beta$ *is hyponormal.*

2) *If* T *is invertible, then* T^{-1} *is hyponormal.*

Indeed, look at the factorization $T^* = KT$ with $\|K\| \leq 1$, see § II.3. Since T and T^* are invertible, $T^{-1} = T^{*-1}K$, whence $(T^{-1})^* = K^*T^{-1}$. This implies T^{-1} hyponormal.

3) *Assume* $\alpha, \beta, \gamma, \delta \in \mathbf{C}$, *so that the operator* $\gamma T + \delta$ *is invertible. Then*

$$(\alpha T + \beta)(\gamma T + \delta)^{-1}$$

is hyponormal.

This follows from 1) and 2).

4) *Let* $A, B \in (X)'$ *be elements in the commutant of* X *in* $L(H)$, *such that* $B^* = B$.

Then the operator

$$X + i(AYA^* + B)$$

is hyponormal.

Indeed, $i[X, AYA^* + B] = iA[X, Y]A^* \geq 0$. In particular, if $f \in L^\infty(\sigma(X))$, then by denoting $f(X) = \int f(\lambda)E(d\lambda)$, where E is the spectral measure of X, the operator

$$X + if(X)Y\bar{f}(X)$$

is still hyponormal.

Obviously similar operations exist for X replaced by Y.

1.1. Definition. Let $T = X + iY \in L(H)$ be a hyponormal operator. The *cut-down* of the operator T, with respect to the Borel set δ of the real axis, is the operator

$$T_\delta = X + iE(\delta)YE(\delta),$$

acting on the space $E(\delta)H$.

Here E stands for the spectral measure of the operator X. In view of observations **4)** above and **5)** below the operator $T_\delta \oplus 0 = E(\delta)TE(\delta)$ is hyponormal.

Analogously, if F denotes the spectral measure of the self-adjoint operator Y, then $T^\sigma = F(\sigma)TF(\sigma)$ is a hyponormal operator acting on the space $F(\sigma)H$, for every Borel subset σ of the imaginary axis. However, in general, $(T_\delta)^\sigma \neq (T^\sigma)_\delta$. The next section is essentially devoted to an important spectral property of the operators T_δ.

Now, let us come back to the list of operations.

5) *The restriction of a hyponormal operator to a closed invariant subspace is still hyponormal.*

Indeed, let K be an invariant subspace of the operator $T \in L(H)$, and denote $P = P_K^H$ and $T' = T|K$. For any vector $\xi \in K$,

$$\langle [T'^*, T']\xi, \xi\rangle = \|T'\xi\|^2 - \|T'^*\xi\|^2 = \|T\xi\|^2 - \|PT^*\xi\| \geq \|T\xi\|^2 - \|T^*\xi\|^2.$$

6) *The compression of a hyponormal operator to a closed invariant subspace of its real or imaginary part is hyponormal.*

This follows from 4).

7) *If* $T = X + iY \in L(H)$ *is hyponormal, and* X *or* Y *are invertible, then* $X^{-1} - iY$ *or* $X - iY^{-1}$ *are hyponormal operators, too.*

The proof consists in a simple algebraic observation:

$$[X^{-1}, Y] = X^{-1}[Y, X]X^{-1}.$$

After the singular integral model of a hyponormal operator will have been discussed in Chapter VII, we shall be able to characterize those real valued functions $\phi, \psi \in L^{\infty}(\mathbf{R})$, with the property that

$$\phi(X) + i\psi(Y)$$

is a hyponormal operator, whenever $X + iY$ has this property (see Exercise VII.1).

8) *The set of all hyponormal operators acting on a Hilbert space is closed in the strong operatorial topology.*

Indeed, if $\text{so}-\lim_n T_n = T$ and T_n are hyponormal operators, then from the inequalities $|\langle T_n^*\xi, \eta\rangle| \leq \|\eta\| \, \|T_n\xi\|$ we infer $\|\langle T^*\xi, \eta\rangle\| \leq \|\eta\| \, \|T\xi\|$ for any vectors ξ, η. Whence T is hyponormal, too.

The next property of the cut-down operation will be needed later.

1.2. Proposition. *Let* $T \in L(H)$ *be a hyponormal operator and let* $\delta \subset \mathbf{R}$ *be a Borel set.*

If T *is pure, then* T_δ *is also pure hyponormal.*

Proof. Let E denote the spectral measure of $X = \mathbf{Re}\, T$, and put $H_\delta = E(\delta)H$. Assume the operator $T_\delta \in L(H_\delta)$ has a normal part, i.e. that there exists a closed subspace $K \subset H_\delta$ which reduces T_δ to a normal operator. Since $T|K = T_\delta|K$ it will be sufficient to prove that the space K reduces T.

Since K reduces the operator $\mathbf{Re}\, T_\delta$ and $\mathbf{Im}\, T_\delta$, we find

$$E(\delta)[T^*, T]E(\delta)\xi = [T_\delta^*, T_\delta]\xi = 0$$

for every $\xi \in K$. Therefore

$$\| D^{\frac{1}{2}}\xi \| = \langle D\xi, \xi\rangle = \langle E(\delta)DE(\delta)\xi, \xi\rangle = 0, \quad \xi \in K.$$

That means $DK = 0$ and, since the subspace K is invariant under X, $DX^nK = \{0\}$ for all $n \geq 0$. But

$$[X^n, Y] = \sum_{p=0}^{n-1} X^{n-1-p}[X, Y]X^p,$$

whence $[X^n, Y]K = 0$ for all $n \geq 0$.

The spectral projection $E(\delta)$ is a strong limit of polynomials in X, so $[E(\delta), Y]K = 0$. Consequently $YK \subset E(\delta)YK \subset K$.

In conclusion we have proved that the subspace K is invariant for both X and Y, and the proof is complete.

2. Spectral mapping results

At this moment a question naturally arises. How does the spectrum of a hyponormal operator transform under the operations listed in the preceding section? The difficult operation from this point of view is the cut-down. Quite specifically, this section is devoted to the computation of the spectrum of a real cut-down of a hyponormal operator. In our approach this result, due to Putnam, will be needed only in establishing some properties of the principal function, see Chapter XI.

The notations for the decomposition of the spectrum are those introduced in §III.1. We begin with a few simple homotopy properties of the residual spectrum.

2.1. Lemma. *Let* $S \in L(H)$ *be an arbitrary operator and* $\gamma : [0,1] \to \mathbf{C}$ *a continuous arc, such that*

$$\gamma([0,1]) \cap \sigma(S) \neq \emptyset, \quad \text{and} \quad \gamma([0,1]) \cap \sigma_{ap}(S) = \emptyset.$$

Then $\gamma([0,1]) \subset \sigma_r(S)$.

Proof. It follows from the inclusion $\partial\sigma(S) \subset \sigma_{ap}(S)$.

The next lemma is a parametric variant of Lemma 2.1 and will be a key argument in what follows.

2.2. Lemma. *Let* $A : [0,1] \to L(H)$ *be a continuous path in the norm topology and let* $\tau : [0,1] \to \mathbf{C}$ *be continuous. Assume that*

(i) $\tau(0) \in \sigma(A(0))$,

(ii) $\tau(t) \in \mathbf{C} \setminus \sigma_{ap}(A(t))$, $t \in [0,1]$.

Then $\tau(t) \in \sigma_r(A(t))$ *for every* $t \in [0,1]$.

Proof. For all $t \in [0,1]$, $A(t) - \tau(t)$ is a semi-Fredholm operator by assumption (ii). Then by the continuity of the Fredholm index, one gets $\mathbf{ind}(A(t) - \tau(t)) = \mathbf{ind}(A(0) - \tau(0))$, $t \in [0,1]$. But $\mathbf{ind}(A(0) - \tau(0)) < 0$ by assumption (i), forcing $\tau(t)$ to be in $\sigma_r(A(t))$ for every $t \in [0,1]$.

A technical consequence of the last lemma is the following.

2.3. Proposition. *Let* $A : [0,1] \to L(H)$ *be continuous in the norm topology, let* σ *be a subset of* $\mathbf{C}$ *and* $\theta : [0,1] \times \sigma \to \mathbf{C}$ *be a function with the properties*

(i) $\theta(\cdot,\lambda) : [0,1] \to \mathbf{C}$ *is continuous for any* $\lambda \in \sigma$,

(ii) $\theta_t = \theta(t,\cdot) : \sigma \to \mathbf{C}$ *is one-to-one for any* $t \in [0,1]$ *and* $\theta_0 | \sigma = \text{id}$.

Assume that

(4) $$\sigma_{ap}(A(t)) \cap \theta_t(\sigma) = \theta_t(\sigma_{ap}(A(0)) \cap \sigma), \quad t \in [0,1].$$

Then for any $t \in [0,1]$ *one has*

$$\sigma_r(A(t)) \cap \theta_t(\sigma) = \theta_t(\sigma_r(A(0)) \cap \sigma).$$

The reader is urged to draw a picture in order to visualize this statement.

Proof. Fix a point $\lambda \in \sigma_r(A(0)) \cap \sigma$ and consider the path $\tau = \theta(\cdot,\lambda)$. By assumption

$$\tau(0) = \theta_0(\lambda) = \lambda \in \sigma(T(0)).$$

Because the map θ_t is one-to-one, assumption (4) implies $\tau(t) \in \mathbf{C} \setminus \sigma_{ap}(A(t))$ for $t \in [0,1]$. By using Lemma 2.3 one finds $\tau(t) \in \sigma_r(A(t))$ for all $t \in [0,1]$. Therefore we have proved the inclusion $\theta_t(\sigma_r(A(0)) \cap \sigma) \subset \sigma_r(A(t)) \cap \theta_t(\sigma)$.

In order to prove the converse inclusion we pick a number $s \in [0,1]$ and a point $\lambda \in \sigma_r(A(s)) \cap \theta_s(\sigma)$. Let us define the paths

$$B(t) = A(s(1 - t)), \quad t \in [0,1],$$

$$\tau(t) = \theta(s(1 - t), \theta_s^{-1}(\lambda)), \quad t \in [0,1].$$

Both are continuous functions and

$$\tau(0) = \theta(s, \theta_s^{-1}(\lambda)) = \lambda \in \sigma(A(s)) = \sigma(B(0)).$$

By the assumptions we infer $\theta_s^{-1}(\lambda) \notin \sigma_{ap}(A(0))$, whence

$$\tau(t) \notin \sigma_{ap}(A(s(1 - t))) = \sigma_{ap}(B(t)).$$

Again in view of Lemma 2.2 we conclude $\tau(t) \in \sigma_r(B(t))$ for all $t \in [0,1]$. In particular, for $t = 1$, one obtains

$$\theta(0, \theta_s^{-1}(\lambda)) \in \sigma_r(A(0)),$$

that is $\theta_s^{-1}(\lambda) \in \sigma_r(A(0))$, and the inclusion $\sigma_r(A(s)) \cap \theta_s(\sigma) \subset \theta_s(\sigma_r(A(0)) \cap \sigma)$ is verified. The proof is complete.

Before making use of the above homotopy principle in the computation of the spectrum of a cut-down operator we need the following result, which in the case of normal operators is immediate.

Let $z = \mathbf{Re}\, z + i\,\mathbf{Im}\, z$ denote the Cartesian coordinates of a complex number $z \in \mathbf{C}$, where $\mathbf{Re}, \mathbf{Im} : \mathbf{C} \to \mathbf{R}$ are the natural projections.

2.4. Theorem. (Putnam). *Let* $T = X + iY \in L(H)$ *be a hyponormal operator decomposed into Cartesian parts. Then*

$$\mathbf{Re}\sigma(T) = \sigma(X), \quad \text{and} \quad \mathrm{Im}\sigma(T) = \sigma(Y).$$

Here $\mathbf{Re}\sigma(T) = \{x \in \mathbf{R} : (\exists)\, y \in \mathbf{R},\ x + iy \in \sigma(T)\}$ and analogously for $\mathbf{Im}\,\sigma(T)$.

Proof. Let $x \in \mathrm{Re}\sigma(T)$. Then there exists a point $y \in \mathbf{R}$ with the property $z = x + iy \in \partial\sigma(T)$. Consequently there exists a sequence (ξ_n) of unit vectors in H, so that $\lim_{n\to\infty} T_z \xi_n = 0$.

But a straightforward computation shows that

$$\text{(5)} \qquad T_z^* T_z = X_x^2 + Y_y^2 + \tfrac{1}{2} D,$$

therefore $\lim_{n\to\infty} X_x^2 \xi_n = 0$. Thus $x \in \sigma_{ap}(X)$ and the inclusion $\mathbf{Re}\sigma(T) \subset \sigma(X)$ is proved.

The proof of the reverse inclusion is more delicate. In fact we can prove a little more, namely.

2.5. Lemma. *For every* $x \in \sigma(X)$ *there exists a point* $y \in \sigma(Y)$ *and a sequence of unit vectors* (ξ_n) *in* H*, so that*

$$\lim_{n\to\infty} X_x \xi_n = \lim_{n\to\infty} Y_y \xi_n = 0.$$

By taking into account relation (5) this suffices for the proof of the inclusion $\sigma(X) \subset \mathbf{Re}\sigma(T)$.

Proof of Lemma 2.5. Take, for $x \in \sigma(X)$ fixed, a sequence of unit vectors (η_n) in H so that $\lim_{n\to\infty} X_x \eta_n = 0$. This is possible because $\sigma(X) = \sigma_{ap}(X)$. Hence $\lim_n Y X_x \eta_n = 0$, and consequently the sequence

$$\| D^{\frac{1}{2}} \eta_n \| = \langle D\eta_n, \eta_n \rangle = \langle 2i[X_x, Y]\eta_n, \eta_n \rangle = \langle 2i\eta_n, YX_x\eta_n \rangle - \langle 2iYX_x\eta_n, \eta_n \rangle$$

converges to zero. Thus $\lim_n D\eta_n = 0$ and consequently

$$\lim_n X_x Y\eta_n = 0.$$

In the case $\lim_n Y\eta_n = 0$ the proof is over. Otherwise the computations can be obviously iterated with $(Y\eta_n)$ instead of (η_n), so that $\lim_n X_x Y^2\eta_n = 0$ and similarly:

$$\lim_n X_x Y^k \eta_n = 0, \quad k \geq 0.$$

By a standard approximation argument we find

$$\lim_n X_x \phi(Y)\eta_n = 0 \tag{6}$$

for any continuous function $\phi \in C(\mathbf{R})$.

Let F denote the spectral measure of the operator Y and $\Delta_0 = [-\|Y\| - 1, \|Y\| + 1]$. Then obviously $\|F(\Delta_0)\eta_n\| = \|\eta_n\| = 1$ for any $n \geq 0$. For at least one, say Δ_1, of the two halves $\Delta = [-\|Y\| - 1, 0]$ or $\Delta = [0, \|Y\| + 1]$ of the interval Δ_0, we find a subsequence (η_n^1) of (η_n), with the property $\|F(\Delta_1)\eta_n^1\| \geq \frac{1}{2}$, $n \geq 0$. An obvious continuation of this process yields a decreasing sequence of intervals

$$\Delta_0 \supset \Delta_1 \supset \ldots, \ \mathbf{length}(\Delta_k) = (\|Y\| + 1)/2^{k-1}, \quad k \geq 0,$$

and of subsequence (η_n^k) of the given sequence (η_n), with the following property:

$$\|F(\Delta_k)\eta_n^k\| \geq 1/2^k \quad \text{for } n \geq 0 \text{ and } k \geq 0.$$

Let us consider, for any k, a continuous function $\phi_k : \mathbf{R} \to [0,1]$, such that $\phi_k | \Delta_k \equiv 1$ and $\mathbf{dist}(\mathbf{C} \setminus \mathbf{supp}(\phi_k), \Delta_k) < 1/k$. It is plain to verify that

$$\|\eta_n^k\| \geq \|\phi_n(Y)\eta_n^k\| \geq \|F(\Delta_k)\eta_n^k\| \geq 1/2^k$$

for any $n \geq 0$.

Denote $\{y\} = \bigcap_{k=0}^{\infty} \Delta_k$ and consider a function $p : \mathbf{N} \to \mathbf{N}$ so that, by relation (6),

$$\|X_x \phi_n(Y)\eta_{p(n)}^n\| < 1/2^{2n}, \quad n \geq 0.$$

Take $\xi_n = \phi_n(Y)\eta_{p(n)}^n / \|\phi_n(Y)\eta_{p(n)}^n\|$, so that $\|\xi_n\| = 1$ and $\lim_n X_x \xi_n = 0$.

On the other hand, from $\sigma(\phi_n(Y)) \subset \mathbf{supp}\, \phi_n$ we infer

$$\|Y_y \xi_n\| \leq \int_{\mathbf{supp}\, \phi_n} |t - y| dt \leq \mathbf{length}(\Delta_n) + 1/n$$

for any $n \geq 0$. Accordingly $\lim_n Y_y \xi_n = 0$ and the proof of Lemma 2.5 is over.

It remains to remark that by replacing the operator T by iT the proof of Theorem 2.4 ends, too.

Now we are prepared to compute the spectrum of cut-down operators. Recall that if $T = X + iY$ is a hyponormal operator and E denotes the spectral measure of $X = \mathbf{Re}\, T$, then the operator $T_\beta = E(\beta)TE(\beta)$ is the cut-down hyponormal operator, acting on the space $H_\beta = E(\beta)H$, where β is a Borel subset of $\mathbf{R}$. We are mainly interested in the behaviour of $\sigma(T_\beta)$ where β is an interval.

2.6. Proposition. *Let* $T \in L(H)$ *be hyponormal and* $\Delta = (a,b)$ *be an open interval of* $\mathbf{R}$. *Then*

$$\sigma_{ap}(T) \cap (\Delta \times \mathbf{R}) = \sigma_{ap}(T_\Delta) \cap (\Delta \times \mathbf{R}).$$

Proof. Take $z = x + iy \in \sigma_{ap}(T)$ with $x \in \Delta$ and consider a sequence of unit vectors (ξ_n) of H with the property $\lim\limits_{n\to\infty} T_z\xi_n = 0$. By relation (5) one obtains $\lim\limits_n X_x\xi_n = 0$ and $\lim\limits_n Y_y\xi_n = 0$. Denote $\delta = \min(x - a, b - x)$, so that for any $\xi \in H$ one gets

$$\| X_x\xi \|^2 \geq \int_{\mathbf{R}\setminus\Delta} (t - x)^2 d\| E(t)\xi \|^2 \geq \delta^2 \| E(\mathbf{R}\setminus\Delta)\xi \|^2.$$

Accordingly $\lim\limits_n E(\mathbf{R}\setminus\Delta)\xi_n = 0$, therefore

$$\lim_n (T_\Delta - z)E(\Delta)\xi_n = \lim_n E(\Delta)T_z\xi_n - \lim_n E(\Delta)T_zE(\mathbf{R}\setminus\Delta)\xi_n = 0.$$

Since $\| E(\Delta)\xi_n \|$ tends to 1, we conclude that $z \in \sigma_{ap}(T_\Delta)$.

For the reverse inclusion assume that $z = x + iy \in \sigma_{ap}(T_\Delta)$, with $x \in \Delta$. Let (ξ_n) be a sequence of unit vectors in H_Δ, so that $\lim\limits_{n\to\infty} (T_\Delta - z)\xi_n = 0$. Again by (5) one obtains $\lim\limits_n X_x\xi_n = 0$ and

(7) $$\lim_n E(\Delta)Y_y\xi_n = 0.$$

Since $[T_\Delta^*, T_\Delta] = E(\Delta)[T^*, T]|H_\Delta$, we infer

$$\lim_n E(\Delta)D\xi_n = 0.$$

But D is a positive operator, and

$$\| D^{\frac{1}{2}}\xi_n \|^2 = \langle D\xi_n, \xi_n\rangle = \langle E(\Delta)D\xi_n, \xi_n\rangle,$$

therefore $\lim\limits_n D\xi_n = 0$. By taking into account that $\lim\limits_n X_x\xi_n = 0$ we find

$$\lim_n X_x Y_y \xi_n = 0.$$

Because $x \in \Delta$ and by (7) we derive by a standard approximation procedure,

$$\lim_n E(\mathbf{R} \setminus \Delta) Y_y \xi_n = 0.$$

In conclusion we have proved that $\lim_n Y_y \xi_n = 0$. That means $z \in \sigma_{ap}(T)$, and the proof is complete.

2.7. Corollary. *With the notation of the proposition, let* $\phi \in L^\infty(\mathbf{R})$, *with* $\phi \mid \Delta = 1$, *and consider the hyponormal operator*

$$\tilde{T} = T + i\phi(X)^* Y \phi(X).$$

Then $\sigma_{ap}(\tilde{T}) \cap (\Delta \times \mathbf{R}) = \sigma_{ap}(T) \cap (\Delta \times \mathbf{R})$.

Proof. $\tilde{T}_\Delta = T_\Delta$.

2.8. Theorem. (Putnam). *Let* $T \in L(H)$ *be a hyponormal operator and let* $\Delta \subset \mathbf{R}$ *be an open interval. Then*

$$\sigma(T) \cap (\Delta \times \mathbf{R}) = \sigma(T_\Delta) \cap (\Delta \times \mathbf{R}).$$

By Theorem 2.4 we conclude from this last result that $\sigma(T_\Delta)$ is $\sigma(T) \cap (\overline{\Delta} \times \mathbf{R})$ union possibly with some other points lying on the lines $\partial \Delta \times \mathbf{R}$. We shall see later (Chapter VI) that this is not the case, so that finally we shall be able to prove the relation $\sigma(T_\Delta) = \sigma(T) \cap (\overline{\Delta} \times \mathbf{R})$.

Proof of Theorem 2.8. In view of Proposition 2.6 it remains to prove the equality

$$\sigma_r(T) \cap (\Delta \times \mathbf{R}) = \sigma_r(T_\Delta) \cap (\Delta \times \mathbf{R}).$$

The essential tool in doing this is the homotopy of Proposition 2.3. Take $\sigma = \Delta \times \mathbf{R}$ and define

$$\theta : [0,1] \times \sigma \to \mathbf{C}; \quad \theta(t,\lambda) = \lambda, \ \lambda \in \sigma, \ t \in [0,1].$$

For every $t \in [0,1]$ we define the map

$$\phi_t : \mathbf{R} \to \mathbf{R}, \qquad \phi_t(x) = \begin{cases} 1 - t, & x \notin \Delta \\ 1, & x \in \Delta \end{cases}$$

Chapter VI

THE BASIC INEQUALITIES

Much relevant information on hyponormal operators is concentrated in a few estimates of the norm or of the trace-norm of their self-commutators. The first quantitative results of this nature were obtained by Kato [3] and Putnam [2]. They led to the famous Putnam inequality [6] which we have already met in the case of subnormal operators. Afterwards, Berger and Shaw [1] discovered, by more elaborated tools (the principal function), a sharper inequality which refers to the trace-class norm of the self-commutator of a hyponormal operator. Thanks to a recent observation by Voiculescu [2], Berger and Shaw's inequality is at hand with simple quasitriangularity methods.

This chapter begins with Voiculescu's proof. Once the basic inequalities established, we confine ourselves to present, far from being exhaustive, a few of their applications. The next chapters rely essentially on these estimates.

1. Berger and Shaw's inequality

First of all we need some elementary facts concerning the multiplicity of an operator. Recall that **Rat**(σ) denotes the algebra of all rational functions of a complex variable, with poles off the set $\sigma \subset \mathbf{C}$.

1.1. Definition. Let $A \in L(H)$. The *rational multiplicity* of the operator A is the smallest cardinal number $m(A)$, with the property that there is a set $\{\xi_i\}_{i=1}^{m(A)}$ of $m(A)$ vectors in H, so that

$$\bigvee \{f(A)\xi_j \,;\, f \in \mathbf{Rat}(\sigma(A)),\ 1 \leq j \leq m(A)\} = H.$$

In case $m(A) = 1$ the operator A is said to be *rationally cyclic*. We have already noted in Chapter I that the smallest rationally A-invariant subspace H_ξ which contains a given vector $\xi \in H$ has the property:

$$\sigma(A \mid H_\xi) \subset \sigma(A).$$

1.2. Lemma. *Suppose that the operators* $A, B \in L(H)$ *have disjoint spectra,* $\sigma(A) \cap \sigma(B) = \emptyset$. *Then*

$$m(A \oplus B) = \mathbf{max}(m(A), m(B)).$$

Proof. The equality follows by remarking that the projections $I \oplus 0$ and $0 \oplus I$ are uniform limits of rational functions of the operator $A \oplus B$, with poles off $\sigma(A \oplus B)$.

The main result of this section is the following.

1.3. Theorem. (Voiculescu). *Let* $T \in L(H)$ *be an operator with* $[T^*, T]_- \in C_1(H)$. *If* $A \in C_2(H)$ *and* $m(T + A) < \infty$, *then*

$$\mathbf{Tr}[T^*, T] \leq (m(T + A)/\pi)\mu(\sigma(T + A)).$$

Here S_- denotes the negative part of a self-adjoint operator S and $C_2(H)$ is the ideal of *Hilbert-Schmidt operators* on H, with the norm $|A|_2^2 = \mathbf{Tr}(A^*A)$.

Proof. We divide the proof into several steps.

I) *If* $S \in L(H)$ *and* $P \in L(H)$ *is a finite rank orthogonal projection, then*

$$\mathbf{Tr}\, P[S^*, S]P \leq |(I - P)SP|_2^2.$$

Indeed, if one denotes

$$B = PSP,\ C = PS(I - P) \text{ and } D = (I - P)SP,$$

then

$$P[S^*, S]P = [B^*, B] + D^*D - CC^*.$$

Since the operator P was chosen to be finite dimensional, $\mathbf{Tr}[B^*, B] = 0$. Therefore

$$\mathbf{Tr}\, P[S^*, S]P = |D|_2^2 - |C|_2^2 \leq |D|_2^2,$$

as desired.

II) *Assume* $S \in L(H)$ *has* $m(S) < \infty$. *Then there exists an increasing sequence of finite dimensional projections* $P_n \in L(H)$, *with the properties:*

a) $P_n \uparrow I$,

b) $\mathbf{rank}(I - P_n)SP_n \leq m(S), \quad n \geq 1$.

The proof of this statement runs as follows. Denote $m = m(S)$, $\sigma = \sigma(S)$ and take an m-tuple of vectors $\xi_1, \ldots, \xi_m$ belonging to H, such that

$$\bigvee\{f(S)\xi_j \; ; \; f \in \mathbf{Rat}(\sigma), \; 1 \leq j \leq m\} = H.$$

Consider a countable dense subset $\{z_1, z_2, \ldots\}$ of $\mathbf{C} \setminus \sigma(S)$ and denote by P_n the orthogonal projection onto the finite dimensional subspace:

$$\bigvee\{S^j(S - z_1)^{-1} \ldots (S - z_n)^{-1}\xi_k \; ; \; 1 \leq j \leq 2n, \; 1 \leq k \leq m\}.$$

Since the operator $S \in L(H)$ maps out of this space at most m vectors, we get the estimate

$$\mathbf{rank}(I - P_n)SP_n \leq m, \quad n \geq 1.$$

But every function $f \in \mathbf{Rat}(\sigma)$ can be expressed as a uniform limit of linear combinations of functions like

$$z^j(z - z_1)^{-1} \ldots (z - z_n)^{-1},$$

with $0 \leq j \leq 2n$. This means that the sequence of projections P_n converges strongly to I.

III) *Let* $S \in L(H)$ *and* $K \in C_2(H)$. *If* $S_- \in C_1(H)$, *then*

$$\mathbf{Tr}[S^*, S] \leq m(S + K) \| S + K \|^2.$$

For the proof we may of course assume that $m(S + K) < \infty$. Apply then Step II) in order to find an increasing sequence of finite rank orthogonal projections $P_n \in L(H)$, with the properties $P_n \uparrow I$ and

$$\mathbf{rank}(I - P_n)(S + K)P_n \leq m(S + K), \quad n \geq 1.$$

Because the operator K is Hilbert-Schmidt,

$$\lim_{n \to \infty} | (I - P_n)KP_n |_2 = 0.$$

According to Step I) we infer

$$\mathbf{Tr}[S^*, S] = \lim_{n \to \infty} \mathrm{Tr}\, P_n[S^*, S]P_n \leq \limsup_{n \to \infty} | (I - P_n)SP_n |_2^2 = \limsup_{n \to \infty} | (I - P_n)(S + K)P_n |_2^2 .$$

But

$$| (I - P_n)(S + K)P_n |_2^2 \leq [\mathbf{rank}(I - P_n)(S + K)P_n] \cdot \| (I - P_n)(S + K)P_n \|^2 \leq$$

$$\leq m(S + K) \| S + K \|^2,$$

and the proof of assertion III) is over.

IV) This is the last step of the proof. With the notations of the statement, put $\rho = \| T + A \|$ and $D = \{z \in \mathbf{C} ; |z| \leq \rho\}$. For any $\varepsilon > 0$, there exists a finite set $D_1, \ldots, D_k$ of closed discs contained in $D \setminus \sigma(T + A)$, such that

$$\mu(D) \leq \mu(\sigma(T + A)) + \sum_{j=1}^{k} \mu(D_j) + \varepsilon.$$

Let λ_j and ρ_j be the center and the radius, respectively, of D_j, $1 \leq j \leq k$. Then one obtains

$$\pi\rho^2 \leq \mu(\sigma(T + A)) + \pi \sum_{j=1}^{k} \rho_j^2 + \varepsilon. \tag{1}$$

Our next aim is to fill every disc D_j with the spectrum of an operator S_j, so that the spectrum of the direct sum operator $(T + A) \oplus S_1 \oplus \ldots \oplus S_k$ should be closer to the disc D and its degree of rational multiplicity should be $m = m(T + A)$. Accordingly, we pick S_j to be the direct sum of m copies of the operator $\lambda_j + \rho_j U_+$. Obviously

$$\sigma(S_j) = D_j, \quad m(S_j) = m \quad \text{and} \quad [S_j^*, S_j]_- = 0$$

for any $j = 1, 2, \ldots, k$.

Let us denote

$$S = T \oplus S_1 \oplus \ldots \oplus S_k, \quad K = A \oplus 0 \oplus \ldots \oplus 0.$$

In view of Lemma 1.2 we find $m(S) = m$, and by Step III) above we have

$$\mathbf{Tr}[S^*, S] \leq m \cdot \rho^2.$$

Since $\mathbf{Tr}[U_+^*, U_+] = 1$ it follows $\mathbf{Tr}[S_j^*, S_j] = m\rho_j^2$, $j = 1, \ldots, m$, whence

$$\mathbf{Tr}[T^*, T] + m \sum_{j=1}^{k} \rho_j^2 \leq m \cdot \rho^2.$$

By taking into account (1) we finally obtain

$$\mathbf{Tr}[T^*, T] < (m/\pi)(\mu(\sigma(T + K)) + \varepsilon).$$

As $\varepsilon > 0$ was arbitrary this ends the proof of the theorem.

1.4. Corollary. (Berger and Shaw's Inequality). *Let* $T \in L(H)$ *be a hyponormal operator. Then*

(2) $$\mathbf{Tr}[T^*, T] \leq (m(T)/\pi)\mu(\sigma(T)).$$

This follows directly from Voiculescu's theorem.

1.5. Corollary. *If a hyponormal operator* $T \in L(H)$ *has finite rational multiplicity, then* $\mathrm{Tr}[T^*, T] < \infty$.

We shall discuss further the extremal cases of Berger and Shaw's Inequality, namely the operators for which (2) is an equality. See for the beginning Exercises 2 and 3.

2. Putnam's inequality

The same inequality which was proved in Theorem I.4.1 for subnormal operators is valid for any hyponormal operator.

2.1. Theorem. (Putnam). *Let* $T \in L(H)$ *be a hyponormal operator. Then*

(3) $$\| [T^*, T] \| \leq (1/\pi)\mu(\sigma(T)).$$

Proof. Denote as usually $D = [T^*, T]$ and take a unit vector $\xi \in H$. We have to prove the inequality

(4) $$\langle D\xi, \xi\rangle \leq (1/\pi)\mu(\sigma(T)).$$

Consider the rationally cyclic subspace H_ξ spanned by the vector ξ:

$$H_\xi = \bigvee \{f(T)\xi \,;\, f \in \mathbf{Rat}(\sigma(T))\}.$$

Then the operator $S = T|H_\xi$ is still hyponormal and it has ξ as a rational cyclic vector. Moreover $\|S^*\xi\| \leq \|T^*\xi\|$, whence

$$\langle D\xi, \xi\rangle = \|T\xi\|^2 - \|T^*\xi\|^2 \leq \|S\xi\|^2 - \|S^*\xi\|^2 =$$

$$= \langle [S^*, S]\xi, \xi\rangle \leq \mathbf{Tr}[S^*, S].$$

Since $\sigma(S) \subset \sigma(T)$, Berger and Shaw's Inequality (2) yields

$$\langle D\xi, \xi\rangle \leq \mathbf{Tr}[S^*, S] \leq (1/\pi)\mu(\sigma(S)),$$

and the estimate (4) is proved.

2.2. Corollary. *A hyponormal operator* $T \in L(H)$ *with* $\mu(\sigma(T)) = 0$ *is necessarily normal.*

More can be said, namely:

2.3. Proposition. *Let* $T \in L(H)$ *be a pure hyponormal operator. If* $\Delta \subset \mathbf{C}$ *is an open disc, then* $\mu(\Delta \cap \sigma(T)) > 0$ *whenever* $\Delta \cap \sigma(T) \neq \emptyset$.

In other terms, this proposition asserts that the spectrum of a pure hyponormal operator is a compact set with *positive planar density* in each of its points. We shall see later that any compact subset of $\mathbf{C}$ which possesses this property may arise as the spectrum of a pure hyponormal operator.

Proof of Proposition 2.3. The proof uses the rectangular cuttings introduced in the preceding chapter. Let $\Delta \subset \mathbf{C}$ be an open disc with $\Delta \cap \sigma(T) \neq \emptyset$, and let $\delta \times \delta' \subset \subset \mathbf{R}^2 = \mathbf{C}$ be an open rectangle with $\delta \times \delta' \subset \Delta$ and $(\delta \times \delta') \cap \sigma(T) \neq \emptyset$. It suffices to prove that the last set has positive measure.

We denote by $T = X + iY$ the cartesian decomposition of T and by dE and dF the spectral measures of the self-adjoint operators X, respectively Y. Take two succesive cuttings of the operator T as follows

$$S = (T_\delta)^{\delta'} = F(\delta')E(\delta)TE(\delta)F(\delta').$$

The operator S is still hyponormal and, in virtue of Theorem V.2.8

$$\sigma(S) \cap (\delta \times \delta') = \sigma(T) \cap (\delta \times \delta'). \tag{5}$$

But obviously

$$\sigma(S) \subset \bar{\delta} \times \bar{\delta}',$$

so that

$$\mu(\sigma(S)) = \mu(\sigma(T) \cap (\delta \times \delta')).$$

By Proposition V.1.2 the operator S is pure hyponormal, hence $\mu(\sigma(S)) \neq \emptyset$ and the proof is complete.

2.4. Corollary. *Let* $T \in L(H)$ *be a pure hyponormal operator and let* δ *and* δ' *be open intervals of the real axis. Then*

$$\sigma((T_\delta)^{\delta'}) = \overline{\sigma(T) \cap (\delta \times \delta')}.$$

Proof. Thanks to the equality (5) the inclusion "$\supseteq$" is obvious. Denote $S = (T_\delta)^{\delta'}$ as in the above proof, and $K = \overline{\sigma(T) \cap (\delta \times \delta')}$.

Assume that there exists a point $\lambda \in \sigma(S) \setminus K$. Then an open disc Δ centered at λ will also be disjoint from the set K. In virtue of (5) we must have

$$\Delta \cap \sigma(S) \subset \Delta \cap \partial(\delta \times \delta').$$

As the boundary $\partial(\delta \times \delta')$ has planar measure zero, it follows that $\mu(\Delta \cap \sigma(S)) = 0$ and this contradicts Proposition 2.3. Hence $\sigma(S) \subset K$ and the proof is finished.

Other applications of Putnam's inequality will be presented later. The original proof of Putnam [6] was prior to Berger and Shaw's paper, hence different from the proof presented above. The main idea of the original proof was the following: first a weaker inequality (6) which will be discussed in the next section was established. Then Corollary 2.4 above was proved and finally, by dividing the spectrum of the operator T in small rectangles, the quantity $\mu(\sigma(T))$ was approximated by putting together the measures of the spectra of the cutted pieces, bounded from below by means of the weak inequalities.

3. Commutators and absolute continuity of self-adjoint operators

A series of concrete mathematical problems of quantum physics led to the observation that pairs of self-adjoint operators which are subject to certain commutation relations have necessarily only absolute continuous spectra. A pioneer in isolating such a phenomenon was C.R. Putnam (cf. Putnam [2], [3] and his book [5] which contains many examples and references). Putnam in his book and Kato in [5] gave a simple and definitive explanation of the phenomenon, by means of a weak form of Putnam's Inequality. From all these contributions did the final form of Putnam's Inequality emerge.

The purpose of this section is to present - a posteriori - this weak inequality and one of its applications to the absolute continuity of self-adjoint operators.

In order to keep the original framework we refer below to pairs of self-adjoint operators rather than to hyponormal operators.

For the convenience of the reader we first recall the definition of absolutely continuous vectors with respect to a self-adjoint operator.

Let $A \in L(H)$, $A = A^*$ and denote by dE the spectral measure of the operator A. To any vector $\xi \in H$ one associates the Borel measure ν_ξ on $\mathbf{R}$, defined by the formula

$$\nu_\xi(\sigma) = \langle E(\sigma)\xi , \xi\rangle, \quad \sigma \in B(\mathbf{R}).$$

The *vector* ξ is said to be *absolutely continuous* with respect to A, if the measure ν_ξ is absolutely continuous with respect to the linear Lebesgue measure μ_1. It turns out by a standard argument, which is left as an exercise to the reader, that the subspace $H_{ac}(A)$ of all A-absolutely continuous vectors is closed in H. Moreover $H_{ac}(A)$ is invariant under A, and its orthogonal $H_s(A) = H_{ac}(A)^{\perp}$ consists of those vectors $\xi \in H$ for which the measure ν_ξ is singular with respect to μ_1.

If $H_s(A) = 0$, then the operator A is called *absolutely continuous.* For details consult Kato [4, § X.2]. The typical example of an absolutely continuous self-adjoint operator is $A = M_t$, acting on the space $L^2([0,1], \mu_1)$, $t \in [0,1]$. The classification of absolutely continuous self-adjoint operators is discussed in Section 5 below.

3.1. Proposition. (Kato, Putnam). *Let* $A, B \in L(H)$ *be a pair of bounded self--adjoint operators, with* $i[A, B] \geq 0$ *or* $i[A, B] \leq 0$. *Then*

(6) $$\|[A, B]\| \leq (\|A\|/\pi)\mu_1(\sigma(B)).$$

Proof. Assume $i[A, B] \geq 0$ and consider the hyponormal operator $T = A + iB$, so that $[T^*, T] = 2i[A, B]$. Because

$$\sigma(T) \subset \sigma(A) \times \sigma(B) \subset [-\|A\|, \|A\|] \times \sigma(B),$$

relation (6) follows from Putnam's Inequality.

The *interaction space* of a pair of self-adjoint operators $A, B \in L(H)$ is the smallest closed reducing subspace for both A and B which contains the range of the commutator $[A, B]$. If $T \in L(H)$ is a hyponormal operator, then we easily recognize the pure space $H_p(T)$ to be the interaction space of **Re** T and **Im** T.

3.2. Proposition. *Let* $A, B \in L(H)$ *be self-adjoint operators with the interaction space* $K \subset H$. *If* $i[A, B] \geq 0$ *or* $i[A, B] \leq 0$, *then* $K \subset H_{ac}(A) \cap H_{ac}(B)$.

Proof. Assume that $i[A, B] \geq 0$. It suffices to prove that $K \subset H_{ac}(A)$, the inclusion $K \subset H_{ac}(B)$ following then by replacing (A,B) with (-B,A).

First of all we prove that

(7) $$\mathbf{Ran}\,[A, B] \subset H_{ac}(A).$$

Let E be the spectral measure of the operator A.

If $\delta \subset \mathbf{R}$ is an open interval, then

$$E(\delta)[A, B]E(\delta) = [E(\delta)AE(\delta), E(\delta)BE(\delta)],$$

By applying (6) one finds

(8) $$\|E(\delta)[A, B]E(\delta)\| \leq (\|B\|/\pi)\mu_1(\delta).$$

Take $C \geq 0$ to be the square-root of the operator $i[A, B]$. Then

$$\|E(\delta)C\|^2 = \|CE(\delta)\|^2 = \|E(\delta)[A, B]E(\delta)\|,$$

so that, for any vector $\xi \in H$, relation (8) becomes

$$\langle E(\delta)C\xi, C\xi\rangle = \|E(\delta)C\xi\|^2 \leq (\|B\|/\pi)\|\xi\|^2\mu_1(\delta).$$

In other terms, for any interval $\delta \subset \mathbf{R}$,

$$\nu_{C\xi}(\delta) \leq \textbf{const.}\,\mu_1(\delta).$$

This implies by well known arguments that the measure $\nu_{C\xi}$ is μ_1-absolutely continuous, whence $C\xi \in H_{ac}(A)$. This proves the inclusion (7).

Next we prove that for any natural number $n \geq 1$,

$$B^n[A, B]H \subset H_{ac}(A). \tag{9}$$

In order to do this, consider a variable t running the interval $(\|B\|, \infty)$. Then the operator $B - t$ is invertible and

$$[A, (B - t)^{-1}] = -(B - t)^{-1}[A, B](B - t)^{-1}.$$

In view of (7) we infer

$$(B - t)^{-1}[A, B]H \subset H_{ac}(A),$$

which of course implies (9).

Since $H_{ac}(A)$ is an invariant subspace of A, relation (9) yields

$$A^m B^n[A, B]H \subset H_{ac}(A), \quad m, n \geq 0.$$

Or, it is obvious that the interaction space K is spanned by the vectors $A^m B^n[A, B]\xi$, $m, n \geq 0$, $\xi \in H$. This completes the proof of Proposition 3.3.

3.3. Corollary. *Let* $T \in L(H)$ *be a hyponormal operator. Then*

$$H_p(T) \subset H_{ac}(\mathbf{Re}\,T) \cap H_{ac}(\mathbf{Im}\,T).$$

In particular, the real and imaginary parts of a pure hyponormal operator are absolutely continuous self-adjoint operators. This fact will be frequently used in the sequel.

4. Kato's inequality

In his fundamental papers [3] and [5] devoted to smooth operators, Kato discovered an estimate which turns out to be strongly related to Putnam's and Berger and Shaw's inequalities. It is the aim of this section to present Kato's inequality in a form which will be appropriate for the next chapter, where some functional models of hyponormal operators will be discussed.

The main objects to be investigated in this section are some integral operators with operatorial valued kernel. The natural framework to develop this subject is provided by the decomposition of a Hilbert space into a direct integral of Hilbert spaces. A few details on this notion are collected in the next section. For more details see Nielsen [1].

To fix some notations, let $E = E_\infty$ denote an infinite dimensional Hilbert space endowed with an increasing sequence of subspaces E_j, which span E and have dimension j, correspondingly. Let σ denote a compact subset of $\mathbf{R}$ and consider a measurable function

$$j : \sigma \to \mathbf{N} \cup \{\infty\}$$

which assigns to every $t \in \sigma$ a subspace $H(t) = E_{j(t)}$. We shall work on the Hilbert space $H = \int_\sigma^\oplus H(t)dt$, see Section 5 for the definition.

An important class of operators acting on a direct integral of Hilbert spaces is that of integral operators with essentially bounded kernel. They are defined as follows. Let $H = \int_\sigma^\oplus H(t)dt$ be as above and consider an operator valued kernel $k \in L^\infty(\sigma\times\sigma, L(E))$. If

$$k(s,t)\,|\,H(t)^\perp = 0 \quad \text{and} \quad k(s,t)^*\,|\,H(s)^\perp = 0$$

for almost all $(s,t) \in \sigma\times\sigma$, then

$$(K\xi)(s) = \int_\sigma k(s,t)\xi(t)dt, \quad \xi \in H,\ s \in \sigma, \tag{10}$$

defines a bounded operator on H, and it is called an *integral operator on the direct integral space* H, with the associated kernel k. From Hölder's inequality one easily finds the estimate

$$\|K\| \le \|k\|_\infty \cdot \mu_1(\sigma).$$

If E stands for the spectral measure of the operator $M_t \in L(H)$, then similarly one gets

(11) $$\|E(\delta)KE(\delta)\| \leq \|K\|_{\infty}\mu_1(\delta)$$

for any Borel subset δ of σ. This follows from the concrete action of the projection $E(\delta)$:

$$(E(\delta)\xi)(s) = \begin{cases} \xi(s), & s \in \delta \\ 0, & s \notin \delta. \end{cases}$$

It turns out that (11) is a characteristic property of the positive integral operators on H. Quite specifically, we have the next theorem.

4.1. Theorem. (Kato). *Let* K *be a positive, bounded operator acting on a direct integral* $H = \int_\sigma^+ H(t)dt$ *of Hilbert spaces. The following assertions are equivalent:*

(i) K *is an integral operator with essentially bounded kernel* k,

(ii) $\sup_\delta \|E(\delta)KE(\delta)\|/\mu_1(\delta) < \infty$, *where* δ *runs over all bounded intervals of* $\mathbf{R}$.

Moreover, in this case

(12) $$\|k\|_{\infty,\sigma\times\sigma} = \sup_\delta \|E(\delta)KE(\delta)\|/\mu_1(\delta).$$

Proof. It remains to prove the implication (ii) $\Rightarrow$ (i). So assume $K \geq 0$ and

(13) $$\|E(\delta)KE(\delta)\| \leq C^2\mu_1(\delta),$$

for any bounded interval $\delta \subset \mathbf{R}$, where $C > 0$.

Take $B = K^{\frac{1}{2}}$ as an operator from H into $X := (\mathbf{Ran}\, K)^-$, so that

$$K = B^*B \quad \text{and} \quad \mathbf{Ker}\, B^* = 0.$$

By assumption (13) one finds

$$\|BE(\delta)\|^2 = \|E(\delta)B^*\|^2 \leq C^2\mu_1(\delta), \quad \delta \subset \mathbf{R}.$$

For a fixed vector $\xi \in X$ this yields

$$\left\|\int_{\sigma\cap\delta}(B^*\xi)(t)dt\right\| \leq \mu_1(\delta)^{\frac{1}{2}}\left(\int_{\sigma\cap\delta}\|B^*\xi(t)\|^2dt\right)^{\frac{1}{2}} =$$

$$= \mu_1(\delta)^{\frac{1}{2}}\|E(\delta\cap\sigma)B^*\xi\| \leq C\|\xi\|\mu_1(\delta).$$

Thus

$$m_\xi(\delta) = \int_{\sigma\cap\delta}(B^*\xi)(t)dt$$

is an E-valued absolutely continuous Borel measure. Its Radon-Nicodym derivative $b_\xi^* = dm_\xi/dt$ is a function $b_\xi^* \in L^\infty(\sigma,E)$ with $\|b_\xi^*\| \le C\|\xi\|$. Since the space E is supposed to be separable, there is a set $\sigma(\xi) \subset \sigma$, with $\mu_1(\sigma(\xi)) = 0$, and

$$b_\xi^*(t) = \lim_{\varepsilon \downarrow 0} m_\xi([t - \varepsilon, t + \varepsilon])/2\varepsilon = (B^*\xi)(t)$$

for any $t \in \sigma \backslash \sigma(\xi)$.

Let $\Xi \subset X$ be a countable dense subset, and take $\sigma_0 = \bigcup_{\xi \in \Xi} \sigma(\xi)$, so that

$$(B^*\xi)(t) = b_\xi(t), \quad \text{for } \xi \in \Xi \text{ and } t \in \sigma \backslash \sigma_0.$$

Let $\eta \in X$ be an arbitrary vector. For any $\varepsilon > 0$ there is an element $\xi \in \Xi$ so that $\|\xi - \eta\| < \varepsilon$, whence

$$\|m_\eta(\delta) - m_\xi(\delta)\| \le C\mu_1(\delta)\cdot\varepsilon$$

for any bounded interval $\delta \subset \mathbf{R}$. For a fixed point $t \in \sigma \backslash \sigma_0$ one has

$$\limsup_{s \downarrow 0} \|m_\eta([t - s, t + s])/2s - b_\xi^*(t)\| \le C\cdot\varepsilon,$$

therefore

$$\lim_{s \downarrow 0} \sup_{s' \downarrow 0} \|m_\eta([t - s, t + s])/2s - m_\eta([t - s', t + s'])/2s'\| \le 2C\varepsilon.$$

As ε was arbitrary we infer from the last estimate that $b_\eta^*(t)$ exists. This enables us to define the measurable function $b^* \in L^\infty(\sigma, L(X,H))$ by the formula

$$b^*(t)\xi = \begin{cases} b_\xi^*(t), & t \in \sigma \backslash \sigma_0 \\ 0, & t \in \sigma_0. \end{cases}$$

Moreover, $\|b^*\| \le C$ and $\mathbf{Ran}\, b^*(t) \subset H(t)$ for any $t \in \sigma$.

Defining then $b \in L^\infty(\sigma, L(H,X))$ by $b(t) = (b^*(t))^*$ one easily finds

$$B\xi = \int_\sigma b(t)\xi(t)dt, \quad \xi \in H.$$

Therefore $K = B^*B$ is an integral operator with the essentially bounded kernel $k(s,t) = b^*(s)b(t)$.

Finally one observes that

$$\|k\|_\infty \le \operatorname*{ess\,sup}_{s,t \in \sigma} \|b^*(s)\| \, \|b(t)\| \le C^2,$$

and the proof is complete.

As a direct application of Theorem 4.1 to hyponormal operators we present the next consequence.

4.2. Corollary. (Kato's inequality). *Let* $T = X + iY$ *be a pure hyponormal operator, and denote by* n_X *the spectral multiplicity function of the self-adjoint operator* $X = \mathbf{Re}\, T$. *Then*

(14) $$\mathbf{Tr}[T^*, T] \leq (1/\pi)\mu_1(\sigma(Y)) \int_{\sigma(X)} n_X(t)dt.$$

Proof. Let E denote the spectral measure of the operator X, represented as M_t on a direct integral of Hilbert spaces. By Putnam's Inequality one obtains

$$\| E(\delta)[T^*, T]E(\delta) \| \leq (1/\pi)\mu_1(\sigma(Y))\mu_1(\delta)$$

for any bounded interval $\delta \subset \mathbf{R}$.

By Kato's theorem $[T^*, T]$ is an integral operator with essentially bounded kernel k, and

$$\| k \|_\infty \leq (1/\pi)\mu_1(\sigma(Y)).$$

Since $k(t,t) \geq 0$ a.e. we get

$$\mathbf{Tr}[T^*, T] = \int_{\sigma(X)} \mathbf{Tr}\, k(t,t)dt.$$

As $\mathbf{Tr}\, k(t,t) \leq \| k \|_\infty n_X(t)$ for any $t \in \sigma(X)$, relation (14) is proved.

Similarly one proves the symmetric inequality, namely

$$\mathbf{Tr}[T^*, T] \leq (1/\pi)\mu_1(\sigma(X)) \int_{\sigma(Y)} n_Y(t)dt.$$

4.3. Corollary. *Let* $T = X + iY$ *be a pure hyponormal operator. Assume that the self-adjoint operator* X *is diagonalized on the direct integral* $H = \int_{\sigma(X)}^{\oplus} H(t)dt$ *and that the self-commutator* $[T^*, T]$ *is represented on* H *by an integral operator with the kernel* $b^*(s)b(t)$, $b(t) \in L(H(t), X)$.

If $\mathbf{Tr}[T^*, T] < \infty$, *then* $b(t) \in C_2(H(t), X)$ *for a.a.* $t \in \sigma(X)$.

This point of view, i.e. to represent the hyponormal operator $T = X + iY$ on the direct integral space which diagonalizes X or Y, will be crucial for the next chapters.

5. SUPPLEMENT: The structure of absolutely continuous self-adjoint operators

The classification problem of self-adjoint operators guided the development of functional analysis from its origins (at the beginning of this century) until nowadays. The solution to this problem was achieved by von Neumann, in a very concise and manageable form (cf. Stone [1] for the historical evolution and Halmos [2] for von Neumann's contribution). A compulsory tool in von Neumann's classification was the notion of decomposition of a Hilbert space into a direct integral Hilbert spaces.

The purpose of this Supplement is to recall, without proofs, some basic facts concerning direct integrals of Hilbert spaces and their application to the diagonalization of self-adjoint operators. In view of the requirements of this book we confine ourselves to presenting only the classification of absolutely continuous, bounded self-adjoint operators.

Let $E = E_\infty$ denote a separable Hilbert space of infinite dimension and fix an increasing sequence of subspaces $E_j \subset E$, $1 \leq j < \infty$, so that

$$\dim E_j = j, \qquad \bigvee_{j \geq 1} E_j = E.$$

Let $j : \sigma \rightarrow \mathbf{N} \cup \{\infty\}$ be a measurable function defined on a compact subset σ of $\mathbf{R}$, and denote

$$H(t) = E_{j(t)}, \quad t \in \sigma.$$

The *direct integral* of the family of subspaces $H(t)$ with respect to the linear Lebesgue measure dt is the following closed subspace of $L^2(\sigma, E)$:

$$H = \{f \in L^2(\sigma, E)\ ;\ f(t) \in H(t) \text{ a.e.}\}.$$

This space is usually denoted by

$$H = \int_\sigma^{\oplus} H(t)dt. \tag{15}$$

Of course, the notion extends to any positive measure on $\mathbf{R}$.

The reader will easily verify that the multiplication operator M_t is self-adjoint on H. Moreover, any vector $\xi \in H$ is absolutely continuous with respect to M_t (see Section 3 for the definition). The converse is also true, as follows from the next.

5.1. Theorem. (von Neumann). *Every bounded, absolutely continuous self-adjoint operator* A, *acting on a separable Hilbert space, is unitarily equivalent to the multiplication operator* M_t, *defined on a direct integral of Hilbert spaces over* $\sigma(A)$, *and with respect to the linear Lebesgue measure.*

For a proof consult Halmos [2] or Helson [1].

This theorem provides also the classification of all bounded, absolutely continuous self-adjoint operators. Namely if the operator M_t in the statement acts on the space (15), then the function

$$n_A(t) = \mathbf{dim}\, H(t), \quad t \in \sigma(A)$$

is well defined and provides a complete unitary invariant for A. It is called the *spectral multiplicity* of the operator A. See again Halmos [2] for details.

By keeping the above notations, let us consider an operator valued essentially bounded function $f \in L^{\infty}(\sigma(A), L(E))$ which satisfies the next constraints:

(16) $$f(t)\,|\,H(t)^{\perp} = 0, \quad f(t)^*\,|\,H(t)^{\perp} = 0, \text{ a.e.}$$

Then, the formula

$$(T\xi)(t) = f(t)\xi(t), \quad t \in \sigma(A)$$

defines a linear bounded operator on H which commutes with M_t. We shall denote for simplicity

(17) $$T = \int_{\sigma(A)} f(t)dt.$$

5.2. Proposition. *The commutant* $(M_t)'$ *of the multiplication operator* M_t *in* $L(H)$ *coincides with the algebra of operators* (17), *where* $f \in L^{\infty}(\sigma(A), L(E))$ *satisfies conditions* (16). *Moreover* $\|T\| = \|f\|_{\infty,\sigma(A)}$.

The proofs of the results listed above resort to an equivalent and more flexible definition of the direct integral of Hilbert spaces. We briefly recall, for later use, this alternative approach.

Let σ be a compact subset of $\mathbf{R}$, and let H be a direct integral of Hilbert spaces as defined above by relation (15). We consider a sequence of not necessarily measurable functions $\phi_n : \sigma \to E$, $n \geq 0$, with the following properties:

(i) $\phi_n(t) \in H(t)$ and $\bigvee_{n=0}^{\infty} \phi_n(t) = H(t), \quad t \in \sigma;$

(ii) $\langle \phi_n(t), \phi_m(t) \rangle_E = 0$ for any $m \neq n$ and $t \in \sigma$;

(iii) the function $t \mapsto \|\phi_n(t)\|$ is the characteristic function of a measurable subset of σ.

To the sequence $(\phi_n)_{n \geq 0}$ one associates a linear space G of those functions $f : \sigma \to E$, which have the following properies:

(iv) the functions $t \to \langle f(t), \phi_n(t)\rangle$ are measurable;

(v) $|||f|||^2 := \sum_{n=0}^{\infty} \int_\sigma |\langle f(t), \phi_n(t)\rangle|^2 dt$ is finite.

Of course, under the previous assumptions *G* becomes a Hilbert space with the norm $|||\cdot|||$. Moreover, the multiplication operator M_t naturally acts on *G* as a bounded, absolutely continuous self-adjoint operator.

5.3. Lemma. *With the above notations, the multiplication operators* $M_t \in L(G)$ *and* $M_t \in L(H)$ *are unitarily equivalent.*

The proof of this basic lemma consists in a succesive transformation of the sequence $(\phi_n)_{n\geq 0}$ into a sequence of functions which applies unitarily on the standard basis of characteristic functions related to the space *H*.

Let us finally remark that the commutant of the operator M_t becomes, on the second description of the direct integral space, the set of those operator valued functions $f : \sigma \to L(E)$ which satisfy condition (16) and such that

$$\langle f(t)\phi_m(t), \ \phi_n(t)\rangle$$

are measurable functions on t, for every $m, n \in \mathbf{N}$. An integral operator, as those investigated in Section 4, can be defined analogously in this setting.

Notes. The original proof of Berger and Shaw's Inequality was much more intricate, involving the theory of the principal function, cf. Berger and Shaw [1]. Afterwards Berger [2] sharpened this inequality. Putnam's inequality discovery was preceeded by quite a lot of weaker similar results, synthetized in Putnam's book [5]. For the history of Proposition 3.2 see also Putnam [5]. A rapid account of the theory of direct integrals decomposition is given by Nielsen [1]. Theorem 4.3 is proved in Kato [5], in connection with the abstract perturbation theory of self-adjoint operators. The absolute continuity phenomenology (as it appeared in Proposition 3.3) is crucial in perturbation and scattering theories, cf. for instance Kato [4], Reed–Simon [3] and more recently Mourre [1].

EXERCISES

1. Let $T \in L(H)$ be an operator with the property $[T^*, T]_- \in C_1(H)$. If $A \in C_2(H)$, $m(T + A) = \infty$ and $\mu(\sigma(T + A)) = 0$, then $\mathbf{Tr}[T^*, T] \leq 0$.

2. Let $T \in L(H)$ be a hyponormal operator with finite rational multiplicity $m(T)$. If Berger and Shaw's Inequality is an equality for T, then $T = \bigoplus_{1}^{m(T)} R$, where $\sigma(R) = \sigma(T)$, $m(R) = 1$ and $\mathbf{Tr}[R^*, R] = \mu(\sigma(R))/\pi$.

3. Assume $T \in L(H)$ has the property $\| [T^*, T] \| \in \sigma_p([T^*, T])$. If Putnam's Inequality is an equality for T, then $T = R \oplus S$, where R is as in Exercise 2.

Hint. Use the eigenvector corresponding to the eigenvalue $\| [T^*, T] \|$ of $[T^*, T]$.

4. Compute the spectrum of an angular cutting of an invertible pure hyponormal operator.

5. Let $U \in L(H)$ be a unitary operator and $A \in L(H)$ be a non-negative operator, such that $A \geq UAU^*$.

a) Show that

$$\| A - UAU^* \| \leq \| A \| \; m(\sigma(U)),$$

where m is the normalized Haar measure on the torus $\mathbf{T}$.

b) Show that

$$\| A - UAU^* \| \leq \mu_1(\sigma(A)).$$

7. Let $T = UA$ be the polar decomposition of an invertible pure hyponormal operator. Prove that U is an absolutely continuous unitary operator, and that A is an absolutely continuous self-adjoint operator.

8. In the rational approximation theory one introduces the notion of *analytic capacity* $\alpha(\sigma)$ of a compact set $\sigma \subset \mathbf{C}$, as follows:

$$\alpha(\sigma) = \sup(\lim_{|z| \to \infty} |zf(z)|),$$

where the supremum is taken over all analityc functions $f \in O(\tilde{\mathbf{C}} \setminus \sigma)$, such that $f(\infty) = 0$ and $\| f \|_{\infty, \tilde{\mathbf{C}} \setminus \sigma} \leq 1$. Recall that $\tilde{\mathbf{C}} = \mathbf{C} \cup \{\infty\}$ denotes the Riemann sphere. The analytic capacity measures the obstruction to extending any bounded analytic function defined on $\tilde{\mathbf{C}} \setminus \sigma$ across the set σ. Ahlfors and Beurling's Inequality (Lemma I.4.2) shows that

$$\mu(\sigma) \leq \pi\alpha(\sigma)^2$$

for any compact set $\sigma \subset \mathbf{C}$.

Prove, by using the contractive function C, the following weak form of

Putnam's Inequality: For any hyponormal operator $T \in L(H)$, one has

$$\|[T^*, T]\| \leq \alpha(\sigma(T))^2.$$

9. Let $A \in L(H)$ be an absolutely continuous self-adjoint operator. Show that between the rational multiplicity $m(A)$ and the spectral multiplicity function n_A the following identity holds:

$$m(A) = \operatorname*{ess-sup}_{t \in \sigma(A)} n_A(t).$$

10. (Putnam). If $T = X + iY$ is a pure hyponormal operator, $X = \operatorname{Re} T$, and the spectral multiplicity n_X is bounded a.e., then $[T^*, T]$ is a compact operator.

11. a) Any cyclic self-adjoint operator which is absolutely continuous is unitarily equivalent with M_t acting on the space $L^2(\sigma, d\mu_1)$, where $\sigma \subset \mathbf{R}$ is compact.

b) Prove Theorem 5.1.

c) Prove Proposition 5.2.

12. (Conway-Putnam [1]). Let $T = X + iY$ be the Cartesian form of a pure hyponormal operator with self-commutator D.

a) If

$$\|D\| \in \sigma_{ess}(D) \quad \text{and} \quad \pi\|D\| = \mu(\sigma(T)),$$

then the self-adjoint operators $aX + bY$, $a > 0$, $b > 0$ have uniformly infinite multiplicity.

b)* There exists an irreducible subnormal operator which satisfies the conditions of **a)**.

Chapter VII

FUNCTIONAL MODELS

Kato's Theorem VI.4.1 shows with a bit extra effort that any hyponormal operator can be represented as a combination between operators of multiplication by essentially bounded functions and the Hilbert transform, all acting on a direct integral of Hilbert spaces. Daoxing Xia [1] discovered this model for the first time in the scalar case of hyponormal operators with one dimensional self-commutator. Then Kato [5] and Pincus [1] pointed out the importance of the singular model for general hyponormal operators. By means of this model the perturbation theoretical methods came into the field of hyponormal operators. Afterwards, the existence of the singular integral model for hyponormal operators became available independently from the basic inequalities, due to an observation by Muhly [1]. For this approach the reader may equally well consult Clancey's book [5].

The present chapter is devoted mainly to the existence of the singular integral model for hyponormal operators. A couple of other functional realizations of hyponormal operators, namely the Toeplitz model and the two-dimensional singular model, are also briefly discussed. A complement on the classical theory of one dimensional singular integral operators ends this chapter.

1. The Hilbert transform of vector valued functions

The reader of this section is supposed to be acquainted with the basic L^2-theory of the Fourier and the Hilbert transforms, on the real line. The facts required are summarized in the complement to this chapter. A more comprehensive treatment can be found for instance in Stein and Weiss [1].

Let E be a separable, complex Hilbert space and consider an increasing chain of subspaces $E_j \subset E$, so that

$$\dim E_j = j \quad \text{and} \quad \bigvee_{j=1}^{\infty} E_j = E, \quad 1 \leq j < \infty.$$

We shall work in the Hilbert space $L^2(\mathbf{R},E)$ of E-valued square summable functions with respect to the linear Lebesgue measure μ_1. It is worthwhile remarking that the isometric isomorphism

$$L^2(\mathbf{R},E) \simeq L^2(\mathbf{R})\tilde{\otimes}E$$

holds, where $\tilde{\otimes}$ stands for the completed Hilbertian tensor product. Consequently each of the spaces $L^2(\mathbf{R})\otimes E$ or $\bigcup_{j=1}^{\infty} D(\mathbf{R},E_j)$ is dense in $L^2(\mathbf{R},E)$.

The *Fourier transform* of a vector valued function $f \in L^2(\mathbf{R}, E)$ is defined as follows:

$$\hat{f}(s) = w\text{-}\lim_{n\to\infty}(1/\sqrt{2\pi})\int_{-n}^{n} e^{-ist}f(t)dt.$$

The weak limit means that both sides of this equality must be evaluated on a vector $\xi \in E$. Due to the fact that the Fourier transform is an isometry on the algebraic tensor product $L^2(\mathbf{R})\otimes E$, it extends up to an isometry on $L^2(\mathbf{R}, E)$. Since

$$\hat{\hat{f}} = \check{f}$$

for any $f \in L^2(\mathbf{R},E)$, the inversion formula still holds. Therefore the Fourier transform is a unitary operator from $L^2(\mathbf{R},E)$ onto $L^2(\mathbf{R},E)$.

Let us denote for simplicity $L = \bigcup_{j=1}^{\infty} L^2(\mathbf{R},E_j)$. For any $\phi \in L$, the principal value

$$(H\phi)(t) = \lim_{\varepsilon\downarrow 0} \int_{|\tau|\geq\varepsilon} (\phi(t+\tau)/\tau)d\tau$$

exists a.e. and defines a function $H\phi \in L$. From the scalar case we infer

(1) $$((1/\pi i)H\phi)^{\wedge}(s) = \mathbf{sgn}(s)\cdot\hat{\phi}(s), \quad s \in \mathbf{R},$$

whence $((1/\pi i)H)^2 = I$ and $\|H\phi\| = \pi\|\phi\|$ for any function $\phi \in L$.

This enables us to define a linear and continuous extension of the operator H on $L^2(\mathbf{R},E)$. This operator, still denoted by H, is the *Hilbert transform* of E-valued functions. We put as usually $S = (1/\pi i)H$.

For any function $\phi \in L$ one defines:

$$(P_{\pm}\phi)(t) = \lim_{\varepsilon\downarrow 0}(1/2\pi i)\int_{\mathbf{R}}((\phi(\tau)/(\tau-(t\pm i\varepsilon)))d\tau.$$

We know from the scalar case (see formula (17) below) that $P_{\pm}\phi \in L$ and that

(2) $$(P_{\pm}\phi)\hat{}(s) = \pm\chi_{[0,\infty)}(\pm s)\hat{\phi}(s), \quad s \in \mathbf{R},\ \phi \in L.$$

Therefore $\|P_{\pm}\phi\| \leq \|\phi\|$ for any $\phi \in L$. Thus we may extend by continuity the operators $P_{\pm}$ to $L^2(\mathbf{R},E)$. Of course the extensions preserve the algebraic relations (see Section 6 below):

(3) $$P_+ + P_- = S \text{ and } P_+ - P_- = I.$$

1.1. Proposition. *For any function* $f \in L^2(\mathbf{R},E)$,

$$(P_{\pm}f)(x) = \lim_{\varepsilon\downarrow 0} (1/2\pi i) \int_{\mathbf{R}} (f(t)/(t - (x \pm i\varepsilon)))dt,$$

and

$$(Hf)(x) = \lim_{\varepsilon\downarrow 0} \int_{|\tau|\geq\varepsilon} (f(x + \tau)/\tau)d\tau,$$

where the limits are taken in the norm of the space $L^2(\mathbf{R},E)$.

Proof. For every $\varepsilon > 0$ consider the operator

$$(P_+^{\varepsilon}f)(x) = (1/2\pi i) = \int_{\mathbf{R}}(f(t)/(t - (x + i\varepsilon)))dt, \quad f \in L^2(\mathbf{R},E).$$

One easily computes the Fourier transform of this transformation:

$$(P_+^{\varepsilon}f)\hat{}(s) = e^{-\varepsilon s}\chi_{[0,\infty)}(s)f(s), \quad s \in \mathbf{R}.$$

In particular this shows that for any $\varepsilon > 0$:

$$\|P_+^{\varepsilon}\| \leq 1.$$

On the other hand we learn from (2) that

$$(P_+f)\hat{}(s) = \chi_{[0,\infty)}(s)f(s)$$

for any $f \in L^2(\mathbf{R},E)$. Therefore

$$\|(P_+^{\varepsilon} - P_+)f\|^2 = \|(P_+^{\varepsilon}f)\hat{} - (P_+f)\hat{}\|^2 = \int_0^{\infty}(e^{-\varepsilon s} - 1)^2 \|f(s)\|_E^2 ds.$$

By Lebesgue's dominated convergence theorem we conclude that

$$\lim_{\varepsilon\downarrow 0} \|P_+^{\varepsilon}f - P_+f\|_{L^2(\mathbf{R},E)} = 0.$$

One deals similarly with the operators P_- and H. This concludes the proof of the proposition.

1.2. Corollary. (Sohotskii-Plemelj-Privalov (SPP) formulae). *Let* $f \in L^2(\mathbf{R},E)$. *There is a set* N(f) *of null Lebesgue measure, such that*

$$(1/\pi i)(Hf)(x) = \lim_{\varepsilon \downarrow 0} (1/\pi i) \int_{\mathbf{R}} (f(t)/(t - (x \pm i\varepsilon))) \mp f(x),$$

for all $x \in \mathbf{R} \setminus N(f)$, *where the limit is taken in* E.

Proof. From (3) we already know that

$$(1/\pi i)H = P_{\pm} \mp I.$$

Therefore it suffices to remark, in view of Proposition 1.1, that

$$\lim_{\varepsilon \downarrow 0} \| (P^{\varepsilon}_{\pm} f)(x) - (P_{\pm} f)(x) \|_E = 0$$

for almost all $x \in \mathbf{R}$.

In fact the only things we really need in this chapter are the algebraic properties of the Hilbert transform as they result from the Fourier transform formula (1). However, later on we shall use the SPP formulae. Proposition 1.1 and Corollary 1.2 still hold for any integrable E-valued function and with respect to any positive measure on $\mathbf{R}$. In this case, however, the proofs are substantially more difficult, see Asano [1]. We also point out that for the Fourier transform method it was essential that E be a Hilbert space rather than a Banach space.

2. The singular integral model

Let $T \in L(H)$ be a hyponormal operator with the Cartesian decomposition $T = X + iY$ and self-commutator $D = [T^*, T]$.

By von Neumann's theorem we may assume that $H = \int^{\oplus}_{\sigma(X)} H(t)dt$ and $X = M_t$. Throughout this section we denote for simplicity $\sigma = \sigma(X)$. Above, all $H(t)$ are subspaces in a fixed separable Hilbert space E.

Kato's Theorem VI.4.1 tells us that the operator D acts as an integral operator with essentially bounded kernel, on the direct integral space H. More precisely, there exist a Hilbert space X and a function $b \in L^{\infty}(\sigma, L(E,X))$, with the properties:

(4) $$(\pi/2)(D\xi)(t) = \int_{\sigma} b^*(t)b(s)\xi(s)ds; \quad \xi \in H,\ t \in \sigma,$$

and

(5) $$\operatorname*{ess\,sup}_{t\in\sigma} \| b(t) \|^2 \leq \mu_1(\sigma(Y))/2.$$

Next we denote by M_t, M_b and $S = (1/\pi i)H$ the corresponding multiplication operators, respectively the normalized Hilbert transform, on $L^2(\sigma,E)$. Since

$$2i([M_t\,, -S]\eta)(x) = -(2i/\pi i)\int_\sigma ((x - t)\eta(t)/(t - x))dt = (2/\pi)\int_\sigma \eta(t)dt,$$

relation (4) yields

$$2i[M_t\,, -M_b^* S M_b] = 2iM_b^*[M_t\,, -S]M_b = D.$$

Therefore $M_t - iM_b^* S M_b$ is a hyponormal operator with the same self-commutator as T. Consequently $[M_t\,, Y + M_b^* S M_b] = 0$, and by Proposition VI.4.2 there is a function $a \in L^\infty(\sigma, L(E))$, such that $Y = M_a - M_b^* S M_b$ and

$$a(t)^* = a(t), \quad a(t)\,|\,H(t)^\perp = 0$$

for almost all $t \in \sigma$.

In conclusion we have proved the next theorem.

2.1. Theorem. *Every hyponormal operator* $T \in L(H)$ *is unitarily equivalent to an operator* T *acting on the direct integral of Hilbert spaces* $H = \int_{\sigma(X)}^{\oplus} H(t)dt$, *which diagonalizes* $X = \mathbf{Re}\,T$, *and which is given by the formula:*

$$(T\xi)(x) = x\xi(x) + i[a(x)\xi(x) - (1/\pi i) \int_{\sigma(X)} (b^*(x)b(t)\xi(t)/(t - x))dt],$$

where $\xi \in H$ *and* $a = a^*$, b *are operator valued, essentially bounded functions on* $\sigma(X)$.

The symmetric statement, with the self-adjoint operator $Y = \mathbf{Im}\,T$ diagonalized instead of X, holds too.

2.2. Corollary. *Let* $T = X + iY$ *be a pure hyponormal operator. If* $X = \mathbf{Re}\,T$ *has a cyclic vector and* $\mathbf{rank}\,[T^*, T] = 1$, *then the operator* T *is unitarily equivalent to the operator* $T : L^2(\sigma(X)) \to L^2(\sigma(X))$:

$$(Sf)(x) = xf(x) + i[a(x)f(x) - (1/\pi i) \int_{\sigma(X)} (\overline{b(x)}b(t)f(t)/(t - x))dt],$$

where $a, b \in L^\infty(\sigma(X))$, *and* $a = \bar{a}$ a.e.

For the proof it suffices to remark from the construction of the singular integral model that the operators b(t) may be chosen with values in the space $X = (\mathbf{Ran}\,[T^*, T])^-$. We already met the singular operators described in Corollary 2.2 in Chapter II.

Our next aim is to characterize the purity of a hyponormal operator in terms of its singular integral model.

2.3. Proposition. *Let* T *be a hyponormal operator with functional model as given by Theorem* 2.1. *The operator* T *is pure hyponormal if and only if*

$$H(x) = \bigvee_{n=0}^{\infty} a(x)^n \mathbf{Ran}\,[b^*(x)]$$

for almost all $x \in \sigma(X)$.

Proof. Let us denote by $K(x)$ the right hand term in the equality asserted by the statement. Put $K = \int^{\oplus} K(x)dx$.

One easily remarks that K is a reducing subspace for the singular operator T (the model of T). Moreover,

$$\mathbf{Ran}\,[T^*, T]^- = \int^{\oplus} \mathbf{Ran}\, b^*(x)^- dx,$$

whence K is the smallest reducing subspace of T which contains the range of the self-commutator $[T^*, T]$. By virtue of Corollary I.1.5, the hyponormal operator T is pure if and only if $H = K$.

2.4. Corollary. *Let* T *be a pure subnormal operator with functional model as given by Theorem* 2.1. *Then*

$$H(x) = (\mathbf{Ran}\, b^*(x))^-$$

for almost all $x \in \sigma(X)$.

Proof. If $N = \begin{bmatrix} T & A \\ 0 & B \end{bmatrix}$ denotes a normal extension of the model T of T, written in matricial form, then the relation $[N^*, N] = 0$ implies:

$$[T^*, T] = AA^* \quad \text{and} \quad T^*A = AB^*.$$

Therefore

$$\mathbf{Ran}\, A = \mathbf{Ran}[T^*, T] = \{b^*(x)\xi, \xi \in X\}^-.$$

But the singular integral expression of the operator T^* shows that the function

$a(x)b^*(x)\xi$ belongs to the subspace **Ran** A for every $\xi \in X$. According to Proposition 2.3, **Ran** $b^*(x)$ is a dense subspace of the fibre $H(x)$, for almost all $x \in \sigma(X)$.

3. The two-dimensional singular integral model

Hyponormal operators can be equally well modelled as singular integral operators on a Hilbert space of functions depending on two real variables. This functional representation contains as an independent and final parameter only the unitary invariant distribution Γ_T, introduced in Chapter II. At this moment, however, this or other two dimensional models cannot supply the deep insight brought in the field by the one dimensional model. For this reason we shall refer only occasionally to the present section.

Let $T \in L(H)$ be a pure hyponormal operator. The contractive operator valued function $C \in L^\infty(\mathbf{C}, L(H))$ was introduced in Section II.1 by the pointwise conditions

$$T_z^* C(z) = D^{\frac{1}{2}}, \quad C(z)^* \,|\, \mathbf{Ker}\, T_z^* = 0; \quad z \in \mathbf{C}.$$

Recall the notations $D = [T^*, T]$, $X = (\mathbf{Ran}\, D)^-$. In Theorem II.4.6 it was proved that the distribution $\Gamma_T = -D^{\frac{1}{2}} \partial C \in D'(\mathbf{C}, L(X))$ is a complete unitary invariant for T. The operator T can be recovered from the function C (and implicitly from Γ_T) as follows.

Consider the Hilbert space H_T obtained as the separate completion of $D(\mathbf{C}, X)$ in the seminorm

$$\|\phi\|_T = \|\int_{\mathbf{C}} C(z)\, \partial\phi(z) d\mu(z)\|, \quad \phi \in D(\mathbf{C}, X).$$

3.1. Theorem. *The pure hyponormal operator* $T \in L(H)$ *is unitarily equivalent to the operator* $\tilde{T} \in L(H_T)$ *given by the formula*

$$\tilde{T}^* f = \bar{z} f, \quad f \in H_T.$$

Proof. The map $U : D(\mathbf{C}, X) \longrightarrow H$:

$$U(\phi) = \int_{\mathbf{C}} C(z)\, \partial\phi(z) d\mu(z), \quad \phi \in D(\mathbf{C}, X),$$

obviously determines an isometry $U : H_T \longrightarrow H$. In view of Proposition II.4.1 U is onto, hence it is unitary. For any function $\phi \in D(\mathbf{C}, X)$,

$$U(\bar{z}\phi) = \int C\, \partial(\bar{z}\phi) d\mu = \int \bar{z} C\, \partial\phi d\mu = T^* \int C\, \partial\phi d\mu = T^* U(\phi).$$

In conclusion U^*TU is the desired operator.

3.2. Corollary. *With the notations of Theorem* 3.1,

$$\tilde{T}\phi = z\phi - (1/\pi)(\Gamma_T\cdot\phi) * (1/\bar{z}),$$

and

$$[\tilde{T}^*, \tilde{T}]\phi = (1/\pi)\int D^{\frac{1}{2}}C\,\partial\phi d\mu,$$

for any function $\phi \in D(\mathbf{C}, X)$.

Proof. Both formulae must be interpreted in H_T, i.e. as the representatives of the two functions in this space.

Let us recall the relation

$$T_z(\partial C)\cdot\phi = \lim_{\varepsilon\downarrow 0} CD^{\frac{1}{2}}\partial C_\varepsilon\phi, \quad \phi \in D(\mathbf{C}, X),$$

proved in Chapter II. Accordingly, one gets

$$TU(\phi) = \int TC(z)\,\partial\phi(z)d\mu(z) = \int C(z)\,\partial(z\phi)(z)d\mu(z) - \langle T_z\,\partial C\cdot\phi\,, \mathbf{1}\rangle =$$

$$= U(z\phi) + \lim_{\varepsilon\downarrow 0}\int C(z)\Gamma_\varepsilon(z)\phi(z)d\mu(z) = U(z\phi) - \pi^{-1}U(\lim_{\varepsilon\downarrow 0}(\Gamma_\varepsilon\phi)*(1/\bar{z})) =$$

$$= U(z\phi - \pi^{-1}(\Gamma\cdot\phi)*(1/\bar{z})).$$

Since $U(\phi) = 0$ for every function $\phi \in D(\mathbf{C}, X)$, which satisfies $\mathbf{supp}\,\phi \cap \sigma(T) = \emptyset$, we have adopted the notational convention

$$U(\psi) = U(\chi\psi), \quad \psi \in E(\mathbf{C}, X),$$

where χ is any function in $D(\mathbf{C})$, such that $\chi = 1$ in a neighbourhood of $\sigma(T)$.

The expression of the operator T can be easily transformed into the next non-distributional form:

$$\tilde{T}\phi = z\phi - D^{\frac{1}{2}}C\phi - \pi^{-1}\int_{\mathbf{C}}(D^{\frac{1}{2}}C(\zeta)\,\partial\phi(\zeta)/(\bar{\zeta} - \bar{z}))d\mu(\zeta).$$

Consequently

$$U^*[T^*, T]U\phi = [\tilde{T}^*, \tilde{T}]\phi = \pi^{-1}\int_{\mathbf{C}} D^{\frac{1}{2}}C(\zeta)\,\partial\phi(\zeta)d\mu(\zeta),$$

for any function $\phi \in D(\mathbf{C}, X)$.

This ends the proof of Corollary 3.2.

The last part of this section is devoted to an abstract characterization of the unitary invariant Γ_T. This characterization points out a minimal set of properties for an operator valued distribution to coincide with some Γ_T with T hyponormal.

3.3. Theorem. *Let X be a separable Hilbert space and let $\Gamma \in E'(\mathbf{C},L(X))$. The following conditions are equivalent:*

(i) *There exists a measurable function* $K : \mathbf{C} \times \mathbf{C} \to L(X)$ *with the properties:*

a) *K is a positive definite kernel;*

b) *there is a constant M, such that*

$$\| K(w,z) \| \leq M(1 + |wz|)^{-1}, \quad (w,z) \in \mathbf{C}^2;$$

c) *for every function* $\phi \in D(\mathbf{C},X)$ *the following relation*

(6) $$(w - z)[\partial_z K(w,z)]\phi(z) = \Gamma(z)\phi(z) - \lim_{\varepsilon \downarrow 0} K(w,z)\Gamma_\varepsilon(z)\phi(z)$$

holds in the weak topology of the space $D'(\mathbf{C}^2,X)$.

(ii) *There exist a pure hyponormal operator T, acting on a Hilbert space H, and an isometry* $U : \mathbf{Ran}\,[T^*, T]^- \to X$, *with the property* $\Gamma = U\Gamma_T U^*$.

Proof. (ii)$\Rightarrow$(i). This implication is a direct consequence of Proposition II.3.6, by remarking that the kernel $K(w,z) = UC(w)^*C(z)U^*$ fulfils conditions a), b), c).

(i) $\Rightarrow$(ii) The proof of this implication is divided in four steps.

I) By Kolmogorov Theorem I.5.1 one finds a factorization

$$K(w,z) = C_1(w)^*C_1(z), \quad (w,z) \in \mathbf{C}^2,$$

where $C_1 : \mathbf{C} \to L(X,M)$ is a measurable function and M is an auxiliary Hilbert space. We choose the space M to be minimal, that is so that $\bigvee_{z \in \mathbf{C}} \mathbf{Ran}\, C_1(z) = M$.

Property b) of the kernel K implies

(7) $$\| C_1(z) \|^2 \leq M(1 + |z|^2)^{-1}, \quad z \in \mathbf{C}.$$

By looking at condition c) off the compact set $\mathbf{supp}(\Gamma)$, one finds

$$(w - z)[\partial_z K(w,z)] = 0, \quad z \notin \mathbf{supp}(\Gamma).$$

Since the distribution $(w - z)[\partial_z K(w,z)]$ cannot be supported by the diagonal without being zero (because of its order), one obtains $\partial_z C_1(z) = 0$ for $z \notin \mathbf{supp}(\Gamma)$. Accordingly, the antianalytic function C_1 has a power series expansion at infinity, which, by taking into account (7), has no free term:

$$C_1(z) = -R/\bar{z} + \sum_{n \geq 1} R_n/\bar{z}^{n+1}, \quad |z| \gg 0.$$

Let us write equation (6) for large values of $|w|$:

$$(w - z)C_1^*(w)\,\partial C_1(z)\phi(z) = \Gamma(z)\phi(z) - C_1^*(w)\lim_{\varepsilon\downarrow 0} C_1(z)\Gamma_\varepsilon(z)\phi(z).$$

By identifying term by term the power series expansion in w of both sides, one obtains

$$\Gamma = -R^*\,\partial C_1.$$

Denote $F = -R^*C_1 \in L^\infty(\mathbf{C})\hat{\otimes}_\varepsilon L(X)$, so that $\Gamma = \partial F$, and

$$\lim_{|z|,|w|\to\infty} w\bar{z}K(w,z) = \lim_{|z|\to\infty} \bar{z}F(z) = R^*R.$$

We put $A = (R^*R)^{\frac{1}{2}} \in L(M)$.

II) Next we shall consider the separate completion Hilbert space $\boldsymbol{H}$ of the space $D(\mathbf{C},X)$ in the seminorm

$$\|\phi\|_{\boldsymbol{H}}^2 = \int\langle K(w,z)\,\partial\phi(z),\ \partial\phi(w)\rangle d\mu(w)d\mu(z).$$

Because the kernel K is antianalytic in z off **supp** (Γ), we obtain $\|\phi\|_{\boldsymbol{H}} = 0$ whenever **supp**(ϕ) is disjoint of **supp**(Γ). Whence the multiplication operator with $\bar{z}$ is bounded in the seminorm $\|\cdot\|_{\boldsymbol{H}}$. Let S^* denote the continuous extension of this operator on $\boldsymbol{H}$. We shall write formally

$$S^*\phi = \bar{z}\phi.$$

Throughout this proof ϕ and ψ denote elements of the space $D(\mathbf{C},X)$.

We claim that the adjoint S of S^* acts by the formula:

(8) $$S\phi = z\phi - \pi^{-1}(\Gamma\cdot\phi)*(1/\bar{z}).$$

Indeed, at the level of distributions equality (8) is equivalent to:

$$\partial(S\phi) = z\,\partial\phi + \phi + \Gamma\cdot\phi = z\,\partial\phi + \phi + \lim_{\varepsilon\downarrow 0}\Gamma_\varepsilon\phi.$$

Accordingly

$$\langle S\phi, \psi\rangle_{\boldsymbol{H}} = \int\langle K(w,z)\,\partial(S\phi)(z),\ \partial\psi(w)\rangle d\mu(z)d\mu(w) =$$

$$= \int\langle K(w,z)z\,\partial\phi(z),\ \partial\psi(w)\rangle d\mu(z)d\mu(w) + \int\langle K(w,z)\phi(z),\ \partial\psi(w)\rangle d\mu(z)d\mu(w) +$$

$$+ \lim_{\varepsilon\downarrow 0}\int\langle K(w,z)\Gamma_\varepsilon(z)\phi(z),\ \partial\psi(w)\rangle d\mu(z)d\mu(w).$$

Here we have implicitly used the existence of the weak limit

$$K(w,z)\Gamma\phi = \lim_{\varepsilon\downarrow 0} K(w,z)\Gamma_\varepsilon(z)\phi(z),$$

asserted by condition c).

By taking into account equation (6), one finds from the above computations

$$<S\phi , \psi> = <\phi , S^*\psi>,$$

as desired.

A non-distributional form of the operator S can be easily derived from (8):

$$S\phi = (z + F)\phi + \pi^{-1}\int (F(\zeta)\partial\phi(\zeta)/(\overline{\zeta} - \overline{z}))d\mu(\zeta).$$

This formula shows that the range of the operator $[S^* , S]$ consists of classes in the space H of identically constant functions in a neighbourhood of **supp** (Γ). If $\chi\otimes x$ denotes a representative of such a function, with $\chi \in D(\mathbf{C})$, $\chi \equiv 1$ on **supp** (Γ) and $x \in X$, then the residue formula at ∞ for the antianalytic function C_1 gives

$$\int K(w,z)\partial\chi(z)d\mu(z) = \pi C_1^*(w)R.$$

Hence

$$<\chi\otimes x , \phi>_H = <x , \pi\int R^* C_1(w)\partial\phi(w)d\mu(w)>,$$

and consequently

$$<[S^* , S]\phi , \phi> = || \int F(z)\partial\phi(z)d\mu(z) ||^2.$$

In conclusion we have proved that S is a hyponormal operator. It will be the desired operator provided we compare the kernels $C(w)^*C(z)$ and $K(w,z)$.

III) By the definition of the scalar product of the space $\boldsymbol{H}$ one finds that

$$V\phi = \int C_1(z)\partial\phi(z)d\mu(z)$$

is an isometry $V : \boldsymbol{H} \to M$. Let us denote

$$T = VSV^*.$$

Therefore T is a hyponormal operator acting on the space M. By construction, the operators T and T^* vanish on the space $M \ominus V\boldsymbol{H}$. Relation (8) and the definition of S^* imply

$$T^*\int C_1\partial\phi d\mu = \int C_1\overline{z}\partial\phi d\mu$$

and

$$[T^* , T]\int C_1\partial\phi d\mu = RR^*\int C_1\partial\phi d\mu = A^2\int C_1\partial\phi d\mu.$$

Accordingly $\partial(T_z^*C_1) = 0$ in the sense of distributions and $[T^* , T]^{\frac{1}{2}} = A$. By looking at the value at ∞ of the anti-entire function $T_z^* C_1$ one gets

(9) $$T_z^* C_1(z) = R, \quad z \in \mathbf{C}.$$

On the other hand, the contractive function attached to the hyponormal operator T satisfies $T_z^* C(z) = A$. If $R^* = UA$ is the polar decomposition of the operator R^*, then relation (9) implies

(10) $$T_z(C_1(z) - C(z)U^*) = 0, \quad z \in \mathbf{C}.$$

IV) It remains to prove that $C_1 = CU^*$. Then the kernels $UC(w)^*C(z)U^*$ and $K(w,z)$ would be equal and the proof would be finished.

Let $M = M_p \oplus M_n$ be the orthogonal decomposition of M into pure and normal parts, with respect to the hyponormal operator T. Notice that the operators $C(z)$, $z \in \mathbf{C}$, take their values only in M_p.

Because H is a separable Hilbert space, the set of those complex numbers z with the property $\mathbf{Ker}\, T_z^* \cap \mathbf{Ker}\, T_z \neq 0$ is at most countable. Consequently, relation (10) yields

$$C_1(z) = C(z)U^* \quad \text{for a.a. } z \in \mathbf{C}.$$

Since $M_p \subset VH$, the subspace VH is generated by vectors of the form $\int (C' - CU^*)\, \partial \phi d\mu$ and $\int CU^* \partial \phi d\mu$. In conclusion we get $M_p = VH$. This shows that the operator $T | VH$ is pure.

Equation (6) implies the next equality of distributions:

(11) $$T_z \partial C' \cdot \phi = -\lim_{\varepsilon \downarrow 0} C' \Gamma_\varepsilon \phi.$$

On the other hand, a similar expression holds for the function CU^*, namely:

(12) $$T_z \partial CU^* \cdot \phi = \lim_{\varepsilon \downarrow 0} CU^* UD^{\frac{1}{2}} \partial C_\varepsilon U^* \phi.$$

By virtue of Corollary II.4.3, the invariant Γ is completely determined by the values of the function $C(z)^*C(z)$ for $|z|$ large. Whence we get

$$\Gamma = -UD^{\frac{1}{2}} \partial CU^*.$$

From (11) and (12) we obtain:

$$T_z \partial (C_1 - CU^*) \cdot \phi = 0.$$

By integrating the last identity as in the proof of Proposition II.4.1 one finds

$$AT^n(C_1 - CU^*) = 0$$

for every non-negative integer n. By relation (9) one derives

$$AT^nT^{*m}(C_1 - CU^*) = 0, \quad n, m \geq 0.$$

Since T is a pure hyponormal operator on the space $V\mathcal{H}$, the last identity implies $C_1 = CU^*$, the equality being considered in the space $L^\infty(\mathbf{C})\hat{\otimes}_\varepsilon L(X)$.

The proof of Theorem 3.3 is complete.

Some applications of Theorem 3.3 will be discussed in the last two chapters of this book.

4. The Toeplitz model

This section is purely informative and is independent of the rest of the book. We reproduce below, after Sz.-Nagy and Foiaş [3], the construction of a Toeplitz type model for hyponormal operators. For an application of this model to operators with one dimensional self-commutators see Clancey [6]. The terminology used in this section is detailed in the monograph of Sz.-Nagy and Foiaş [1].

Recall that the *Toeplitz operator* T_ϕ with symbol $\phi \in L^\infty(\mathbf{T})$ and acting on the Hardy space $H^2(\mathbf{T})$ is defined by the formula

$$T_\phi f = P(\phi f), \quad f \in H^2(\mathbf{T}),$$

where P is the orthogonal projection of $L^2(\mathbf{T})$ onto $H^2(\mathbf{T})$. The class of Toeplitz operators is remarkable by its properties, a large amount of work being devoted to it.

Let U_+ stand for the unilateral shift on $H^2(\mathbf{T})$. It is a Toeplitz operator $U_+ = T_z = M_z$, where $z \in \mathbf{T}$ denotes the coordinate. A straightforward computation shows that

$$U_+^* T_\phi U_+ = T_\phi, \quad \phi \in L^\infty(\mathbf{T}).$$

It was this relation which was generalized by Sz.-Nagy and Foiaş as follows. Fix a Hilbert space H and a contraction $K \in L(H)$. An operator $T \in L(H)$ is said to be *K-Toeplitz* if

$$K^*TK = T.$$

Then the operator T has a "symbol" acting on the minimal isometric dilation space of K, which is naturally related to T, exactly as in the classical case.

Our interest for these topics is motivated by the fact that hyponormal operators fit into this scheme. More precisely, let $T \in L(H)$ be hyponormal and let $K \in L(H)$ be the unique contraction subject to the relations:

$$T^* = KT \quad \text{and} \quad K|\mathbf{Ker}\, T^* = 0,$$

see §II.3. Then

$$T = T^*K^* = KTK^*,$$

so that T is a K^*-Toeplitz operator. This simple fact has the following consequences:

4.1. Theorem. (Sz.-Nagy and Foiaş). *Let* $T \in L(H)$ *be a hyponormal operator. There exists a Hilbert space* K, *a normal operator* N, *a unitary* U *on* K, *and a contraction* $R \in L(H,K)$ *with the following properties:*

1) $T = R^*NR$ *(i.e.* T *is* R*-Toeplitz),*
2) $\|N\| = \|T\|$,
3) $NU = UN = N^*$ *(i.e.* N *is* U*-Toeplitz),*
4) $\|R^*U\xi\| \leq \|R^*\xi\|, \quad \xi \in K$,
5) $K = \bigvee_{n=1}^{\infty} U^n RH$, *and*
6) $\sigma(\lambda N + \mu N^*) \subset \sigma_{ap}(\lambda T + \mu T^*)$, *for any scalars* $\lambda, \mu \in \mathbf{C}$.

The idea of the proof is the following. Consider the minimal isometric dilation V of the contraction K, acting on the Hilbert space

$$L = \bigvee_{n=0}^{\infty} V^n H.$$

The unitary part U of V acts on the space

$$K = \bigcap_{n=0}^{\infty} V^n L \subset L.$$

The operator R in the statement is the restriction to H of the orthogonal projection of L onto K. Since

$$T = K^n T K^{*n}, \quad n \geq 0,$$

the following strong operatorial limit exists

$$A' = \text{so-}\lim_{n\to\infty} V^n T K^{*n}.$$

Then the operator $N \in L(K)$ is defined by the continuous extension of

$$NU^nA = U^nA', \quad n \geq 0.$$

For a proof of the spectral inclusion 6) consult Sz.-Nagy and Foiaş [3].

The process can be reversed, in order to derive hyponormal operators from a normal one. More exactly, let $N \in L(K)$ be a normal operator and let V be any isometry on K which satisfies

$$N^* = VN.$$

Take a V^*-invariant subspace H of K and denote by $P = P_H^K$ the orthogonal projection. Then

$$T = PN|H$$

is a hyponormal operator, and $K = PV|H$ is a contraction which satisfies $T^* = KT$. The normal symbol of the operator T is then the compression of N to the minimal isometric dilation space of K:

$$\bigcap_{n \geq 0} \bigvee_{m \geq n} V^m H,$$

regarded as a subspace of K.

5. SUPPLEMENT: One dimensional singular integral operators

a) The Cauchy singular integral

Let $\Omega \subset \mathbf{C}$ be a bounded domain with smooth boundary $\gamma = \partial\Omega$. The boundary behaviour of the analytic function F defined on $\mathbf{C} \setminus \Omega$ by

$$F(z) = \int_\gamma (f(\zeta)/(\zeta - z))d\zeta, \quad z \notin \gamma,$$

where f is for instance an integrable function on γ, has been recognized and investigated as one of the basic problems of complex analysis for more than one hundred years. It is the aim of the present complement to review some of the basic facts and their ramifications concerning this subject.

Assume that the principal value of the singular integral

$$(13) \qquad \lim_{\varepsilon \downarrow 0} \int_{\substack{\zeta \in \gamma \\ |\zeta - z| \geq \varepsilon}} (f(\zeta)/(\zeta - z))d\zeta = g(z)$$

exists, for a scalar function $f : \gamma \to \mathbf{C}$ and a point $z \in \gamma$. Then it is called the *Cauchy singular integral* of f at z, and it is simply denoted by $g(z) = \int_\gamma (f(\zeta)/(\zeta - z))d\zeta$.

The simplest class of functions f for which the above limit exists is offered by the next.

5.1. Lemma. *Assume that* $f \in \mathbf{Lip}(\alpha,\gamma)$, *for an order* $0 < \alpha < 1$. *Then the Cauchy singular integral of* f *exists.*

Proof. In order to recall for the reader some standard techniques, we reproduce the whole proof of this simple lemma.

Let $t \in \gamma$ be fixed and denote $\gamma_\varepsilon = \{\zeta \in \gamma\,;\ |\zeta - t| \geq \varepsilon\}$. Then

$$\int_{\gamma_\varepsilon} (f(\zeta)/(\zeta - t))d\zeta = \int_{\gamma_\varepsilon} ((f(\zeta) - f(t))/(\zeta - t))d\zeta + f(t) \int_{\gamma_\varepsilon} d\zeta/(\zeta - t).$$

The first term in the right hand side obviously converges to $\int_\gamma ((f(\zeta) - f(t))/(\zeta - t))d\zeta$, while the second integral can be easily computed as follows:

$$\int_{\gamma_\varepsilon} d\zeta/(\zeta - t) = \mathbf{ln}(\zeta - t)\big|_{\gamma_\varepsilon} = \mathbf{ln}(|\zeta_2 - t|/|\zeta_1 - t|) + i\,\mathbf{arg}(\zeta - t)\big|_{\gamma_\varepsilon},$$

where ζ_1, ζ_2 are the boundaries of the arc γ_ε. Since $|\zeta_1 - t| = |\zeta_2 - t| = \varepsilon$ and $\lim\limits_{\varepsilon\downarrow 0} \mathbf{arg}(\zeta - t)\big|_{\gamma_\varepsilon} = \pi$, one obtains

$$\int_\gamma d\zeta/(\zeta - t) = \lim_{\varepsilon\downarrow 0} \int_{\gamma_\varepsilon} d\zeta/(\zeta - t) = \pi i.$$

Thus, we get the equality

$$(14)\qquad (1/\pi i)\int_\gamma (f(\zeta)/(\zeta - t))d\zeta = (1/\pi i)\int_\gamma ((f(\zeta) - f(t))/(\zeta - t))d\zeta + f(t),$$

and the proof is complete.

5.2. Definition. The *Cauchy singular integral operator* is defined as

$$(Sf)(t) = (1/\pi i)\int_\gamma (f(\zeta)/(\zeta - t))d\zeta, \quad t \in \gamma.$$

For the moment it has a good meaning for any function $f \in \mathbf{Lip}(\alpha,\gamma)$, $0 < \alpha < 1$, but in this case Sf could be, a priori, a worse function. Similarly, with a fixed linear measure on γ, one proves that Sf exists for any function $f \in L^p(\gamma)$, $1 < p < \infty$. The behaviour of S on every of these classes is settled by the next classical result.

5.3. Theorem. *On any of the spaces* $X = \mathbf{Lip}(\alpha,\gamma)$, $0 < \alpha < 1$ *or* $X = L^p(\gamma)$, $1 < p < \infty$, $S(X) \subset X$ *and* $S : X \to X$ *is a continuous operator.*

For a proof the reader can consult Mikhlin and Prössdorf [1] and Privalov [1].

b. The Sohotskii - Plemelj - Privalov formulae

Again $\Omega \subset \mathbf{C}$ is a bounded domain with smooth boundary $\gamma = \partial\Omega$. One denotes $\Omega_+ = \Omega$ and $\Omega_- = \mathbf{C} \setminus \Omega$ or $\Omega_- = \tilde{\mathbf{C}} \setminus \Omega = (\mathbf{C} \setminus \Omega) \cup \{\infty\}$, depending on the context.

If $f : \gamma \to \mathbf{C}$ is a sufficiently good function, then there is a remarkable relation between the boundary values, when $z \to t \in \gamma$, of the analytic function

$$F(z) = (1/2\pi i)\int_\gamma (f(\zeta)/(\zeta - t))d\zeta, \quad z \notin \gamma,$$

and the singular integral $(Sf)(t)$. This subject has a long history from which we recall two moments.

5.4. Theorem. (Sohotskii, 1873). *Assume* $f \in \mathbf{Lip}(\alpha,\gamma)$, *where* $0 < \alpha < 1$. *Then the analytic function* $F \in O(\Omega_+ \cup \Omega_-)$ *has non-tangential limits* $F_\pm(t) = \lim_{z \to t,\, z \in \Omega_\pm} F(z)$ *at every point* $t \in \gamma$, *and*

(15) $$F_\pm(t) = \pm\frac{1}{2}f(t) + \frac{1}{2}(Sf)(t), \quad t \in \gamma.$$

The theorem was proved by Yu. V. Sohotskii in his dissertation at the Petersburg University, and later independently discovered by Plemelj in connection with some concrete problems of function theory. A proof can be found in Mikhlin and Prössdorf book [1].

The similar result for measurable functions has the following statement.

5.5. Theorem. (Privalov, 1919). *Assume* $f \in L^p(\gamma)$, *where* $1 < p < \infty$. *Then the function* $F \in O(\Omega_+ \cup \Omega_-)$ *has non-tangential limits* $F_\pm(t) = \lim_{z \to t,\, z \in \Omega_\pm} F(z)$ *at almost every point* $t \in \gamma$, *and*

(15)' $$F_\pm(t) = \pm\frac{1}{2}f(t) + \frac{1}{2}(Sf)(t), \text{ a. e.}$$

The idea of the proof (see Privalov [1] for details) is as follows: one can assume by Lebesgue's theorem that, off a subset of γ of linear measure zero,

$$f(t) = \lim_{\varepsilon \downarrow 0} (1/2\varepsilon) \int_{\gamma \setminus \gamma_\varepsilon} f(\zeta)d\zeta,$$

where $\gamma_\varepsilon = \{\zeta \in \gamma\,;\ |\zeta - t| \geq \varepsilon\}$. Fix such a point $t \in \gamma$. Since the statement is obvious for constant functions, we may assume

$$\lim_{\varepsilon \downarrow 0} (1/2\varepsilon) \int_{\gamma \setminus \gamma_\varepsilon} f(\zeta)d\zeta = 0.$$

Decompose then the difference

$$\int_{\gamma} (f(\zeta)/(\zeta - t))d\zeta - \int_{\gamma_\varepsilon} (f(\zeta)/(\zeta - t))d\zeta$$

into

$$(t - z) \int_{\gamma_\varepsilon} (f(\zeta)/(\zeta - z)(\zeta - t))d\zeta + \int_{\gamma\setminus\gamma_\varepsilon} (f(\zeta)/(\zeta - t))d\zeta,$$

and show that each term converges uniformly to zero when z belongs to a non-tangential sector of $\Omega_\pm$, at $t \in \gamma$.

The pair of formulae (15) or (15)' are usually called the *Sohotskii - Plemelj - - Privalov* (or in short SPP) *formulae.*

Form (15) and (15)' one easily deduces that

$$(P_\pm f)(t) = \lim_{z\to t,\, z\in\Omega_\pm} F(z), \quad t \in \gamma,$$

are linear bounded operators on $\mathbf{Lip}(\alpha,\gamma)$, respectively on $L^p(\gamma)$. An important interpretation of these operators appears in the next.

5.6. Corollary. *Assume* $f \in \mathbf{Lip}(\alpha,\gamma)$, $0 < \alpha < 1$.

a) *The function* f *extends continuously to an analytic function on* Ω_+, *if and only if* $P_- f = 0$.

b) *The function* f *extends continuously to an analytic function on* Ω_-, *vanishing at* ∞, *if and only if* $P_+ f = 0$.

Proof. Apply the SPP - formulae.

This corollary shows in particular, by Liouville's theorem, that

(16) $$P_+P_- = P_-P_+ = 0,$$

when $P_\pm$ are regarded as operators on $\mathbf{Lip}(\alpha,\gamma)$. Since this is a dense subspace in any $L^p(\gamma)$, $1 < p < \infty$, the same vanishing relation holds on $L^p(\gamma)$. But the SPP formulae can be equivalently written:

(17) $$P_+ + P_- = S, \quad P_+ - P_- = I.$$

Therefore, by (16), P_+ and $-P_-$ are projections:

$$P_+^2 = P_+, \quad P_-^2 = -P_-,$$

whence

(18) $$S^2 = I.$$

This identity is known as the *Poincaré - Bertrand formula.*

In other terms the Cauchy singular integral operator is a symmetry on every of the Banach spaces **Lip**(α,γ), $0 < \alpha < 1$ or $L^p(\gamma)$, $1 < p < \infty$.

c. The Hilbert transform on $L^2(\mathbf{R})$

The existence of the Cauchy singular integral Sf for $f \in L^2(\gamma)$ is obviously a local question. In this case the Fourier transform on $L^2(\mathbf{R})$ can be successfully used in the analysis of the operator S.

The Fourier transform of a function $\phi \in D(\mathbf{R})$ is taken to be

$$\hat{\phi}(t) = (1/\sqrt{2\pi}) \int_{\mathbf{R}} e^{-itx}\phi(x)dx,$$

and $(\cdot)^\wedge$ extends up to a unitary operator on $L^2(\mathbf{R})$. As usually one denotes

$$\check{\phi}(x) = \phi(-x), \quad \phi \in D(\mathbf{R}).$$

Let us consider on $\mathbf{R}$ the principal value distribution v.p. $(1/x)$, defined by

$$(\text{v.p.}(1/x), \phi) = \lim_{\varepsilon \downarrow 0} \int_{|x| \geq \varepsilon} (\phi(x)/x)dx, \quad \phi \in D(\mathbf{R}).$$

It is a temperate distribution of order 1, which satisfies the identities:

$$x\cdot(\text{v.p.}(1/x)) = \mathbf{1}, \quad (\text{v.p.}(1/x))^\vee = -\text{v.p.}(1/x).$$

Let $u = (\text{v.p.}(1/x))^\wedge$ be its Fourier transform. Then

$$u' = -2\pi i\,\delta \quad \text{and} \quad \check{u} = -u.$$

The last equations have a unique solution, namely $u = -\pi i \operatorname{sgn}(x)$. In particular this shows that $u \in L^\infty(\mathbf{R})$ and $((1/\pi i)u)^2 = \mathbf{1}$.

Consequently the convolution operator

$$H\phi = -(\text{v.p.}(1/x)) * \phi$$

is bounded on $L^2(\mathbf{R})$ and $((1/\pi i)H)^2 = I$. But

$$(H\phi)(x) = -[(\text{v.p.}(1/x)) * \phi](x) = \lim_{\varepsilon \downarrow 0} \int_{|t-x| \geq \varepsilon} (\phi(t)/(t - x))dt,$$

hence, with our previous notation

$$H = \pi i S.$$

The operator $H : L^2(\mathbf{R}) \longrightarrow L^2(\mathbf{R})$ is usually called *the Hilbert transform.* We have already proved the formula

$$(19) \qquad (H\phi)^{\wedge}(x) = \pi i \operatorname{sgn}(x))\hat{\phi}(x), \quad \phi \in L^2(\mathbf{R}).$$

The SPP formulae on $L^2(\mathbf{R})$ can be easily derived in this way, by computing the Fourier transform of the functions

$$\int_{\mathbf{R}}(\phi(t)/(t \pm i\varepsilon))dt \quad \text{and} \quad \int_{|t| \geq \varepsilon} (\phi(t)/t)dt; \quad \varepsilon > 0, \ \phi \in \mathcal{D}(\mathbf{R}).$$

The advantage of the Fourier transform method lies in the fact that it extends almost automatically to vector valued functions, see §VII.1.

We end this section with the following simple remark.

5.7. Lemma. *Let* $\phi \in \mathcal{D}(\mathbf{R})$. *Then the commutator* $[M_\phi, H]$ *is a compact operator on* $L^2(\mathbf{R})$.

Indeed, it is well known that an operator like

$$[M_\phi, H]f(x) = \int_{\mathbf{R}}((\phi(x) - \phi(t))/(t - x))f(t)dt, \quad f \in L^2(\mathbf{R}),$$

is compact because the kernel $(\phi(x) - \phi(t))/(t - x)$ is continuous and compactly supported on $\mathbf{R}^2$.

d. Singular integral operators with continuous symbol

Throughout this section Ω is a simply connected domain of the Riemann sphere $\tilde{\mathbf{C}} = \mathbf{C} \cup \{\infty\}$, with smooth boundary $\gamma = \partial\Omega$, and such that either $\infty \notin \overline{\Omega}$ or $\Omega = \mathbf{C}_+$ is the upper half plane. This extension is necessary because we need to include in our discussion the unbounded domain $\mathbf{C}_+$, with the boundary $\tilde{\mathbf{R}} = \mathbf{R} \cup \{\infty\}$. Anyhow, if $\infty \in \gamma$ or $\infty \notin \gamma$, we endow the curve γ with the linear metric induced on $\gamma \setminus \{\infty\}$ by the Lebesgue measure on $\mathbf{C}$. For instance $\mathbf{R} = \tilde{\mathbf{R}} \setminus \{\infty\}$ is endowed with the usual translation invariant measure and not with the induced measure from the embedding $\mathbf{R} \subset \tilde{\mathbf{C}}$. This approach unifies the apparently quite different two cases.

The adjoint S^* of the Cauchy singular integral operator S acting on $L^2(\gamma)$ is given by the formula

$$(S^*f)(z) = (\pi i)^{-1}\int_{\gamma} e^{i(d(\xi)-d(z))}(\overline{\xi} - \overline{z})^{-1}f(\xi)d\overline{\xi},$$

where $f \in L^2(\gamma)$ and $d(\xi)$ denotes the argument of the oriented tangent of the curve γ at the point ξ.

By comparing the kernel of the operator S^* to that of S one easily finds:

$$S^* = S + \text{compact}, \tag{20}$$

and

$$JS^*J = -S + \text{compact}, \tag{21}$$

where J is the involution $Jf = \bar{f}$, $f \in L^2(\gamma)$.

Next we denote by $C(\gamma)$ the unital C^*-algebra of continuous functions on the compact curve γ. It is worthwhile remarking that any function $\phi \in C(\gamma)$ gives a bounded multiplication operator M_ϕ on $L^2(\gamma)$.

5.8. Lemma. *Let* $a, b \in C(\gamma)$. *The operator* $M_a + M_bS$ *is compact on* $L^2(\gamma)$ *if and only if* $a = b = 0$.

Proof. Assume that the operator $M_a + M_bS$ is compact. According to relations (20) and (21) we find that the operators $M_{\bar{a}} \pm M_{\bar{b}}S$ are also compact. Then $M_{\bar{a}}$ is a compact operator, whence $a = 0$. Similarly from $M_{\bar{b}}S$ compact is follows that $M_{\bar{b}} = (M_{\bar{b}}S)S$ is also a compact operator, and $b = 0$.

The *algebra* $A(\gamma)$ *of singular integral operators with continuous symbol* is by definition:

$$A(\gamma) = \{M_a + M_bS + K;\quad a, b \in C(\gamma),\quad K \in K(L^2(\gamma))\}.$$

By virtue of Lemma 5.7 one easily proves that $A(\gamma)$ is a unital subalgebra of $L(L^2(\gamma))$. Moreover, Lemma 5.8 shows that the operator $M_a + M_bS + K$ uniquely determines the entries a, b and K. In a more formal setting this fact can be stated as follows:

Let $\mathbf{C}[\varepsilon]$ be the $\mathbf{C}$-algebra with the generator ε, such that $\varepsilon^2 = 1$.

5.9. Lemma. *The exact sequence of* $\mathbf{C}$*-algebras*

$$0 \to K(L^2(\gamma)) \to A(\gamma) \xrightarrow{\Phi} C(\gamma) \otimes \mathbf{C}[\varepsilon] \to 0$$

holds, where $\Phi(M_a + M_bS + K) = a + b\varepsilon;\quad a, b \in C(\gamma)$.

The map Φ, called the *symbol morphism*, exists because $S^2 = I$.

In order to prove that $A(\gamma)$ is a uniformly closed subalgebra of $L(L^2(\gamma))$, we need the following technical lemma.

5.10. Lemma. *For* $a \in C(\gamma)$ *one has* $M_a = A + K$, *where* A *is an invertible oper-*

ator and K *is compact (both on* $L^2(\gamma)$*) if and only if* $a(\xi) \neq 0$ *for every* $\xi \in \gamma$.

Proof. Assume $M_a = A + K$ as in the statement. Then M_a is a Fredholm operator and in particular **dim Ker** M_a is finite. Thus the set $\{\xi \in \gamma;\ a(\xi) = 0\}$ has null linear measure, whence **dim Ker** $M_a = 0$. Therefore M_a is a one-to-one operator with closed range.

The invariance of the Fredholm index under compact perturbations yields

$$\mathbf{ind}\, M_a = \mathbf{ind}\, A = 0.$$

Thus M_a is an invertible operator on $L^2(\gamma)$. Then it is well known that the function a has no zeros on γ.

5.11. Theorem. (Gohberg [1]). *The algebra* $A(\gamma)$ *of singular integral operators with continuous symbol is a* C^**-subalgebra of* $L(L^2(\gamma))$.

Proof. The algebra $A(\gamma)$ is a $*$-subalgebra of $L(L^2(\gamma))$ by relation (20). It remains to be proved that $A(\gamma)$ is closed in the norm topology.

Let $(M_{a_n} + M_{b_n} S + K_n)$ be a fundamental sequence in $L(L^2(\gamma))$. In virtue of (20) and (21), there are compact operators K'_n, such that $(M_{a_n} - M_{b_n} S + K'_n)$ is also a fundamental sequence. Accordingly $(M_{a_n} + C_n)$ is a fundamental sequence, where $C_n = (1/2)(K_n + K'_n)$.

Fix a positive ε and assume that

$$\| M_{a_n} - M_{a_m} + C_n - C_m \| < \varepsilon,$$

for $n, m \geq N_\varepsilon$. Under these assumptions, the operator $A = \lambda - (M_{a_n} - M_{a_m} + C_n - C_m)$ is invertible, whenever $\lambda \in \mathbf{C}$ and $|\lambda| \geq \varepsilon$. By applying Lemma 5.10, one finds that the function $\lambda - a_n(\xi) + a_m(\xi)$ is not vanishing for $\xi \in \gamma$ and $n, m \geq N_\varepsilon$. Since λ was chosen arbitrary with the property $|\lambda| \geq \varepsilon$, it follows that

$$\sup_{\xi \in \gamma} | a_n(\xi) - a_m(\xi) | < \varepsilon ,$$

whenever $n, m \geq N_\varepsilon$. This proves the uniform convergence of the sequence (a_n) in $C(\gamma)$.

Similarly one proves the convergence of (b_n) in $C(\gamma)$. This completes the proof of Theorem 5.11.

The principal result concerning the algebra $A(\gamma)$ of singular integral operators goes back to F. Noether [1]. More precisely, the sufficiency implication and the index formula in the next theorem was proved in 1921 by F. Noether [1], while the necessity implication was obtained in 1952 by I. Gohberg [1]. Their theorem constituted the basis for the Atiyah - Singer index theory and it also furnishes an important example and motivation for the recent theory of extensions of C^*-algebras.

5.12. Theorem. (F. Noether; I. Gohberg). *An operator* T *is Fredholm if and only if its symbol* $\phi(T) = a + b\varepsilon$ *is invertible in* $C(\gamma)\otimes\mathbf{C}[\varepsilon]$. *In that case*

(23)
$$\mathbf{ind}\, T = (1/2\pi i)\int_\gamma d[\ln((a(t)) - b(t)) - b(t))/(a(t) + b(t))] = $$
$$= (1/2\pi)\mathbf{arg}((a - b)/(a + b))\,\Big|_\gamma .$$

Proof. The exact sequence (22) shows that the image of the algebra $A(\gamma)$ in the Calkin algebra $L(L^2(\gamma))/K(L^2(\gamma))$ is isomorphic to the symbols algebra $C(\gamma)\otimes\mathbf{C}[\varepsilon]$.

A straightforward computation proves that the symbol $a + b\varepsilon$ is invertible if and only if its conjugate $a - b\varepsilon$ is invertible, and both conditions are equivalent to the non-vanishing of the function $a^2 - b^2$ on γ.

Thus, if $a^2 - b^2$ does not vanish on γ, the operator $M_a + M_bS \in A(\gamma)$ is Fredholm.

Conversely, assume that the operator $M_a + M_bS$ is Fredholm on $L^2(\gamma)$. If would exist a point $\xi \in \gamma$ such that $a^2(\xi) = b^2(\xi)$, then the multiplication operator by $a + b\varepsilon$ on $C(\gamma)\otimes\mathbf{C}[\varepsilon]$ would be either not one-to-one or would not have closed range. On the other hand, the multiplication with the class of $M_a + M_bS$ on the Calkin algebra has closed range and is one-to-one (in fact this operator is invertible in the Calkin algebra). This fact contradicts the isomorphism established at the begining of this proof.

Without going into full details, the idea of proving the index formula in the statement is the following.

Let $M_a + M_bS \in A(\gamma)$ be a Fredholm operator. By the invariance of the index under small perturbations and the smoothness of the curve γ, we may assume that a and b are restrictions on γ of rational functions, with the property that $a^2 - b^2$ has not poles or zeros on γ.

Let us denote $\sigma = a - b$ and $\tau = a + b$. An easy algebraic computation shows that the map $\alpha : C(\gamma)\otimes C(\gamma) \rightarrow C(\gamma)\otimes\mathbf{C}[\varepsilon]$:

$$\alpha(\sigma, \tau) = (\sigma + \tau)/2 - (\sigma - \tau)\varepsilon/2$$

is multiplicative. By the additivity of the Fredholm index one gets

$$\mathbf{ind}\, T(\sigma_1\sigma_2, \tau_1\tau_2) = \mathbf{ind}\, T(\sigma_1,\tau_1) + \mathbf{ind}\, T(\sigma_2,\tau_2),$$

where $T(\sigma,\tau)$ denotes the singular integral operator whose index equals $\alpha(\sigma,\tau)$.

By decomposing the rational functions σ and τ into factors and by noticing that the right hand side term of formula (23) also depends additively on σ and τ, we are reduced to prove that

$$\mathbf{ind}\, T(\sigma, \tau) = 1, \tag{24}$$

where $\sigma(\zeta) = \zeta - \lambda$, $\tau(\xi) = 1$ and λ is an interior point of Ω.

The last equality follows by a direct computation, by taking into account the Sohotskii - Plemelj - Privalov formulae (Corollary 5.6 above):

$$S(\zeta - \lambda)^m = \mathbf{sgn}(m)(\zeta - \lambda)^m, \quad m \in \mathbf{Z}.$$

Indeed,

$$T(\xi - \lambda,1)(\xi - \lambda)^m = (1/2)[((\xi - \lambda) + 1) + \mathbf{sgn}(m)(1 - (\xi - \lambda))](\xi - \lambda)^m,$$

whence

$$T(\xi - \lambda,1)f_m = \begin{cases} f_m, & m \geq 0 \\ f_{m+1}, & m < 0. \end{cases}$$

Because $f_m(\xi) = (\xi - \lambda)^m$, $m \in \mathbf{Z}$, form a basis (not necessarily orthogonal) of $L^2(\gamma)$, relation (24) follows.

Although we have stated Theorem 5.12 only for integral operators on $L^2(\gamma)$, it holds on every space $L^p(\gamma)$, $1 < p < \infty$ or even on $\mathbf{Lip}(\alpha,\gamma)$, $0 < \alpha < 1$. In the last case the symbol must be taken in the Banach algebra $\mathbf{Lip}(\alpha,\gamma)\otimes\mathbf{C}[\varepsilon]$.

e. The Riemann - Hilbert problem

Under the assumptions of the preceding section on the curve γ, the *Riemann - - Hilbert problem* can be formulated as follows.

Fix an order α, $0 < \alpha < 1$. Given the invertible function $G \in \mathbf{Lip}(\alpha,\gamma)$ and given $g \in \mathbf{Lip}(\alpha,\gamma)$, find all pairs of analytic functions $F_\pm \in O(\Omega_\pm) \cap \mathbf{Lip}(\alpha,\overline{\Omega}_\pm)$, such that

$$F_+(t) = G(t)F_-(t) + g(t), \quad t \in \gamma. \tag{25}$$

It turns out by elementary consideration, cf. Muskhelishvili [1], that the problem has a finite dimensional linear variety of solutions.

By the SPP formulae, the Riemann - Hilbert problem is equivalent to the following singular integral equation on the curve γ:

$$[(1 + G) + (1 - G)S]f = 2g,$$

with the function $f \in \mathbf{Lip}(\alpha,\gamma)$ as unknown.

The given datum g and the index of the operator $T = (1 + G) + (1 - G)S$ provide full information about the number of the solutions to problem (25). Notice that, in view of Noether's theorem the operator T is Fredholm on the Banach space $\mathbf{Lip}(\alpha,\gamma)$, and

$$\mathbf{ind}\, T = (1/2\pi)\mathbf{arg}\, G \,|_{\gamma}.$$

Similarly, the Riemann - Hilbert problem, can be stated for analytic functions $F_{\pm} \in O(\Omega_{\pm})$ with boundary values belonging to $L^2(\gamma)$. In that case the function G must be taken to be continuous on γ, while g could be any element of $L^2(\gamma)$.

Conversely, to every singular integral equation

$$(M_a + M_b S)f = h$$

corresponds, in the case when $a^2 - b^2 \neq 0$ on γ, the Riemann-Hilbert problem with these parameters

$$G = (a - b)/(a + b) \text{ and } g = h/(a + b).$$

Multi-dimensional Riemann - Hilbert problems, that is when $F_{\pm}$, g are vector valued functions and G is a matrix of continuous functions on γ, are also classical, cf. Muskhelishvili [1].

Notes. The solution of operator commutator equations by singular integral operators goes back to Friedrichs [1]. The representation of the basis of self-adjoint operators which fulfil the hyponormality commutator condition as Cauchy singular integral operators is essentially due to Kato [5]. Then Carey, Pincus, Daoxing Xia and other authors have settled and exploited this representation in the framework of the theory of hyponormal operators. Corollary 2.4 appears in Carey - Pincus [9].

The investigation of the properties of the Hilbert transform on spaces of vector valued functions or on exotic spaces of scalar functions is not yet finished, cf. Dynkin [3].

Section 3 is reproduced after Martin and Putinar [1]. Other two dimensional singular integral models for hyponormal or subnormal operators have been recently discussed by Pincus - Xia - Xia [1] and D. Xia [5]. For an application of the Toeplitz model see Clancey [6].

The theory of one dimensional singular integral operators was mainly developed by the Russian school, cf. Mikhlin [1], Muskhelishvili [1], Gohberg - Krupnik [1]. This domain remains very active by its wide applications.

EXERCISES

1. Let $T \in L(H)$ be a hyponormal operator with cartesian decomposition $T = X + iY$. Let $\phi \in L^{\infty}_{\mathbf{R}}(\sigma(X))$ and $\psi \in L^{\infty}_{\mathbf{R}}(\sigma(Y))$. Prove that the operator $\phi(X) + i\psi(Y)$ is hyponormal if the kernels

$$(\phi(t) - \phi(x))/(t - x), \quad (\psi(t) - \psi(x))/(t - x).$$

are positive definite on $\mathbf{R}^2 \setminus$ diagonal.

Hint. Use the one dimensional singular integral model.

2. Compute the Fredholm index of the singular integral operator $M_t + i(M_a - M_b^*SM_b)$ acting on $L^2(\mathbf{R})$, where $a, b \in C_o(\mathbf{R})$, $a = \bar{a}$ and $a(0) \pm |b(0)|^2 \neq 0$.

3. Find the Riemann - Hilbert problem associated to the operator $M_t + i(M_a - M_{|b|^2}S)$, where a and b are as in Exercise 2.

4. Classify, up to unitary equivalence, all pure hyponormal operators $T \in L(H)$, with $\mathbf{rank}[T^*, T] = 1$ and $m(\mathbf{Re}\, T) = 1$.

Hint. After representing T as $M_t + i(M_a - M_b^*SM_b)$, show that the pair of real functions $(a, |b|)$ classifies T.

5. a) If $n \in \mathbf{N}$ is odd, then the principal value of the singular integral $\int_{-1}^{1}(dt/t^n)$ exists.

b) If $\phi \in \mathbf{Lip}(\alpha,[-1,1])$ and $\alpha > n - 1$, then v.p. $\int_{-1}^{1}(f(t)/t^n)dt$ exists.

6. Let $\Omega \subset \mathbf{C}$ be a bounded simply connected domain, with smooth boundary $\gamma = \partial\Omega$.

a) Assume $f : \gamma \to \mathbf{R}$ is continuous and that $\int_{\gamma}(f(\zeta)/(\zeta - z)) = 0$ for any $z \in \Omega_-$. Then $f = 0$.

b) Assume $f \in L^1(\gamma)$. Then the non-tangential limits $P_{\pm}f$ exist a.e. on γ if and only if the Cauchy singular integral $(Sf)(t)$ exists for almost all $t \in \gamma$.

c) If $f \in L^1(\gamma)$, then the non-tangential limits P_+f exist and are equal to f a.e. if and only if $\int_{\gamma} f(\zeta)\zeta^n d\zeta = 0$ for all $n \in \mathbf{N}$.

7. a) The Cauchy singular integral operator S acts on trigonometric polynomials defined on **T** by the formula:

$$S\left(\sum_{k=-n}^{n} a_k e^{ikt}\right) = \sum_{k=-n}^{n} \operatorname{sgn}(k) a_k e^{ikt}.$$

b)* (Hunt - Muckenhoupt - Wheeden). Let $w \in L^1(\mathbf{T},dt)$, $w \geq 0$ a.e. The operator S is bounded on $L^p(\mathbf{T},wdt)$, $1 < p < \infty$, if and only if

$$\sup_I\left[\left(|I|^{-1}\int_I wdt\right)^{1/p}\cdot\left(|I|^{-1}\int_I w^{-p'/p}dt\right)^{1/p'}\right] < \infty,$$

where the supremum is taken over all non-trivial arcs I of **T**, and $1/p' + 1/p = 1$.

8. Let S be a pure subnormal operator. Prove that the spectral multiplicities of the self--adjoint operators **Re** S and **Im** S do not exceed **rank** $[S^*, S]$.

9*(D. Xia [3]). An operator $T \in L(H)$ is called *semi-hyponormal* if $(T^*T)^{\frac{1}{2}} \geq (TT^*)^{\frac{1}{2}}$.

If T is an invertible semi-hyponormal operator, then it is unitarily equivalent to a singular integral operator T of the form

$$(Tf)(z) = (za(z)^*/2\pi i)\int_{\mathbf{T}}(a(\zeta)f(\zeta)/(\zeta - z))d\zeta + zb(z)f(z), \quad z \in \mathbf{T},$$

acting on a direct integral of Hilbert spaces over the one dimensional torus **T**, where a and b are measurable essentially bounded functions and $b \geq 0$.

Chapter VIII

METHODS OF PERTURBATION THEORY

Though not every pair of bounded self-adjoint operators $X, Y \in L(H)$, satisfying $[X, Y] \in C_1$ (and possibly $i[X, Y] \geq 0$) fits into the classical scheme of perturbation theory, namely $X - Y$ is not necessarily trace-class, the existence of the singular integral model for the couple (X, Y) offers some analogies and even an effective way of applying perturbation theoretical methods to the study of such a pair of operators. This idea, due to Carey and Pincus, will be developed starting with the next chapter.

Our goal is to collect in the present chapter a few of the specific methods and the main results referring to trace-class perturbations of self-adjoint operators. We select from this wide subject only those topics which will be later related to hyponormal operators. Further details may be found by the interested reader in the excellent monographs of Kato [4], Reed and Simon [3], and Baumgärtel and Wollenberg [1].

The basic facts are gradually presented below with essentially complete proofs. The construction of the phase shift and the proof of the trace formula are new, both due to Voiculescu [7].

1. The phase shift

The phase shift is a real function on $\mathbf{R}$, which synthetically expresses the differences between the spectra of a self-adjoint operator and a trace-class self-adjoint perturbation. This function appeared in the works of the Russian school of mathematical physics, and it was M.G. Krein [1] who founded its mathematical theory.

In the present section we reproduce after Voiculescu [7] a new (and quick) method of constructing the phase shift. This approach relies on the classical Weyl – von Neumann theorem.

Let $A \in L(H)$ be self-adjoint, $A = A^*$. The main problem to deal with in this chapter is the relationship between the spectral behaviour of A and that of a "thin" perturbation of it $A + V$. By thin we mean here that the self-adjoint operator V belongs to an ideal of compact operators, like $C_1(H)$. One of the oldest and most important results in this direction is the next result.

1.1. Theorem. (Weyl - von Neumann). *Let* A *be a bounded self-adjoint operator on the separable Hilbert space H. For every* $\varepsilon > 0$ *there exists a Hilbert - Schmidt self--adjoint operator* V, *with* $\|V\|_2 < \varepsilon$ *and such that the operator* $A + V$ *has only point spectrum.*

In other words the operator A can be diagonalized after a perturbation, small in the Hilbert - Schmidt norm. Recall that

$$\|V\|_2^2 = \mathbf{Tr}(V^*V).$$

Proof. The operator A decomposes into a countable direct sum of cyclic self--adjoint operators:

$$A = \bigoplus_{i=1}^{\infty} A_i .$$

Some of the operators A_i may be zero.

The statement reduces to any of the terms A_i, so we may assume for the moment that A is cyclic. Let $\xi \in H$ be a cyclic vector. Then A is unitarily equivalent with the multiplication operator M_t on $L^2(\nu)$, where $d\nu = \langle dE\cdot\xi, \xi\rangle$ and E is the spectral measure of A.

Choose a finite Borel partition $(\delta_j)_{j=1}^n$ of $\sigma(A)$, such that

$$\mathbf{diam}(\delta_j) \leq a/n,$$

with $a = \mathbf{length}(\sigma(A))$.

Let P_1 be the orthogonal projection of H onto the finite dimensional subspace

$$P_1 H = \bigvee_{j=1}^{n} E(\delta_j)\xi.$$

We define $V_1 = -(I - P_1)AP_1 - P_1A(I - P_1)$, so that the operator $A + V_1$ is reduced by P_1. To draw a picture of this operation, we may look at the matricial decomposition of A with respect to the decomposition $H = P_1 H \oplus (I - P_1)H$. Then $A + V_1$ is obtained from A by deleting the marked corners

$$A = \begin{bmatrix} * & \boxed{*} \\ \boxed{*} & * \end{bmatrix}.$$

The Hilbert - Schmidt norm of V_1 is at hand, because

$$\|V_1\|_2 = 2\|(I - P_1)AP_1\| = \|[A, P_1]\|_2$$

and

$$[A, P_1] = \sum_{j=1}^{n} E(\delta_j)[A, P_1]E(\delta_j).$$

Accordingly,

$$\|V_1\|_2^2 \leq n \cdot \sum_{j=1}^{n} \|E(\delta_j)[A, P_j]E(\delta_j)\|_2^2 \leq n \cdot (4a^2/n^2) = 4a^2/n,$$

because **rank** $E(\delta_j)[A, P_j]E(\delta_j) \leq 2$ and $\|E(\delta_j)[A, P_j]E(\delta_j)\| \leq 2\,\mathbf{diam}(\delta_j)$, $1 \leq j \leq n$. All these estimates can be easily deduced from the functional realization of A.

By repeating the procedure for the cyclic direct summands of the operator $(A + V_1)|(I - P_1)H$, one finds a projection $P_2 \leq I - P_1$ and an operator V_2 with $\|V_2\|_2$ arbitrary small, so that $A + V_1 + V_2$ is reduced by $P_1 + P_2$, and so on.

Finally one obtains an increasing sequence of finite dimensional projections $\sum_{j=1}^{m} P_j$, strongly convergent to I, and the operator $V = \sum_{j=1}^{\infty} V_j$, with the properties

$$\|V\|_2 < \varepsilon,$$

and

$$[P_j, A + V] = 0, \quad j \geq 1.$$

This shows that the self-adjoint operator A + V is diagonalizable.

1.2. Corollary. *If the operator* $A \in L(H)$, $A = A^*$ *has only singular spectrum, then there is an increasing sequence* $(P_n)_{n=1}^{\infty}$ *of finite rank orthogonal projections so that* $\lim_{n\to\infty} \|[P_n, A]\|_1 = 0$.

Proof. We may assume without loss of generality that the operator A has a cyclic vector ξ. Let us denote by E the spectral measure of A.

Choose an increasing sequence of coverings $(\delta_k^{(i)})_{k=0}^{\infty}$, $i \in \mathbf{N}$, of **supp**$\langle dE\xi,\xi\rangle$, with disjoint intervals, such that

$$\sum_{k=0}^{\infty} \mathbf{diam}(\delta_k^{(i)}) < 2^{-i}, \quad i \in \mathbf{N}.$$

For every $i \in \mathbf{N}$, choose a sufficiently large m_i, with the property

$$\|\sum_{k=0}^{m_i} E(\delta_k^{(i)})\xi - \xi\| \leq 2^{-i}.$$

Then the finite dimensional projection P_i onto the subspace

$$\bigvee_{k=0}^{m_i} E(\delta_k^{(i)})\xi,$$

satisfies:

$$[P_i, A] = \sum_{k=0}^{m_i} E(\delta_k^{(i)})[P_i, A]E(\delta_k^{(i)}).$$

Since the operators $E(\delta_k^{(i)})[P_i, A]E(\delta_k^{(i)})$ have the rank at most equal to 2 and $\| E(\delta_k^{(i)})[P_i, A]E(\delta_k^{(i)}) \| \leq 2\,\mathbf{diam}(\delta_k^{(i)})$, one finally finds:

$$\|[P_i, A\|_1 \leq \sum_{k=0}^{m_i} \| E(\delta_k^{(i)})[P_i, A]E(\delta_k^{(i)}) \|_1 \leq 4 \sum_{k=0}^{m_i} \mathbf{diam}(\delta_k^{(i)}) \leq 4\cdot 2^{-i}.$$

On the other hand the sequence of finite dimensional projections P_i is increasing, hence there exists $Q = \text{so-}\lim_{i\to\infty} P_i$. Because $Q\xi = \xi$ and $[Q, A] = 0$, one gets $Q = I$.

The proof is complete.

There are several generalizations of the Weyl - von Neumann theorem; in particular one can replace the Hilbert - Schmidt norm $\|\cdot\|_2$ by any Schatten - von Neumann norm $\|\cdot\|_p$, with $1 < p \leq 2$. The case $p = 1$ of trace-class perturbation is completely different, as we shall see later.

The next lemmas prepare the definition of the phase shift.

1.3. Lemma. *Let* A *be a self-adjoint operator acting on a finite dimensional Hilbert space, and let* V *be a non-negative operator of rank one.*

There exists a function $\phi \in L^{\infty}(\mathbf{R})$, *such that*

(i) $\mathbf{supp}(\phi) \subset [-M, M]$, *where* $M = \|A\| + \|V\|$,

(ii) $\mathbf{Tr}(p(A+V) - p(A)) = \int_{\mathbf{R}} p'(t)\phi(t)dt$, *for every polynomial* p.

Moreover, the function ϕ *takes only the value* 0 *and* 1, *and* $\|\phi\|_1 \leq \|V\|_1$.

Proof. Let $\lambda_1 \leq \lambda_2 \leq \ldots \leq \lambda_n$ denote the eigenvalues of the operator A, counted with multiplicities. Similarly, let $\lambda'_1 \leq \lambda'_2 \leq \ldots \leq \lambda'_n$ denote the eigenvalues of $A + V$.

In virtue of Courant's minimax principle, see for instance Dunford and Schwartz [1], we have

$$\lambda_j = \min_{\dim E = j} \max_{\substack{\xi \in E \\ \|\xi\| = 1}} \langle A\xi, \xi \rangle,$$

where E runs over all j-dimensional subspaces of H.

Since $V \geq 0$, obviously we get

$$\lambda_j \leq \lambda_j', \quad 1 \leq j \leq n.$$

We shall prove that

$$\lambda_j' \leq \lambda_{j+1}, \quad 1 \leq j \leq n.$$

In order to do this, let E be a (j+1)-dimensional space, such that

$$\max_{\substack{\xi \in E \\ \|\xi\|=1}} \langle A\xi , \xi \rangle = \lambda_{j+1}.$$

There exists a j-dimensional space $E' \subset E$, such that $E' \subset E \cap \mathbf{Ker}\, V$, so that

$$\lambda_j' \leq \max_{\substack{\xi \in E' \\ \|\xi\|=1}} \langle (A + V)\xi , \xi \rangle = \max_{\substack{\xi \in E' \\ \|\xi\|=1}} \langle A\xi , \xi \rangle \leq \lambda_{j+1}.$$

One defines ϕ as the following combination of characteristic functions:

$$\phi = \chi_{[\lambda_1,\lambda_1']} + \chi_{[\lambda_2,\lambda_2']} + \dots + \chi_{[\lambda_n,\lambda_n']}.$$

Then

$$\mathbf{Tr}(p(A + V) - p(A)) = \sum_j (p(\lambda_j') - p(\lambda_j)) = \int_{\mathbf{R}} p'(t)\phi(t)dt,$$

for every polynomial $p \in \mathbf{C}[t]$.

The other properties of ϕ are obvious.

1.4. Lemma. *Let* A *be a self-adjoint operator on the Hilbert space* H *and let* V *be a non-negative operator of rank one on* H.

There exists a function $\phi \in L^\infty(\mathbf{R})$, *such that*

(i) $\mathbf{supp}\,\phi \subset [-M,M]$, *where* $M = \|A\| + \|V\|$,

(ii) $\mathbf{Tr}(p(A + V) - p(A)) = \int_{\mathbf{R}} p'(t)\phi(t)dt$ *for every polynomial* p, *and*

(iii) $\|\phi\|_1 \leq \|V\|_1$.

Proof. By the Weyl von Neumann theorem these exists an increasing sequence of finite dimensional projections $P_n \uparrow I$, such that $\|[A , P_n]\|_2 \xrightarrow[n]{} 0$.

Fix a natural number m. Then

$$|\mathbf{Tr}((P_nAP_n)^m - A^mP_n)| = |\mathbf{Tr} \sum_{k=1}^{m-1} P_nA^k(I - P_n)(AP_n)^{m-k}| \leq$$

$$\leq \mathbf{const.} \| (I - P_n)AP_n \|_2 (\sup_k \| P_n A^k (I - P_n) \|_2).$$

Since for every k, $\lim_n \| P_n A^k (I - P_n) \|_2 = 0$, we have

$$\mathbf{Tr}((A + V)^m - A^m) = \lim_n \mathbf{Tr}((A + V)^m P_n - A^m P_n) = \lim_n \mathbf{Tr}((P_n(A + V)P_n)^m - (P_n AP_n)^m).$$

Let, by Lemma 1.3, ϕ_n be the function associated to the finite dimensional perturbation $P_n AP_n \rightarrow P_n AP_n + P_n VP_n$. Then $\| \phi_n \|_1 \leq \| V \|_1$ and $0 \leq \phi_n \leq 1$. Therefore the sequence (ϕ_n) is relatively compact in the weak topology of $L^\infty[-M,M]$. Let ϕ be a weak limit of ϕ_n. Then

$$\mathbf{Tr}(p(A + K) - p(A)) = \int_{-M}^{M} p'(t)\phi(t)dt,$$

for every polynomial p.

But the polynomial functions are dense in the predual $L^1[-M,M]$ of $L^\infty[-M,M]$, hence ϕ is unique.

1.5. Theorem. *Let* A *be a self-adjoint operator on a separable Hilbert space H and let K be a trace class self-adjoint operator on H.*

There exists a unique function $\phi \in L^1(\mathbf{R})$, *such that*

(i) $\mathbf{supp}\, \phi \subset [-M,M]$, *where* $M = \| A \| + \| K \|$,

(ii) $\mathbf{Tr}(p(A + K) - p(A)) = \int_{\mathbf{R}} p'(t)\phi(t)dt$, *for every polynomial* p, *and*

(iii) $\| \phi \|_1 \leq \| K \|_1$.

Proof. Write $K = K_+ - K_-$, with $K_\pm \geq 0$, then decompose $K_\pm = \sum_{j=1}^{\infty} K_{\pm,j}$, with $\mathbf{rank}\, K_{\pm,j} \leq 1$, and finally apply successively Lemma 1.4.

The function ϕ in the preceding theorem is called the *phase shift* of the perturbation problem $A \rightarrow A + K$ and it will be denoted by $\phi(A \rightarrow A + K)$ or $\phi_{(A \rightarrow A+K)}(t)$.

1.6. Proposition. *If the self-adjoint operator* A *is purely singular and* $K = K^*$ *is trace class, then the phase shift* $\phi_{(A \rightarrow A+K)}$ *is integer valued.*

Proof. By Corollary 1.2 there exists an increasing sequence of orthogonal projections $P_n \uparrow I$, such that $\lim_n \| [P_n, A] \|_1 = 0$.

With a fixed $\varepsilon > 0$, we may assume that

$$\| [P_n, A] \|_1 < \varepsilon \cdot 2^{-n+1}, \text{ and } \| [P_n, K] \|_1 < \varepsilon \cdot 2^{-n+1}.$$

Let $A_1 = \sum_{n\geq 0} (P_{n+1} - P_n)A(P_{n+1} - P_n)$ and $K_1 = \sum_{n\geq 0} (P_{n+1} - P_n)K(P_{n+1} - P_n)$. A straightforward computation shows that $\| A - A_1 \|_1 < \varepsilon$ and $\| K - K_1 \|_1 < \varepsilon$.

But

$$\phi(A \longrightarrow A + K) = \phi(A \longrightarrow A_1) + \phi(A_1 \longrightarrow A_1 + K_1) + \phi(A_1 + K_1, A + K),$$

and

$$\phi(A_1 \longrightarrow A_1 + K_1) = \sum_{n\geq 0} \phi((P_{n+1} - P_n)A(P_{n+1} - P_n) \longrightarrow (P_{n+1} - P_n)(A + K)(P_{n+1} - P_n)).$$

The last function is integer valued by Lemma 1.3, and $\| \phi(A \longrightarrow A_1) \|_1 + \| \phi(A_1 + K_1 \longrightarrow A + K)) \|_1 < 2\varepsilon$.

An immediate property of the phase shift, derived directly from Lemma 1.3 is the following.

1.7. Proposition. *The phase shift of a finite rank perturbation* $A \longrightarrow A + K$ *satisfies*

$$| \phi_{(A\to A+K)}(t) | \leq \mathbf{rank}\, K, \quad t \in \mathbf{R}.$$

Other properties of the phase shift will be discussed at the end of this chapter.

Originally, the definition of the phase shift was related to some infinite determinants naturally attached by Krein to a trace-class perturbation, see Krein [2] or Gohberg and Krein [1]. In order to present this initial approach we firstly recall some facts concerning *infinite determinants.*

Let $K \in C_1(H)$, with $\| K \| < 1$. Then the series

$$\mathbf{log}(I + K) = \sum_{j=0}^{\infty} (-1)^j (K^{j+1}/j+1)$$

is absolutely convergent. We put by definition

$$\mathbf{det}(I + K) = \mathbf{exp}(\mathbf{Tr}\,\mathbf{log}(I + K)),$$

by noticing that the series $\sum_{j=0}^{\infty} (-1)^j \mathbf{Tr}(K^j/(j + 1))$ is absolutely convergent. The reader will easily check that the above definition is compatible with the usual one in the case $\mathbf{rank}\, K < \infty$.

For an arbitrary trace-class operator $K \in C_1(H)$, it turns out that the analytic function $\lambda \longmapsto \mathbf{det}\,(1 + \lambda K)$, defined initially for $|\lambda| < \| K \|^{-1}$, has an analytic extension over the whole complex plane. In particular $\mathbf{det}(I + K)$ has a good meaning, see Gohberg and Krein [1]. We mention that

$$\mathbf{det}\,(I + K) = \lim_{n\to\infty} \mathbf{det}\,(I + K_n),$$

whenever $K_n \in C_1(H)$ and $\lim_{n\to\infty} \| K - K_n \|_1 = 0$. Consequently

$$\mathbf{det}(I + TK) = \mathbf{det}(I + KT)$$

for any operators $T \in L(H)$, $K \in C_1(H)$.

The *perturbation determinant* of the self-adjoint, trace-class perturbation $A \longrightarrow A + K$ is the analytic function

$$\Delta_{(A\to A+K)}(z) = \mathbf{det}(I + K(A - z)^{-1}) = \mathbf{det}((A + K - z)(A - z)^{-1}),$$

defined off $\sigma(A)$.

The perturbation determinant and the phase shift are connected by an exponential formula, derived by a classical theorem of Verblunsky, see the supplement to this chapter.

1.8. Theorem. (M.G. Krein). *Let* A, A' *be self-adjoint operators on a Hilbert space, with* A - A' *trace-class. Then*

(1) $$\Delta_{(A\to A')}(z) = \exp\left(\int_{\mathbf{R}} (\phi_{(A\to A')}(t)/(t - z))dt\right),$$

for any $z \in \mathbf{C} \setminus \mathbf{R}$.

Proof. Set $M = \| A \| + \| A' - A \|$. Both functions in equality (1) are analytic off the real interval $[-M, M]$. So it suffices to prove (1) for large values of $|z|$.

Denote $K = A' - A$ and take $|z| > M$. Then $\| K(A - z)^{-1} \| < 1$, whence by the definition of infinite determinants

$$\Delta_{(A\to A')}(z) = \mathbf{det}(I + K(A - z)^{-1}) = \exp(\mathbf{Tr}\,\mathbf{log}(I + K(A - z)^{-1})) =$$

$$= \exp(\mathbf{Tr}\,\mathbf{log}((I - (A'/z))(I - (A/z))^{-1})).$$

Provided $|z|$ is large enough, the Hausdorff series

$$\mathbf{log}(\mathbf{exp}\,U\,\mathbf{exp}\,V) - U - V = \tfrac{1}{2}[U, V] + (1/12)[U, [U, V]] + \ldots$$

is convergent in the trace-class norm, see II.§7 in Bourbaki [1]. Since in our case all commutators have zero trace, one finds

$$\Delta_{(A\to A')}(z) = \exp(\mathbf{Tr}\,\mathbf{log}(I - (A'/z)) - \mathbf{Tr}\,\mathbf{log}\,(I - (A/z))) =$$

$$= \exp(-\mathbf{Tr} \sum_{j=0}^{\infty} (((A')^{j+1} - A^{j+1})/(j+1))\cdot z^{-j-1}) =$$

$$= \exp(- \sum_{j=0}^{\infty} (\int_{\mathbf{R}} \phi_{(A\to A')}(t) t^m dt) z^{-j-1}) = \exp \int_{\mathbf{R}} (\phi_{(A\to A')}(t)/(t-z))dt$$

and the proof is complete.

The supplement to this chapter is devoted to the function theoretic relation $\Delta \leftrightarrow \phi$ insured by the exponential representation (1).

2. Abstract symbols and Friedrichs operations

The importance of the symbol of a singular integral operator has already been revealed by F. Noether's index theorem, Chapter VII.§ 5.d. The basic rôle of symbols in the theory of linear partial differential equations is also well known. A very general object in abstract operator theory unifies both notions and turns out to be a powerful tool in physical applications.

It is the aim of this section to bring into discussion the abstract operatorial symbols as they appeared in the work of Friedrichs [1].

Let $X \in L(H)$ be a self-adjoint operator, $X = X^*$, with the associated group of unitaries e^{-itX}, $t \in \mathbf{R}$. For any operator $T \in L(H)$ we introduce the functions

$$\Gamma_X(T;\lambda) = i\int_0^\lambda e^{itX} T e^{-itX} dt, \tag{2}$$

$$S_X(T;\lambda) = e^{i\lambda X} T e^{-i\lambda X}, \tag{3}$$

depending on $\lambda \in \mathbf{R}$.

Let us denote by $D(\Gamma_X^\pm)$ the set of those operators $T \in L(H)$ for which the strong-operator limits exist

$$\Gamma_X^\pm(T) = \text{so-}\lim_{\lambda\to\pm\infty} \Gamma_X(T;\lambda), \tag{4}$$

correspondingly.

It is important to note that the domains $D(\Gamma_X^+)$ and $D(\Gamma_X^-)$ are in general distinct. It is immediate that $D(\Gamma_X^\pm)$ are vector spaces. The linear maps

$$\Gamma_X^\pm : D(\Gamma_X^\pm) \longrightarrow L(H)$$

are called the *Friedrichs operations*.

Of course the limits exist only under some additional assumptions. If $T \in D(\Gamma_X^\pm)$, we shall write in short

$$i \int_0^{\pm\infty} e^{itX} T e^{-itX} dt = \Gamma_X^{\pm}(T).$$

Analogously one denotes by $D(S_X^{\pm})$ the sets of those operators for which the following strong limits exist:

$$S_X^{\pm}(T) = \underset{\lambda \to \pm\infty}{\text{so-}\mathbf{lim}}\, S_X(T;\lambda). \tag{5}$$

For instance every element T in the commutant of X belongs to $D(S_X^{\pm})$.

The linear maps

$$S_X^{\pm} : D(S_X^{\pm}) \longrightarrow L(H)$$

are called the *symbols* with respect to X.

Next we shall be interested in conditions which insure the existence of the symbols $S_X^{\pm}(T)$. A first connection between Friedrichs operations, symbols and commutators appears in the next lemma.

2.1. Lemma. *For any* $T \in L(H)$, $T \in D(S_X^{\pm})$ *if and only if* $[T, X] \in D(\Gamma_X^{\pm})$. *In that case*

$$T = S_X^{\pm}(T) + \Gamma_X^{\pm}([T, X]). \tag{6}$$

Proof. The proof easily follows from the relation:

$$(d/dt)(e^{itX} T e^{-itX}) = -ie^{itX}(TX - XT)e^{-itX}.$$

Indeed, by (2) and (3) one derives

$$S_X(T;\lambda) = T - \Gamma_X([T, X] ; \lambda)$$

and this identity ends the proof.

2.2. Proposition. *Assume that* $T \in D(\Gamma_X^{\pm})$. *Then*

(i) $T = [\Gamma_X^{\pm}(T), X]$,

(ii) $T \in D(S_X^{\pm})$ *and* $S_X^{\pm}(T) = 0$,

(iii) $\Gamma_X^{\pm}(T) \in D(S_X^{\pm})$ *and* $S_X^{\pm}(\Gamma_X^{\pm}(T)) = 0$.

Of course the statements for Γ_X^{+} or Γ_X are meant to be distinct.

Proof. For a real λ we have

$$e^{i\lambda X}\Gamma_X^+(T)e^{-i\lambda X} = i\int_\lambda^\infty e^{itX}Te^{-itX}dt.$$

Denote by $I(\lambda)$ the last expression. It is plain that

$$e^{i\lambda X}I(0) = I(\lambda)e^{i\lambda X},$$

whence, by taking derivatives one obtains

$$ie^{i\lambda X}X\cdot I(0) = -ie^{i\lambda X}T + iI(\lambda)e^{i\lambda X}\cdot X.$$

In particular, for $\lambda = 0$, one finds

$$X\Gamma_X^+(T) = -T + \Gamma_X^+(T)X.$$

A similar computation for Γ_X^- ends the proof of (i).

The statement (ii) follows directly from (i), Lemma 2.1 and the observation that, if $T \in D(\Gamma_X^\pm)$, then $[T, X] \in D(\Gamma_X^\pm)$ and in that case

$$\Gamma_X^\pm([T, X]) = [\Gamma_X^\pm(T), X].$$

Now assertion (iii) follows easily from Lemma 2.1.

The most imporatant properties of the symbols are collected in the next proposition.

2.3. Proposition. *Assume all symbols below exist. Then:*

(i) $[S_X^\pm(T), X] = 0$,

(ii) $S_X^\pm(T)$ *depends linearly and multiplicatively on* T,

(iii) $S_X^\pm(T^*) = S_X^\pm(T)^*$,

(iv) $\|S_X^\pm(T)\| \leq \|T\|$,

(v) $D(S_X^\pm)$ *is closed in the operatorial norm topology of* $L(H)$.

All statements are quite immediate, their proofs being left to the reader.

2.4. Corollary. *Assume* $T, T^* \in D(S_X^\pm)$. *Then the unital* C^**-algebra* $C^*(T)$ *generated by* T *in* $L(H)$ *is contained in* $D(S_X^\pm)$ *and*

$$S_X^\pm : C^*(T) \to (X)'$$

are unital $*$-morphisms.

We conclude the first part of this section devoted to abstract symbols by a useful technical lemma.

2.5. Lemma. *If* $T \in D(\Gamma_X^{\pm})$, *then*

(7) $$\Gamma_X^{\pm}(T) = \text{so-}\lim_{\varepsilon \downarrow 0} i\int_0^{\pm\infty} e^{\mp\varepsilon t} e^{itX} T e^{-itX} dt.$$

Proof. In virtue of Proposition 2.2.(i),

$$ie^{itX} T e^{-itX} = -(d/dt)(e^{itX} \Gamma_X^{+}(T) e^{-itX}).$$

Therefore, for any $\varepsilon > 0$, one obtains

$$i\int_0^{\infty} e^{-\varepsilon t} e^{itX} T e^{-itX} dt = -\int_0^{\infty} e^{-\varepsilon t} (e^{itX} \Gamma_X^{+}(T) e^{-itX})' dt =$$

$$= -(e^{-\varepsilon t} e^{itX} \Gamma_X^{+}(T) e^{-itX}) \Big|_0^{\infty} - \varepsilon \int_0^{\infty} e^{-\varepsilon t} e^{itX} \Gamma_X^{+}(T) e^{-itX} dt.$$

By Proposition 2.2(iii)

$$\text{so-}\lim_{t \to \infty} e^{itX} \Gamma_X^{+}(T) e^{-itX} = 0.$$

Accordingly

$$\text{so-}\lim_{\varepsilon \downarrow 0} (e^{-\varepsilon t} e^{itX} \Gamma_X^{+}(T) e^{-itX}) \Big|_0^{\infty} = -\Gamma_X^{+}(T).$$

and

$$\text{so-}\lim_{\varepsilon \downarrow 0} \varepsilon \int_0^{\infty} e^{-\varepsilon t} e^{itX} \Gamma_X^{+}(T) e^{-itX} dt = 0.$$

This proves relation (7). The statement for Γ_X^{-} follows similarly.

At this moment we may turn our attention to some examples and applications. First we shall derive a fundamental property of hyponormal operators.

2.6. Proposition. *Let* $T = X + iY$ *be the cartesian decomposition of a hyponormal operator. The symbols* $s_X^{\pm}(T)$, $s_X^{\pm}(T^*)$, $s_Y^{\pm}(T)$, $s_Y^{\pm}(T^*)$ *exists.*

Proof. Denote as usually $D = [T^*, T] = 2i[X, Y]$. Since $D \geq 0$ and $(d/dt)(e^{itX} Y e^{-itX}) = \frac{1}{2} e^{itX} D e^{-itX}$, the function $t \to e^{itX} Y e^{-itX}$ of self-adjoint operators is bounded and increasing. Hence the strong limits $S^{\pm}_{X}(Y)$, see (5), exist. Consequently $S^{\pm}_{X}(T) = X + iS^{\pm}_{X}(Y)$ and $S^{\pm}_{X}(T^*) = X - iS^{\pm}_{X}(Y)$ exist.

Similarly one proves that the operators $S^{\pm}_{Y}(X)$ exists.

By Lemma 2.1 the Friedrichs operations $\Gamma^{\pm}_{X}(D)$ and $\Gamma^{\pm}_{Y}(D)$ exist too.

In virtue of Corollary 2.4 the symbols $S^{\pm}_{X}$ provide two unital $*$-morphisms of $C^*(T)$ into the commutant $(X)'$. Accordingly

$$\sigma(S^{\pm}_{X}(T)) \subset \sigma(T)$$

and $S^{\pm}_{X}(T)$ is a normal operator with the same real part as T. Since the equality

$$\sigma(\mathbf{Re}\, S^{\pm}_{X}(T)) = \mathrm{proj.}_1\sigma(S^{\pm}_{X}(T))$$

holds by the normality of the symbols, we infer the inclusion

$$\sigma(X) \subset \mathrm{proj.}_1\sigma(T).$$

This non-trivial fact has already been asserted by Putnam's Theorem V.2.4.

Our next aim is to compute explicitly the symbols of a vector valued singular integral operator. In order to do this, let us recall some notations.

Fix $H = \int_\sigma^{\oplus} H(t)dt$ a decomposition of the Hilbert space H into a direct integral which diagonalizes a self-adjoint operator X, and consider the singular integral operator

$$(Y\xi)(t) = a(t)\xi(t) - (1/\pi i)\int_\sigma (b^*(t)b(s)/(s-t))\xi(s)ds,$$

acting on $\xi \in H$, $t \in \sigma$. Here $a(t) = a(t)^* \in L(H(t))$ and $b(t) \in L(H(t), X)$ are essentially bounded measurable functions on σ, see Chapter VII. Recall that X is an auxiliary Hilbert space.

It is known by Theorem VII.2.1 that any pure hyponormal operator T can be represented like $T = X + iY$ with X and Y as above. By Proposition 2.6 we know that the symbols $S^{\pm}_{X}(Y)$ exist. In order to compute them it suffices to know $\Gamma^{\pm}_{X}([Y, X])$.

But for any $\xi \in H$ and $t \in \sigma$,

$$([Y, X]\xi)(t) = -(1/\pi i)\int_\sigma b^*(t)b(s)\xi(s)\,ds.$$

By Lemma 2.5,

$$(\Gamma^{+}_{X}([Y, X])\xi)(t) = \lim_{\varepsilon \downarrow 0} i\int_0^\infty e^{-\varepsilon s}(e^{isX}[Y, X]e^{-isX}\xi)(t)\,ds =$$

$$= -(1/\pi)\lim_{\varepsilon\downarrow 0} \int_0^\infty e^{-\varepsilon s}\Big(\int_\sigma e^{ist} b^*(t)b(u)e^{-isu}\xi(u)du\Big)\, ds =$$

$$= -(1/\pi)\lim_{\varepsilon\downarrow 0} \int_\sigma b^*(t)b(u)\xi(u)\Big(\int_0^\infty e^{-is(u-t-i\varepsilon)}ds\Big) =$$

$$= -(1/\pi i)\lim_{\varepsilon\downarrow 0} \int_\sigma (b^*(t)b(u)\xi(u)/(u - (t + i\varepsilon)))\, du.$$

The order in the double integral can be reserved because $b^*(t)b(u)$ is a bounded kernel.

Finally, the vector valued (SPP) formulae, see Corollary VII.1.2, yield

$$(\Gamma_X^+([Y, X])\xi)(t) = -b^*(t)b(t)\xi(t) - (1/\pi i)\int (b^*(t)b(s)/(s - t))\xi(s)\, ds.$$

Similarly one computes

$$(\Gamma_X^-([Y, X])\xi)(t) = b^*(t)b(t)\xi(t) - (1/\pi i)\int (b^*(t)b(s)/(s - t))\xi(s)\, ds.$$

By putting together the pieces we conclude with the next formulae

$$(S_X^{\pm}(Y)\xi)(t) = (a(t) \pm b^*(t)b(t))\xi(t),$$

for any $\xi \in H$ and $t \in \sigma$.

In other terms the operators $S_X^{\pm}(Y)$ are diagonalizable with respect to the direct integral decomposition of H, and

(8) $$S_X^{\pm}(Y) = \int_{\sigma(X)}^{\oplus} (a(t) \pm b^*(t)b(t))\, dt.$$

Compare these symbols with those introduced in the scalar case, in §VII.5.d.

If one diagonalizes the imaginary part Y of the hyponormal operator $T = X + iY$ on $H = \int_\sigma^\oplus H(t)dt$, then X would take the form

$$(X\xi)(t) = \alpha(t)\xi(t) + (1/\pi i)\int (\beta^*(t)\beta(s)/(s - t))\xi(s)ds.$$

The same computations give in that case

(8)' $$S_Y^{\pm}(X) = \int_{\sigma(Y)}^{\oplus} (\alpha(t) \mp \beta^*(t)\beta(t))\, dt.$$

3. The Birman - Kato - Rosenblum scattering theory

The aim of this section is to present a deep theorem and some of its consequences, all belonging to the mathematical theory of scattering.

Throughout this section A and B are two self-adjoint operators acting on the same Hilbert space H. One of the problems to deal with below is to compare the spectra of A and B when A - B is trace-class. We shall denote by $H_{ac}(A)$ the subspace of absolute continuity of A, and by $P_{ac}(A)$ the orthogonal projection of H onto $H_{ac}(A)$.

With regard to our purpose it is useful to look at the *generalized wave operators* $W^{\pm}(B,A)$ induced by

$$W^{\pm}(B,A) = \text{so-}\lim_{t\to\pm\infty} e^{itB}e^{-itA}P_{ac}(A),$$

whenever they exist.

3.1. Proposition. *If* $W^+ = W^+(B,A)$ *exists, then:*

(i) W^+ *is a partial isometry with initial space* $H_{ac}(A)$ *and final space* W^+H *reducing* B,

(ii) $BW^+ = W^+AP_{ac}(A)$,

(iii) $W^+H \subset H_{ac}(B)$.

In particular the proposition asserts that the unitary equivalence

$$A\,|\,H_{ac}(A) \simeq B\,|\,W^+H$$

holds. A similar result is valid for $W^- = W^-(B,A)$.

Proof. If $\xi \in H_{ac}(A)^{\perp}$, then $W^+\xi = 0$. If $\xi \in H_{ac}(A)$, then

$$\|e^{itB}e^{-itA}\xi\| = \|\xi\|, \quad t \in \mathbf{R}.$$

Consequently $\|W^+\xi\| = \|\xi\|$.

Let $s \in \mathbf{R}$ be fixed. Then

$$\text{so-}\lim_{t\to\infty} e^{itB}e^{-itA}P_{ac}(A) = \text{so-}\lim_{t\to\infty} e^{i(t+s)B}e^{-i(t+s)A}P_{ac}(A),$$

so that

$$W^+ = e^{isB}W^+e^{-isA}P_{ac}(A),$$

and

$$e^{-isB}W^+ = W^+e^{-isA}P_{ac}(A).$$

By taking derivatives at s = 0 one gets

$$BW^+ = W^+AP_{ac}(A).$$

This identity proves (i) and (ii). Then the restriction $B|W^+H$ is unitarily equivalent with $A|H_{ac}(A)$ and the proof is complete.

3.2. Lemma. (The chain rule). *Assume that* A, B, C *are self-adjoint operators and* $W^+(B,A)$, $W^+(C,B)$ *exist. Then* $W^+(C,A)$ *exists and*

$$W^+(C,A) = W^+(C,B)W^+(B,A).$$

Similarly for W^-.

Proof. Since the operator valued functions

$$e^{itC}e^{-itB}P_{ac}(B), \qquad e^{itB}e^{-itA}P_{ac}(A)$$

are uniformly bounded for $t \in \mathbf{R}$, we can multiply their strong limits. Therefore

$$W^+(C,B)W^+(B,A) = \text{so-}\lim_{t\to\infty} e^{itC}P_{ac}(B)e^{-itA}P_{ac}(A).$$

Thus the proof will be complete if we are able to prove that

$$\text{so-}\lim_{t\to\infty} e^{itC}(I - P_{ac}(B))e^{-itA}P_{ac}(A) = 0.$$

Since e^{itC} and e^{itB} are unitaries, it suffices to prove that

$$\text{so-}\lim_{t\to\infty} e^{itB}(I - P_{ac}(B))e^{-itA}P_{ac}(A) = 0.$$

As the operators e^{itB} and $P_{ac}(B)$ commute, the last equality is equivalent to

$$(I - P_{ac}(B))(\text{so-}\lim_{t\to\infty} e^{itB}e^{-itA}P_{ac}(A)) = 0,$$

which follows by Proposition 3.1. (iii).

3.3. Corollary. *If* $W^+(B,A)$ *and* $W^+(A,B)$ *exist, then* $W^+(B, A)H = H_{ac}(B)$.

Proof. It suffices to remark that

$$W^+(B,A)\cdot W^+(A,B) = W^+(B,B) = P_{ac}(B).$$

As a first criterion for the existence of the wave operators we state the next theorem.

3.4. Theorem. (Kato - Rosenblum). *Let* $A, B \in L(H)$ *be self-adjoint operators* *with* A - B *trace class. Then the generalized wave operators* $W^{\pm}(B,A)$ *exist, and* $W^{\pm}(B,A)H = H_{ac}(B)$.

By Corollary 3.3 if suffices to prove only the existence part. The theorem asserts in particular that the operators $A|H_{ac}(A)$ and $B|H_{ac}(B)$ are unitarily equivalent. Thus absolutely continuous spectra are invariant under trace-class perturbations.

In the next chapters we shall make use of a more general result, namely.

3.5. Theorem. (Birman - Kato - Rosenblum). *Let* $A, B \in L(H)$ *be self-adjoint operators and let* $T \in L(H)$ *be such that* $BT - TA$ *is trace-class. Then the strong operatorial limits*

$$\Omega^{\pm}_{B,A}(T) = \text{so-}\lim_{t\to\pm\infty} e^{itB}Te^{-itA}$$

exist.

It is quite clear that Theorem 3.4 follows from Theorem 3.5. An important application of Birman - Kato - Rosenblum's theorem is the following.

3.6. Corollary. *If* $A \in L(H)$ *is self-adjoint and* $T \in L(H)$ *satisfies* $[T, A] \in C_1(H)$, *then the symbols* $S^{\pm}_A(T)$ *exist.*

Proof of Theorem 3.5. Let E denote the spectral measure of a self-adjoint operator $A \in L(H)$. A vector $\xi \in H$ is called A - *smooth* if

$$\|\xi\|_A^2 = \sup_I \|E(I)\xi\|^2/|I| = \sup_I (1/|I|)\int_I \langle dE(t)\xi, \xi\rangle$$

is finite. The supremum is taken over all nontrivial finite intervals of $\mathbf{R}$.

The set of all A-smooth vectors is denoted by $H_\infty(A)$. Clearly $H_\infty(A) \subset H_{ac}(A)$, and $H_\infty(A)$ is a vector space. Moreover, an approximation by truncations shows that $H_\infty(A)$ is a dense subspace of $H_{ac}(A)$.

3.7. Lemma. *Assume that* $\xi \in H_\infty(A)$. *Then*

(i) $\int_{\mathbf{R}} |\langle e^{-itA}\xi, \eta\rangle|^2 dt \le 2\pi \|\xi\|_A^2 \|\eta\|^2, \quad \eta \in H;$

(ii) *If* $V \in C_2(H)$, *then*

$$\int_{\mathbf{R}} \|Ve^{-itA}\xi\|^2 dt \le 2\pi \|\xi\|_A^2 \cdot \|V\|_2^2;$$

(iii) *If* $K \in L(H)$ *is compact, then*

$$\lim_{t \to \pm \infty} \| K e^{-itA} \xi \| = 0.$$

Proof. Let P denote the orthogonal projection onto the cyclic subspace of A generated by the vector ξ. Since $\xi \in H_{ac}(A)$, there is a positive measurable function $\rho \in L^1(\mathbf{R})$, such that

$$\langle dE(t)\xi , \xi \rangle = \rho(t)^2 dt.$$

The assumption $\xi \in H_{\infty}(A)$ is equivalent to $\rho \in L^{\infty}(\mathbf{R})$. Moreover $\|\rho\|_{\infty} = \|\xi\|_A$.

The Hilbert space PH can be identified with $L^2(\mathbf{R}, \rho^2 dt)$ and under this isomorphism ξ corresponds to the function identically equal to 1 and e^{-isA} becomes the multiplication operator by e^{-ist}. Let $\eta \in PH$ be a fixed vector, represented on $L^2(\mathbf{R}, \rho^2 dt)$ by a function ψ. Then

$$\langle e^{-isA}\xi , \eta \rangle = \int_{\mathbf{R}} \psi(t)\rho(t)^2 e^{-ist} dt \tag{9}$$

and by the Plancherel Theorem

$$\int_{\mathbf{R}} |\langle e^{-isA}\xi , \eta\rangle|^2 ds = 2\pi \int |\psi(t)|^2 \rho(t)^4 dt \le 2\pi \|\rho\|_{\infty}^2 \int |\psi(t)|^2 \rho(t)^2 dt = 2\pi \|\rho\|_{\infty}^2 \|\eta\|^2.$$

This proves the point (i).

The assertion (ii) easily follows from (i). Indeed, let

$$V = \sum_{j=1}^{\infty} \lambda_j \theta_j \otimes \eta_j$$

be Hilbert-Schmidt. Here (θ_j), (η_j) are orthonormal systems in H and the sequence of complex numbers (λ_j) satisfies $\sum_{j=1}^{\infty} |\lambda_j|^2 < \infty$.

Then

$$\int_{\mathbf{R}} \|Ve^{-isA}\xi\|^2 ds = \sum_{j=1}^{\infty} |\lambda_j|^2 \int |\langle e^{-isA}\xi , \eta_j\rangle|^2 ds \le$$

$$\le 2\pi \|\xi\|_A^2 \sum_{j=1}^{\infty} |\lambda_j|^2 = 2\pi \|\xi\|_A^2 \|V\|_2^2.$$

In order to prove (iii), let $\eta \in H$ be a fixed vector, represented as above by the function $\psi \in L^2(\mathbf{R}, \rho^2 dt)$. Since $\langle e^{-isA}\xi , \eta\rangle$ is by (9) the Fourier transform of an L^1-function the familiar Riemann - Lebesgue Theorem yields

$$\lim_{s\to\pm\infty} \langle e^{-isA}\xi , \eta\rangle = 0.$$

Thus assertion (iii) is valid for any finite rank operator K. Because the function $e^{-isA}\xi$ is uniformly bounded on $s \in \mathbf{R}$, (iii) follows by an approximation argument, and the proof of the lemma is complete.

Next we take two Hilbert-Schmidt operators $V_1, V_2 \in C_2(H)$ and consider the function

$$\Gamma_A(V_2^*V_1 ; t) = i\int_0^t e^{isA}V_2^*V_1e^{-isA}ds, \quad t \in \mathbf{R}.$$

3.8. Lemma. *Let* $\xi \in H_\infty(A)$, $s > 0$ *and* $t \in \mathbf{R}$. *Then*

$$|\langle e^{itA}\Gamma_A(V_2^*V_1 ; s)e^{-itA}\xi , \xi\rangle| \leq \left(\int_t^\infty \|V_1e^{-iuA}\xi\|^2du\right)^{\frac{1}{2}}\left(\int_t^\infty \|V_2e^{-iuA}\xi\|^2du\right)^{\frac{1}{2}}.$$

Proof. For the proof it suffices to remark that

$$|\langle e^{itA}\Gamma_A(V_2^*V_1 ; s)e^{-itA}\xi,\xi\rangle| \leq \int_t^\infty |\langle V_1e^{-itA}\xi , V_2e^{-inA}\xi\rangle| du$$

and to apply the Cauchy–Schwarz inequality.

Now we can proceed to the proof of Theorem 3.5. Let $A, B \in L(H)$ be self- -adjoint, let $T \in L(H)$ be such that $BT - TA \in C_1(H)$ and denote

$$\Omega(t) = e^{itB}Te^{-itA}, \quad t \in \mathbf{R}.$$

We have to prove that $\text{so-}\lim_{t\to\pm\infty} \Omega(t)P_{ac}(A)$ exist. Obviously it suffices to consider the limit for $t \to \infty$.

Since $\|\Omega(t)\| \leq \|T\|$ for every $t \in \mathbf{R}$, it is enough to check that

$$\lim_{\min(s,t)\to\infty} \|(\Omega(t) - \Omega(s))\xi\| = 0 \tag{10}$$

for any vector $\xi \in H_\infty(A)$.

Let us introduce the notation

$$K = BT - TA,$$

$$L(t , s) = [\Omega(t)^*\Omega(s) , A].$$

Notice the identities

$$\Omega(t)A = B\Omega(t) - e^{itB}Ke^{-itA},$$

$$A\Omega(t)^* = \Omega(t)^*B - e^{itA}K^*e^{-itB},$$

and

(11) $$L(t,s) = -e^{itA}T^{*}e^{i(s-t)B}Ke^{-isA} + e^{itA}K^{*}e^{i(s-t)B}Te^{-isA}\ ; \quad s, t \in \mathbf{R},$$

which are straightforward.

We may assume $s - t = u > 0$, so that (11) becomes

(12) $$L(t, t+u) = e^{itA}(T^{*}e^{iuB}Ke^{-iuA} - e^{iuA}K^{*}e^{-iuB}T)e^{-itA}.$$

Take the polar factorization of K, $K = V|K|$ and put $H = |K|^{\frac{1}{2}}$ so that the operator H is Hilbert–Schmidt. An application of (12), Lemmas 3.7 and 3.8 yields the estimate

(13) $$|\langle \Gamma_A(L(t,s); v)\xi, \xi \rangle| \leq 2(2\pi)^{\frac{1}{2}} \|H\|_2 \|T\| \left(\int_t^{\infty} \|He^{-ixA}\xi\|^2 dx\right)^{\frac{1}{2}},$$

where $t < s$ and $v \geq 0$.

On the other hand, one knows by the proof of Lemma 2.1 that for any operator S,

$$S = e^{ixA}Se^{-ixA} - \Gamma_A([S, A]\ ; x).$$

The last identity will be applied to the operator $S = D(t,s)^{*}D(t,s)$, where we put by notation

$$D(t,s) = \Omega(t) - \Omega(s).$$

Thus

$$D(t,s)^{*}D(t,s) = e^{ixA}D(t,s)^{*}D(t,s)e^{-ixA} + \Gamma_A(L(t,s) - L(t,t) - L(s,s) + L(s,t)\ ; x).$$

From (13) one gets

$$\|D(t,s)\xi\|^2 \leq \|D(t,s)e^{-ivA}\xi\|^2 + 8(2\pi)^{\frac{1}{2}} \|H\|_2 \|T\| \left(\int_t^{\infty} \|He^{-ixA}\xi\|^2 dx\right)^{\frac{1}{2}},$$

for $t < s$ and any $v \geq 0$.

But the operator $D(t,s)$ is compact since

$$D(t,s) = i\int_t^s e^{ixB}Ke^{-ixA}dx.$$

Lemma 3.7.(iii) shows that

$$\lim_{v \to \infty} \|D(t,s)e^{-ivA}\xi\| = 0.$$

Consequently

$$\|D(t,s)\xi\|^2 \leq 8(2\pi)^{\frac{1}{2}} \|H\|_2 \|T\| (\int_t^\infty \|He^{-ixA}\xi\|^2 dx)^{\frac{1}{2}}$$

for $t < s$.

But $\|He^{-ixA}\xi\|^2$ is an integrable function on x, see Lemma 3.7(ii) therefore (10) holds and the proof of Theorem 3.5 is complete.

We conclude this section with a few facts concerning smooth vectors. Let $A \in L(H)$ be an absolutely continuous self-adjoint operator. We have already remarked that the space $H_\infty(A)$ of smooth vectors is then dense in H, and that

$$Ve^{-itA}\xi \in L^2(\mathbf{R},H),$$

whenever $\xi \in H_\infty(A)$ and $V \in C_2(H)$, see Lemma 3.7(ii). This observation has, by a Fourier transform argument, an interesting interpretation, as follows.

Let us denote in short by

$$R(z) = (A - z)^{-1}, \quad z \in \rho(A),$$

the resolvent function of A, and fix a $\xi \in H_\infty(A)$, $V \in C_2(H)$ and an $\varepsilon > 0$. One easily checks that the function on $x \in \mathbf{R}$, $VR(x + i\varepsilon)\xi$ is the inverse Fourier transform of the function on t

$$(2\pi)^{\frac{1}{2}} ie^{-\varepsilon t}\chi_{[0,\infty]}(t)Ve^{-itA}\xi.$$

Accordingly, Plancherel's Theorem yields

$$\int_{\mathbf{R}} \|VR(x + i\varepsilon)\xi\|^2 dx = 2\pi\int_0^\infty e^{-2\varepsilon t} \|Ve^{-itA}\xi\|^2 dt. \tag{14}$$

In particular, the last identity shows that the analytic function $VR(z)\xi$ belongs to the Hardy space $H^2(\mathbf{C}_+,H)$, where $\mathbf{C}_+ = \{z \in \mathbf{C} \mid \mathbf{Im}\, z > 0\}$. Thus the nontangential limits $\lim_{z \to x} VR(z)\xi$ exist for almost all $x \in \mathbf{R}$. Let us denote symbolically this limit by $VR(x + i0)\xi$. In particular

$$VR(x + i0)\xi = \lim_{\varepsilon \downarrow 0} VR(x + i\varepsilon)\xi,$$

almost everywhere or in the topology of the space $L^2(\mathbf{R},H)$.

Analogously, with the same notations,

$$\int_{\mathbf{R}} \|VR(x - i\varepsilon)\xi\|^2 dx = 2\pi \int_{-\infty}^0 e^{2\varepsilon t} \|Ve^{-itA}\xi\|^2 dt \tag{15}$$

and $VR(x - i0)\xi$ exists.

3.9. Proposition. *Let* $A \in L(H)$ *be self-adjoint and* $T \in D(\Gamma_A^+) \cap C_1(H)$. *Then*

$$\lim_{\varepsilon \downarrow 0} \varepsilon \langle R(x - i\varepsilon)\Gamma_A^+(T)R(x + i\varepsilon)\xi, \xi\rangle = 0$$

for any smooth vector $\xi \in H_\infty(A)$ *and for a.e.* $x \in \mathbf{R}$.

Proof. We factor $T = V_2^* V_1$, where $V_1, V_2 \in C_2(H)$. Then (15) yields

$$-2\varepsilon i \langle R(x - i\varepsilon)\Gamma_A^+(T)R(x + i\varepsilon)\xi, \xi\rangle = 2\varepsilon\int_0^\infty \langle V_1 e^{-itA}R(x + i\varepsilon)\xi, V_2 e^{-itA}R(x + i\varepsilon)\xi\rangle dt =$$

$$= (\varepsilon/\pi)\int_{\mathbf{R}} \langle V_1 R(x + i\varepsilon)R(t + i0)\xi, V_2 R(x + i\varepsilon)R(t + i0)\xi\rangle dt =$$

$$= \int_{\mathbf{R}} \delta_\varepsilon(x - t)\langle V_1 R(x + i\varepsilon)\xi - V_1 R(t + i0)\xi, V_2 R(x + i\varepsilon)\xi - V_2 R(t + i0)\xi\rangle dt,$$

where

$$\delta_\varepsilon(s) = (\varepsilon/\pi)\cdot(1/(s^2 + \varepsilon^2)) = (1/2\pi i)(1/(s - i\varepsilon) - 1/(s + i\varepsilon))$$

is the Poisson kernel.

Since for every integrable function $f \in L^1(\mathbf{R})$ the convolution $\delta_\varepsilon * f$ converges almost everywhere to f, by Fatou's theorem, the proof of the proposition is over.

By a familiar polarization argument, the preceding proposition is still true for a couple of vectors $\xi, \xi' \in H_\infty(A)$ instead of ξ.

4. Boundary behaviour of compressed resolvents

In order to complete the report on scattering theory we collect in this section some technical lemmas concerning the differentiability of the spectral measure of a self-adjoint operator related to the limit behaviour of its resolvent near the real axis. In our presentation we follow closely Dirman and Entina [1].

Let H be a separable complex Hilbert space and denote by C one of the norm-ideals $C_1(H)$ or $C_2(H)$.

4.1. Lemma. *Assume* $T_n \in C$, $T_n \xrightarrow[n]{wo} T_0$, $T_0 \in L(H)$, *and* $\sup_n \|T_n\|_C < \infty$. *Then* $T_0 \in C$ *and for every pair of compact operators* B, C *we have*

$$\lim_n \|B(T_n - T_0)C\|_C = 0.$$

Proof. Let $T_n : K(H) \to C$ be the linear operator

$$\mathcal{T}_n(C) = B(T_n - T_0)C, \quad C \in K.$$

We have to prove that so-$\lim_n \mathcal{T}_n = 0$.

Since the sequence $\mathcal{T}_n$ is bounded,

$$\|\mathcal{T}_n\| \leq \|B\| \, \|T_n - T_0\|_C \leq 2\sup_n \|T_n\|_C \cdot \|B\|,$$

so that it suffices to prove that

$$\lim_n \|\mathcal{T}_n(C)\| = 0$$

for every rank-one operator $C = (\cdot\,, f)g$.

But $\|BT_n - BT_0 g\|_C \underset{n}{\rightarrow} 0$ by our assumption, hence

$$\|\mathcal{T}_n(\cdot\,,f)g)\| = \|f\| \, \|B(T_n - T_0)g\| \underset{n}{\rightarrow} 0.$$

Let us denote, for the rest of this section, by A a bounded self-adjoint operator on H, with spectral measure E and resolvent function $R(z) = (A - z)^{-1}$, $z \in \rho(A)$. Also, D, D_1 and D_2 will be Hilbert-Schmidt operators on H.

4.2. Lemma. *For almost every* $t \in \mathbf{R}$ *there exists the limit*

$$d(D_1 E_t D_2)/dt$$

in the C_1*-norm.*

Proof. First we prove that the operator valued function $\Theta(t) = D_1 E_t D_2$ has bounded variation in C_1. Let $\Delta_k = [t_k, t_{k+1})$ be a finite partition of $\mathbf{R}$. Then

$$\sum_k \|\Theta(t_{k+1}) - \Theta(t_k)\|_1 = \sum_k \|D_1 E(\Delta_k) D_2\|_1 \leq \sum_k \|D_1 E(\Delta_k)\|_2 \cdot \|E(\Delta_k) D_2\|_2 \leq$$

$$\leq \left(\sum_k \|D_1 E(\Delta_k)\|_2^2\right)^{\frac{1}{2}} \left(\sum_k \|E(\Delta_k) D_1\|_2^2\right)^{\frac{1}{2}} = \|D_1\|_2 \cdot \|D_2\|_2.$$

The operators $\Theta(t) \in C_1$ define a family of continuous functionals on $L(H)$:

$$Y \longmapsto \mathbf{Tr}\, Y^* \Theta(t), \quad Y \in L(H).$$

By a theorem of Gelfand (see Gelfand and Shilov [1]) it follows that $\Theta(t)$ is a weakly differentiable function on t, a.e., that is for almost every $t \in \mathbf{R}$ there is an operator $K(t) \in C_1$, so that

$$d(\mathbf{Tr}\, Y^* \Theta(t))/dt = \mathbf{Tr}\, Y^* K(t),$$

for every $Y \in L(H)$. Consequently the quotients

$$(\Theta(t + \delta) - \Theta(t))/\delta$$

are a.e. bounded in the C_1-norm when $\delta \to 0$.

Let $Y = \langle \cdot, \xi \rangle \eta$ be a one-dimensional operator, so that

$$d\langle \Theta(t)\xi, \eta\rangle/dt = \lim_{\delta \to 0} \langle((\Theta(t + \delta) - \Theta(t))/\delta)\xi, \eta\rangle = \langle K(t)\xi, \eta\rangle, \text{ a.e.} \tag{16}$$

But the operators D_1, D_2 always factor as $D_1 = D_1'D_1''$ and $D_2 = D_2''D_2'$, where $D_i' \in K$ and $D_i'' \in C_2$, $i = 1, 2$. By applying Lemma 2.1 to the product $D_1'(D_1''(E_{t+\delta} - E_t)D_2''/\delta)D_2'$ one obtains the conclusion.

Let us remark that the convergence holds almost everywhere, but independently from the vectors ξ and η.

4.3. Lemma. *For almost every* $t \in \mathbf{R}$ *there exist in the* C_2*-norm the radial limits*

$$M_t^{\pm} = \lim_{\varepsilon \downarrow 0} D_1 R(t + i\varepsilon)D_2,$$

and they are related by

$$M_t^{+} - M_t^{-} = 2\pi i \, d(D_1E_tD_2)/dt.$$

Proof. Let $\xi, \eta \in H$ be fixed. The scalar analytic function

$$\langle D_1R(z)D_2\xi, \eta\rangle = \int_{\mathbf{R}} d\langle E_sD_2\xi, D_1^*\eta\rangle/(s - z)$$

has, in view of the preceding lemma, non-tangential limits, a.e. on the real axis. The difficulty lies in proving that the exceptional (negligible) set on which the limits do not exist is independent from ξ and η. If the norm $\|D_1R_zD_2\|_2$ were bounded a.e. on t, when $z \to t \pm i0$, then we could pick ξ, η in a countable subset of H, and the lemma would be proved by Lemma 4.1, similarly to the proof of Lemma 4.2.

The boundedness of the values $\|D_1R_zD_2\|_2$ when $z \to t \pm i0$ runs as follows. We may assume $\mathbf{Im}\, z > 0$. By the identity

$$4D_1R(z)D_2 = (D_1 + D_2)^*R(z)(D_1 + D_2) - (D_1 - D_2)^*R(z)(D_1 - D_2) -$$
$$- i(D_1^* + iD_2)^*R(z)(D_1^* + iD_2) + i(D_1^* - iD_2)^*R(z)(D_1^* - iD_2)$$

we are reduced to expressions of the form

$$M_z = D^*R(z)D,$$

with $D \in C_2$.

But $\mathbf{Im}\, M_z \geq 0$ for $\mathbf{Im}\, z > 0$, so that

$$\|M_z\|_2^2 = \mathbf{Tr}\, M_z^* M_z \leq \mathbf{det}\,(I + M_z^* M_z) \leq \mathbf{det}\,(I + iM_z^* - iM_z + M_z^* M_z) =$$

$$= \mathbf{det}\,[(I - iM_z^*)(I - iM_z)] = |\mathbf{det}\,(I - iM_z)|^2.$$

Notice also that

$$1 \leq \mathbf{det}\,(I + M_z^* M_z) \leq |\mathbf{det}\,(I - iM_z)|^2.$$

By Fatou's theorem the nontangential limits $\lim\limits_{z \to t+i0} \mathbf{det}(I - iM_z)$ exist almost everywhere on t, hence $\sup\limits_{z \to t+i0} \|M_z\|_2^2 < \infty$, a.e. on t.

By taking adjoints one obtains the same statement for the lower half-plane limits, and the proof is complete.

5. SUPPLEMENT : Integral representations for a class of analytic functions defined in the upper half-plane

For the more detailed study of perturbation determinants and phase shifts we review in the present complement some classical representations of analytic functions in the upper-half plane with non-negative imaginary part. The main representation theorem was established by R. Nevalinna in connection with the moment problem on the line, see the Appendix to the book Krein - Nudelman [1]. The applications to perturbation theory are comprehensively treated in the memoir Aronszajn - Donoghue [1].

Most of the proofs below are only sketched, the details being left to the reader, as more or less routine exercises.

Let $A \in L(H)$ be a self-adjoint operator with the associated spectral measure E. One easily computes the perturbation determinant $\Delta = \Delta_{(A \to A + \xi \otimes \xi)}$ of a non-negative rank-one perturbation of A, $\xi \in H$ (see §1). It is the analytic function

$$\Delta(z) = 1 + \langle (A - z)^{-1} \xi, \xi \rangle,$$

defined for non-real z.

If one denotes $\nu = \langle dE \cdot \xi, \xi \rangle$, then ν is a positive Borel measure compactly supported on $\mathbf{R}$, and

(17) $$\Delta(z) = 1 + \int_{\mathbf{R}} d\nu(t)/(t - z), \quad \mathbf{Im}\, z \neq 0.$$

From (17) it immediatly follows that

$$\mathbf{Im}\,\Delta(z) = (\mathbf{Im}\,z)\int_{\mathbf{R}} d\nu(t)/|t - z|^2, \quad \mathbf{Im}\,z \neq 0.$$

Whence the function $f = \Delta$ has the next properties:

i) f is analytic for $\mathbf{Im}\,z \neq 0$ and for large $|z|$;

ii) $(\mathbf{Im}\,f(z))(\mathbf{Im}\,z) \geq 0$ for $\mathbf{Im}\,z \neq 0$;

iii) $\lim_{|z|\to\infty} f(z) = 1$.

We define ad hoc R to be the set of those analytic functions on $\mathbf{C} \setminus \mathbf{R}$ which fulfil i), ii) and iii).

By using properties of the Poisson kernel of the unit disk, one can prove that any function $f \in R$ is of the form (17).

Indeed, let $f \in R$ and consider the function

$$g(\zeta) = -i[f(i((1 + \zeta)/(1 - \zeta))) - 1].$$

Straightforward computations show that g is an analytic function in the unit disk $\mathbf{D} = \{\zeta \in \mathbf{C}\ ;\ |\zeta| < 1\}$, and

$$u(\zeta) := \mathbf{Re}\,g(\zeta) \geq 0, \quad |\zeta| < 1.$$

Moreover, conditions ii) and iii) imply that g extends analytically across a neighbourhood of the point $-i \in \partial\mathbf{D}$, and $g(-i) = -i(f(\infty) - 1) = 0$.

Fix a point $\zeta \in \mathbf{D}$. Then for every $|\zeta| < r < 1$ we obtain by Poisson's formula

$$u(\zeta) = (1/2\pi)\int_0^{2\pi} \mathbf{Re}((e^{i\theta} + \zeta)/(e^{i\theta} - \zeta))u(re^{i\theta})d\theta.$$

Since the family of positive measures $(u(re^{i\theta})d\theta)_r$ is uniformly bounded on r, Helly's theorem insures the existence of a vague limit of it, say $d\sigma$. Hence, by passing to the limit $r\uparrow 1$, one gets

$$u(\xi) = (1/2\pi)\int_0^{2\pi} \mathbf{Re}((e^{i\theta} + \zeta)/(e^{i\theta} - \zeta))d\sigma(\theta), \quad \zeta \in \mathbf{D}.$$

By transporting back this representation via Cayley's transform, one easily finds that any function $f \in R$ admits the representation

$$f(z) = 1 + \int_{\mathbf{R}} d\nu(t)/(t - z) \tag{18}$$

for a suitable positive Borel measure ν, compactly supported on $\mathbf{R}$. Obviously, the measure ν with this property is unique.

Our next aim is to list a dictionary between some properties of $f \in R$ and those of the associated measure ν.

In virtue of Fatou's Theorem, in every point $x \in \mathbf{R}$ of absolute continuity of ν with respect to the linear Lebesgue measure dt one finds that:

1) The non-tangential limits exist and are equal to

$$\lim_{\substack{\zeta \to x \\ \pm \mathbf{Im}\, \zeta > 0}} (1/\pi)\mathbf{Im}\, f(\zeta) = (d\nu/dt)(x), \quad (\text{a.e. on } x \in \mathbf{R}).$$

Moreover,

2) If $(d\nu/dt)(x)$ and $\text{v.p.}\int d\nu(t)/(t-x)$ exist, then

$$f(x \pm i0) = \lim_{\substack{\zeta \to x \\ \pm \mathbf{Im}\, \zeta > 0}} f(\zeta) = 1 + \text{v.p.}\int d\nu(t)/(t-x) \pm \pi i(d\nu/dt)(x).$$

For the proof use Exercise VII.6. Consequently:

3) The measure ν is singular on the interval $(a,b) \subset \mathbf{R}$ if and only if $f(x \pm i0) \in \mathbf{R}$ for almost all $x \in (a,b)$.

By the inversion formula for the Stieltjes integral one finds:

4) $$\int_a^b d\nu = \lim_{\varepsilon \downarrow 0} (1/\pi)\int_a^b \mathbf{Im}\, f(x + i\varepsilon)dx,$$

for every interval (a,b) such that $\nu(\{a,b\}) = 0$. As a corollary we derive:

5) The restriction of the measure ν to the interval (a,b) is zero if and only if the function f extends analytically across (a,b).

In that case $f(x) \in \mathbf{R}$ for every $x \in (a,b)$, as follows from assumption (ii). Formula 4) also implies:

6) If $\mathbf{Im}\, f(z) \leq M < \infty$ for $\mathbf{Im}\, z > 0$, then the measure ν is absolutely continuous, and $d\nu/dt \leq M/\pi$. In that case representation (18) follows directly from the (SPP) formulae.

Just by decomposing the measure ν into absolute continuous, atomic and continuous singular parts one obtains:

7) The next non-tangential limits exist and are equal to

$$\lim_{\substack{\zeta \to x \\ \pm \mathbf{Im}\, \zeta > 0}} (x - \zeta)f(\zeta) = \nu(\{x\}),$$

for every $x \in \mathbf{R}$.

Throughout the rest of this section $\log(\zeta)$ denotes the branch of the logarithm function defined on $\mathbf{C} \setminus (-\infty, 0]i$ and normalized so that $\log(i) = (\pi/2)i$.

Let $f \in R \setminus \{1\}$ be fixed. Then the function $\log f(z)$ is analytic and $\operatorname{Im}(\log f(z)) \in (0,\pi)$ for $\operatorname{Im} z > 0$. Define

$$F(z) = \begin{cases} \log f(z), & \operatorname{Im} z > 0, \\ \overline{\log f(\overline{z})}, & \operatorname{Im} z < 0. \end{cases}$$

Since $f(x) > 0$ for $|x|$ large enough, the function F analitically extends across $\mathbf{R} \setminus [-M,M]$ for a suitable $M > 0$. Therefore $1 + F \in R$, and consequently the function $1 + F$ can be represented like (18). Due to the fact that $|\operatorname{Im} F| \leq \pi$ and by 6), the corresponding measure $d\nu$ is absolutely continuous with bounded weight $d\nu/dt \leq 1$. In conclusion we have proved the following statement which is a particular case of a theorem of Verblunsky, see Aronszajn - Donoghue [1].

8) Each function $f \in R$ has the exponential representation

$$f(z) = \exp\left(\int_{\mathbf{R}} (\phi(t)/(t - z))dt\right), \quad \operatorname{Im} z \neq 0,$$

where ϕ is a measurable function on $\mathbf{R}$, compactly supported and satisfying $0 \leq \phi \leq 1$.

By the (SPP) formulae (Chapter VII) one easily obtains :

9) The following nontangential limit exists and

$$\lim_{\substack{\zeta \to x \\ \operatorname{Im} \zeta > 0}} (1/\pi)\arg f(\zeta) = \phi(x), \text{ a.e. on } x \in \mathbf{R},$$

where $\arg z \in (0, \pi)$ for z in the upper half-plane. In particular this formula shows that the class of the function ϕ in $L^1(\mathbf{R})$ is uniquely determined by $f \in R$.

By 3) we infer :

10) The measure ν is singular on the interval (a, b) if and olny if ϕ takes only the values 0 and 1, almost everywhere on (a, b).

Thus, if the measure is atomic (or equivalently if it is concentrated on a finite number of points of $\mathbf{R}$), then the function f is rational with poles on $\mathbf{R}$ and the associated function ϕ is the characteristic function of some intervals with endpoints at the poles of f; and conversely.

The last property to be mentioned in our brief review is a criterion in terms of

ϕ, for the corresponding measure ν to possess a point mass at $x \in \mathbf{R}$. Recall by 7) that $\nu(\{x\}) \neq 0$ if and only if

$$\lim_{\varepsilon \downarrow 0} \varepsilon f(x + i\varepsilon) \neq 0.$$

By taking into account the exponential representation of f, this condition is in turn equivalent with

$$\liminf_{\varepsilon \downarrow 0}(\log \varepsilon + \mathbf{Re} \int_{\mathbf{R}} (\phi(t)/(t - x - i\varepsilon))dt) > -\infty.$$

But

$$\mathbf{Re} \int (\phi(t)/(t - x))dt = \int (t\phi(t)/((t - x)^2 + \varepsilon^2))dt,$$

and

$$\log \varepsilon = -\int_0^1 (t/(t^2 + \varepsilon^2))dt + \tfrac{1}{2}\log(1 + \varepsilon^2).$$

After a translation we may take x = 0. Hence it suffices to prove that

$$\liminf_{\varepsilon \downarrow 0} (\int_0^1 (t(\phi(t) - 1)/(t^2 + \varepsilon^2))dt + \int_{-1}^0 (t\phi(t)/(t^2 + \varepsilon^2))dt) > -\infty$$

or equivalently that

$$\limsup_{\varepsilon \downarrow 0} (\int_0^1 (t(1 - \phi)/(t^2 + \varepsilon^2))dt - \int_{-1}^0 (t\phi/(t^2 + \varepsilon^2))dt) < \infty. \tag{19}$$

Since $0 \leq \phi \leq 1$, the monotone convergence theorem for integrals shows that (19) is equivalent to

$$\int_0^1 ((1 - \phi)/t)dt - \int_{-1}^0 (\phi/t)dt < \infty.$$

In conclusion,

11) The measure ν has a point mass at $x \in \mathbf{R}$ if and only if

$$\int_x^{x+1} ((1 - \phi(t))/(t - x))dt - \int_{x-1}^x (\phi(t)/(t - x))dt < \infty.$$

For a continuation of this dictionary the reader may consult Aronszajn and Donoghue [1].

Notes. Perturbation theory of self-adjoint operators is an old and still active branch of operator theory, having many intersections with mathematical physics. The phase shift was introduced in the memoir of Krein [1]. There exists an extensive literature devoted to phase shifts and trace formulae (to quote only a few recent items: Krein [3], Adamjan - Pavlov [1], Birman - Solomjak [1]).

The philosophy and the origins of the Friedrichs stationary method are explained in Friedrichs [2]. For applications to linear differential operators see Kato [4] and Dunford and Schwartz [3]. The connection between boundary values of analytic functions and perturbation theory was intensively exploited by the Russian school of operator theory and mathematical physics, cf. Birman - Entina [1] and Gohberg - Krein [1].

Some excellent bibliographical and historical notes on scattering theory are contained in Kato [4] and in Reed and Simon [3]. The fundamental theorem of Birman - Kato - Rosenblum appears, as stated in Section 3, in Reed and Simon [3].

EXERCISES

1. Let $A, B \in L(H)$ be absolutely continuous self-adjoint, and denote $W(t) = e^{itB}e^{-itA}$, $t \in \mathbf{R}$.

a) For every $s < t$ one has

$$W(s)^{-1} - W(t)^{-1} = -i\int_s^t e^{itA}(B - A)e^{-itB}dt.$$

b) If $W^+ = \text{so-}\lim_{t\to\infty} W(t)$ exists, then

$$W^+ - I = i\int_0^\infty e^{itA}(B - A)e^{-itB}W^+dt,$$

where the improper integral exists as a strong limit.

c) If W^+ exists, then $\Gamma_A^+((B - A)W^+)$ exists and

$$W^+ = I + \Gamma_A^+((B - A)W^+).$$

d) Let $V \in L(H)$ satisfy $(B - A)V \in D(\Gamma_A^+)$ and $V = I + \Gamma_A^+((B - A)V)$. Then W^+ exists and $V = W^+$.

Hint. Prove that $[V, A] = (B - A)V$.

2. Let $X + iY$ be a hyponormal operator in Cartesian decomposition. Prove that every vector of the form $\xi = [X, Y]\eta$, $\eta \in H$, is simultaneously X- and Y- smooth.

3. A pure hyponormal operator $X + iY$, $X = \mathbf{Re}(X + iY)$, is completely determined, up to unitary equivalence, by the commuting self-adjoint operators X and $S_X^{\pm}(Y)$, or respectively Y and $S_Y^{\pm}(X)$.

Hint. Use the singular integral model.

4. Prove that there is a one-to-one correspondence between Borel measures ν compactly supported on $\mathbf{R}$ and functions $\phi \in L^1_{comp}(\mathbf{R})$, $0 \leq \phi \leq 1$, established by the relation

$$1 + \int_{\mathbf{R}} d\nu(t)/(t - z) = \mathbf{exp}\left(\int_{\mathbf{R}} (\phi(t)/(t - z))dt\right), \quad \mathbf{Im}\, z \neq 0.$$

5. Let, with the notations of §5, $f_1, f_2 \in R$ with associated functions ϕ_1, ϕ_2. Prove that $f_1 f_2 \in R$ if and only if $\phi_1 + \phi_2 \leq 1$ a.e.

6. a) The phase shift ϕ of a non-negative rank-one perturbation $A \to A + \xi \otimes \xi$ is almost everywhere integer valued if and only if $\xi \perp H_{ac}(A)$.

b) Prove consequently Proposition 1.6.

7. Any function $\phi \in L^1_{comp}(\mathbf{R})$, $0 \leq \phi \leq 1$ is the phase shift of a rank-one perturbation.

8. Let $A, B \in L(H)$ be self-adjoint operators with cyclic vectors ξ, respectively η. Prove that A and B are unitarily equivalent if and only if

$$\phi_{A \to A+\xi\otimes\xi} = \phi_{B \to B+\eta\otimes\eta} \quad \text{a.e.}\,.$$

9. Let $A \in L(H)$ be self-adjoint, $\xi \in H$, $\xi \neq 0$, and denote $\phi = \phi_{A\to A+\xi\otimes\xi}$. Prove that ξ is an eigenvector of A, corresponding to the eigenvalue $x \in \mathbf{R}$, if and only if

$$\int_x^{x+1} ((1 - \phi(t))/(t - x))dt - \int_{x-1}^{x} (\phi(t)/(t - x))dt < \infty.$$

10. Assume that A and B are self-adjoint operators on the same Hilbert space, and that A - B is trace class.

a) Prove the following identities:

$$S_A^+(W^+(B,A)) = P_{ac}(A),$$

$$S_A^+(W^-(B,A)) = W^+(B,A)^* W^-(B,A).$$

b) If $P_{ac}(A)W^+(B,A) = P_{ac}W^-(B,A)$, then $W^+(B,A) = W^-(B,A)$.

c) Show that for every operator T which has the property that $[T, A]$ is trace-class, one has

$$S_B^{\pm}(T)W^{\pm}(B,A) = W^{\pm}(B,A)S_A^{\pm}(T).$$

d) If P is an orthogonal projection, $P \leq P_{ac}(A)$, then

$$W^{\pm}(PAP,PBP) = PW^{\pm}(A,B)P.$$

11. Let P(x) and Q(x) be two polynomials with real coefficients, whose roots alternate, and which satisfy $\mathbf{sgn}(P(x)Q(x)) = 1$ for $|x| \gg 0$.

Prove the following integral representation formula

$$P(z)/Q(z) = \mathbf{exp}\,(\tfrac{1}{2}\int_{\mathbf{R}}((\mathrm{sgn}(P(x)Q(x))-1)/(x-z))dx), \quad \text{for } z \in \mathbf{C} \setminus \mathbf{R}.$$

12*. (M.G. Krein [4]). Let A be a cyclic self-adjoint operator and let B be a trace-class perturbation of it, $B^* = B$.

Prove that the operator $S = W_+(B,A)^{-1}W_-(B,A)$ commutes with A and it is represented on the functional model of A as the multiplication with the function $\mathbf{exp}(-2\pi i\phi(A \to B))$.

Chapter IX

MOSAICS

Mosaics are refined complete unitary invariants for pure hyponormal operators with trace-class self-commutator. The Cartesian decomposition $T = X + iY$ of such an operator gives rise to the non-negative self-adjoint perturbations $S_X^-(Y) \to S_X^+(Y)$ and respectively $S_Y^+(X) \to S_Y^-(X)$. Roughly, the two corresponding mosaics are some operatorial analogous of the phase shifts of these perturbation problems.

In the first part of the present chapter we introduce after R. W. Carey [1] an operatorial phase shift. The rest of the material belongs exclusively to Carey and Pincus [4], [6], [9].

This chapter relies almost entirely on the main technical results obtained in the previous chapters.

1. The phase operator

The phase shift of a non-negative trace-class perturbation problem for self-adjoint operators was generalized by R.W. Carey [1] to an operator valued function, called in what follows the phase operator. Although this object is no longer additive and is less computational than the phase shift, it gains in its turn an important property, namely it is a complete unitary invariant of the initial perturbation problem.

We begin with a few technical results.

1.1. Lemma. *Let* σ *be a compact subset of* $\mathbf{R}$ *and let* $f \in O(\mathbf{C} \setminus \sigma)$ *be an analytic function with the following properties:*

a) $f(\bar{z}) = \overline{f(z)}$, *and*

b) $0 \leq \mathbf{Im}\, f(z) \leq \pi$, *for* $\mathbf{Im}\, z > 0$

c) $\lim_{|z| \to \infty} f(z) = 0$.

Then there exists an unique element $\phi \in L^\infty(\sigma)$, $0 \leq \phi \leq 1$, *such that*

$$f(z) = \int_\sigma (\phi(t)/(t - z))dt, \quad z \in \mathbf{C} \setminus \sigma.$$

Proof. By Fatou's theorem the limits

$$\lim_{\varepsilon \downarrow 0} (1/\pi)\mathbf{Im}\, f(x + i\varepsilon) =: \phi(x)$$

exist a.e. on $x \in \mathbf{R}$, $\phi \in L^\infty(\mathbf{R})$, $0 \leq \phi \leq 1$ and $\mathbf{supp}\, \phi \subset \sigma$.

Let $R > 0$ and $z \in \mathbf{C}_+$, $\mathbf{Im}\, z > \varepsilon$ be fixed. An application of Cauchy's theorem to two half-disks centered at $\pm i\varepsilon$, with radii equal to R, yields

$$f(z) = (1/\pi) \int_{-R}^{R} (\mathbf{Im} f(t + i\varepsilon)/(t - z))dt + (1/2\pi i)\int_0^\pi (f(R e^{i\theta} + i\varepsilon)e^{i\theta}/(R e^{i\theta} - z) +$$

$$+ f(R e^{i\theta} - i\varepsilon)e^{-i\theta}/(R e^{-i\theta} - z))d\theta.$$

Then one takes limits $\varepsilon \downarrow 0$ and then $R \to \infty$. By Lebesgue's dominated convergence theorem and by assumption c) one derives the desired integral representation.

1.2. Proposition. *Let X be a separable Hibert space and* $\sigma \subset \mathbf{R}$ *be compact. Given an L(X)-valued analytic function* Φ *on* $\mathbf{C} \setminus \sigma$ *with the following properties:*

(i) $\Phi(z)^* = \Phi(\bar{z})$, *and*

(ii) $0 \leq \mathbf{Im}\, \Phi(z) \leq \pi$, *for* $\mathbf{Im}\, z > 0$.

(iii) $\lim_{|z| \to \infty} \Phi(z) = 0$,

there exists a unique class $B \in L^\infty(\sigma; L(X))$, *such that* $0 \leq B(t) \leq I_X$ *a.e. and*

(1) $$\Phi(z) = \int_\sigma (B(t)/(t - z))dt, \quad \textit{for } z \in \mathbf{C} \setminus \sigma.$$

Proof. Let $\xi \in X$ be fixed and consider the analytic function

$$F_\xi(z) = \langle \Phi(z)\xi, \xi \rangle, \quad z \in \mathbf{C} \setminus \sigma.$$

Assume $\Phi(z)\xi$ is not identically zero. Remark that the function $1 + F_\xi$ almost belongs to the class *R* studied in Section VIII.5. An application of Lemma 1.1 shows that indeed $1 + F_\xi \in R$ and gives an element $\phi_\xi \in L^\infty(\sigma)$, $0 \leq \phi_\xi \leq 1$, such that

$$F_\xi(z) = \int_{\mathbf{R}} (\phi_\xi(t)/(t - z))dt,$$

and

$$\phi_\xi(x) = \lim_{\varepsilon \downarrow 0} (1/\pi)\mathbf{Im}\, F_\xi(x + i\varepsilon), \text{ for a.a. } x \in \sigma.$$

If $\Phi(z)\xi \equiv 0$ we put $\phi_\xi = 0$.

Next pick a dense, countable **Q**-vector subspace L of X. Correspondingly, there exists a subset $\delta \subset \sigma$ of Lebesgue measure zero, such that

$$\xi \mapsto \phi_\xi(t), \quad t \in \sigma \setminus \delta$$

is a uniformly bounded family of positive definite quadratic forms on L. There exists then a unique measurable family of quadratic forms on X, indexed over σ, which extends $\phi_\xi(t)$. It defines a measurable function $B(t)$, $t \in \sigma$, of self-adjoint operators on X, such that

$$\phi_\xi(t) = \langle B(t)\xi , \xi\rangle$$

for every $\xi \in X$ and almost all $t \in \sigma$.

Since the properties of the class $B \in L^\infty(\sigma, L(X))$ easily follows from those of the forms $\phi_\xi(t)$, the proof of Proposition 1.1 is over.

We point out that the converse of Proposition 1.1 is also true. More precisely, let $B \in L^\infty(\sigma, L(X))$ be such that $0 \leq B(t) \leq I_X$ a.e. Then the function

$$\Phi(z) = \int_\sigma (B(t)/(t - z))dt$$

satisfies conditions (i), (ii) and (iii) above.

Indeed, it suffices to remark that

$$\text{(2)} \qquad \mathbf{Im}\ \Phi(z) = (\mathbf{Im}\ z)\int_\sigma (B(t)/((t - \mathbf{Re}\ z)^2 + (\mathbf{Im}\ z)^2))dt.$$

In what follows $\mathbf{log} : \mathbf{C} \setminus (-\infty,0]i \to \mathbf{C}$ represents the branch of the logarithm normalized so that $\mathbf{log}\ i = (\pi/2)i$.

Next we prepare some examples of functions as in the statement of Proposition 1.2.

1.3. Lemma. *Let X and H be two Hilbert spaces. Assume that $A \in L(H)$, $A^* = A$, and $K \in L(X,H)$. Then the analytic function*

$$\Psi(z) = I + K^*(A - z)^{-1}K, \quad \mathbf{Im}\ z > 0,$$

satisfies:

a) $\mathbf{Im}\ \Psi(z) \geq 0$, *if* $\mathbf{Im}\ z > 0$,

b) $\Psi(z)$ *is invertible for* $\mathbf{Im}\ z > 0$.

Proof. The resolvent equation yields

$$\mathbf{Im}\ \Psi(z) = (1/2i)K^*[(A - z)^{-1} - (A - \bar{z})^{-1}]K - (\mathbf{Im}\ z)K^*(A - z)^{-1}(A - \bar{z})^{-1}K,$$

whence a) follows.

The inverse of $\Psi(z)$ is

$$\Psi(z)^{-1} = I - K^*(A + KK^* - z)^{-1}K, \quad \mathbf{Im}\, z > 0,$$

as one easily computes.

By sticking to the same assumptions as in Lemma 1.3 we introduce the function

$$\Phi(z) = \mathbf{log}\, \Psi(z), \quad z \in \mathbf{C}_+.$$

This has a good meaning because $0 \notin \sigma(\Psi(z))$ and $\sigma(\Psi(z)) \subset \bar{\mathbf{C}}_+$ for any $z \in \mathbf{C}_+$.

If one denotes $m = \inf\{t ; t \in \sigma(A)\}$ and $M = \sup\{t ; t \in \sigma(A + KK^*)\}$, then $\Psi(t)$ exists and $\Psi(t) = \Psi(t)^*$, for every $t \in \mathbf{R} \setminus [m , M]$. Thus we may extend analytically the function Φ by

$$\Phi(z) = \begin{cases} \Phi(z), & \mathbf{Im}\, z > 0, \\ \mathbf{log}\, \Psi(t), & z = t \in \mathbf{R} \setminus [m , M], \\ \Phi(\bar{z})^*, & \mathbf{Im}\, z < 0. \end{cases}$$

Let $\varepsilon > 0$. By applying Kato's Theorem 5.1 to the functions $\Psi + i\varepsilon$ and by passing to the limit $\varepsilon \downarrow 0$, one obtains

$$0 \leq \mathbf{Im}\, \Phi(z) \leq \pi$$

for any $z \in \mathbf{C}_+$. Accordingly, the function Φ fulfils conditions (i), (ii) and (iii) in Proposition 1.2. In conclusion we have proved the next theorem.

1.4. Theorem. (R.W. Carey). *Let* $A \in L(H)$ *and* $K \in L(X,H)$ *with the Hilbert spaces* H, X *separable and* $A = A^*$. *There exists a unique element* $B \in L^\infty(\mathbf{R};L(X))$, *with* **supp** B *compact and* $0 \leq B \leq I$, *such that*

$$\text{(3)} \qquad I + K^*(A - z)^{-1}K = \mathbf{exp}\Big(\int_{\mathbf{R}} (B(t)/(t - z))dt\Big), \quad z \in \mathbf{C} \setminus \mathbf{R}.$$

The operator valued function B described above is called the *phase operator* of the pair (A,K).

Associated to the pair (A,K) we have the perturbation problem $A \rightarrow A + KK^*$. By taking residues at ∞ in (3) one finds

$$K^*K = \int_{\mathbf{R}} B(t)dt.$$

Thus, if $K \in C_2(H)$, then $\mathbf{Tr}\, K^*K < \infty$, whence $\int_{\mathbf{R}} \mathbf{Tr}\, B(t)dt < \infty$ and

$$\mathbf{Tr}\, K^*K = \int_{\mathbf{R}} \mathbf{Tr}\, B(t)dt.$$

Therefore by taking determinants in (3) we obtain

$$\mathbf{det}(I + KK^*(A - z)^{-1}) = \mathbf{det}(I + K^*(A - z)^{-1}K) = \mathbf{exp}(\int(\mathbf{Tr}\, B(t)/(t - z))dt).$$

According to Krein's theorem VIII.1.8, we conclude

$$\mathbf{Tr}\, B = \phi_{A \to A+KK^*},$$

where the last function is the phase shift, cf. §VIII.1.

The phase operator still reflects the spectral properties of the perturbation problem. We illustrate this relationship by a single example. For further details see Carey [1].

1.5. Lemma. *Let* A *be an invertible self-adjoint operator, with spectral measure* E. *Then*

$$\mathbf{Im}\, \mathbf{log}\, A = \pi E((-\infty, 0]).$$

The proof of this lemma is obvious.

1.6. Proposition. *Let* $A \in L(H)$ *be a purely singular self-adjoint operator and let* $K \in L(X,H)$ *be a Hilbert–Schmidt operator.*

Then for the phase operator B *of the pair* (A,K), $B(t)$ *are projections for almost all* t.

Proof. By Lemma VIII.4.3, the nontangential limits

$$L^{\pm}(t) = \lim_{\varepsilon \downarrow 0} (I + K^*(A - t \mp i\varepsilon)^{-1}K)$$

exist almost everywhere in the Hilbert-Schmidt norm. Since the operator A is purely singular

$$L^+(t) = L^-(t) = L^+(t)^* = L^-(t)^* \quad \text{a.e.}$$

On the other hand, because the phase operator B belongs to $L^2(\mathbf{R}, C_2(X))$, the vector valued (SPP) formulae, see §VII.1, give

$$\lim_{\varepsilon \downarrow 0} \int_{\mathbf{R}} (B(x)/(x - t \pm i\varepsilon))dx = \text{v.p.}\int_{\mathbf{R}} (B(x)/(x - t))dx \pm \pi i B(t), \quad \text{a.e.},$$

in the Hilbert–Schmidt norm.

In view of the proof of Lemma 1.3, the operator $L^+(t)$ is invertible almost everywhere. Consequently

$$\text{v.p.}\int(B(x)/(x - t))dx + \pi i B(t) = \mathbf{log}\, L^+(t), \quad \text{a.e.}$$

But Lemma 1.5 yields $\mathbf{Im\,log}\, L^+(t) = \pi B(t)$, whence $B(t)$ is an orthogonal projection for almost all $t \in \mathbf{R}$. This concludes the proof of Proposition 1.6.

Our final purpose is to perform a construction inverse to that in Carey's theorem.

1.7. Theorem. *Let X be a separable Hilbert space and let* $B \in L^\infty_{comp}(\mathbf{R}, L(X))$, *such that* $0 \le B \le I$.

Then there exist a Hilbert space H and operators $A \in L(H)$, $K \in L(X,H)$, $A = A^*$, *with the property*

$$I + K^*(A - z)^{-1}K = \mathbf{exp}\left(\int_{\mathbf{R}}(B(t)/(t - z))dt\right), \quad z \in \mathbf{C} \setminus \mathbf{R}.$$

Proof. Let δ be a compact interval of $\mathbf{R}$, whose interior contains $\mathbf{supp}\, B$. Consider the function

$$\Theta(z) = \mathbf{exp}\left(\int(B(t)/(t - z))dt\right) - I, \tag{4}$$

which is analytic off δ.

Since $0 \le \mathbf{Im}\int(B(t)/(t - z))dt \le \pi$ for $z \in \mathbf{C}_+$ by (2), Kato's Theorem 6.1 shows that $\mathbf{Im}\,\Theta(z) \ge 0$ for any $z \in \mathbf{C}_+$. Moreover, for a unit vector $\xi \in H$,

- $\mathbf{Im}\langle\Theta(z)\xi, \xi\rangle \ge 0$, $\mathbf{Im}\, z > 0$,
- $\langle\Theta(z)\xi, \xi\rangle$ is real and analytic for $z \in \mathbf{R}$, $|z| \gg 0$,
- $\lim_{|z|\to\infty} \langle\Theta(z)\xi, \xi\rangle = 0$.

Thus the function

$$f(z) = 1 + \langle\Theta(z)\xi, \xi\rangle, \quad z \in \mathbf{C} \setminus \mathbf{R},$$

belongs to the class R introduced in § VIII.5.

Consequently, there exists a positive measure ν_ξ, supported on the interval δ, so that

$$\langle\Theta(z)\xi, \xi\rangle = \int_{\mathbf{R}}(d\nu_\xi(t)/(t - z)), \quad z \in \mathbf{C} \setminus \overline{\delta}, \tag{5}$$

see § VIII.5.

By taking residues at ∞ in (4) and (5), respectively, one finds

$$\int_{\mathbf{R}} d\nu_\xi(t) \le \left\langle\left(\int_{\mathbf{R}} B(t)dt\right)\xi, \xi\right\rangle, \quad \|\xi\| = 1.$$

This inequality enables us to define, by a familiar polarization argument, a positive operator valued measure dQ, with the properties:

- $\mathbf{supp}(dQ) \subset\subset \delta$,
- $\langle (dQ)\xi, \xi \rangle = \|\xi\|^2 d\nu_\xi$, $\xi \in H$.
- $\int_{\mathbf{R}} dQ(t) \leq \int_{\mathbf{R}} B(t)dt$.

Next consider a positive constant C, such that

$$\int_{\mathbf{R}} B(t)dt \leq C.$$

Let $a \in \mathbf{R}$ be chosen such that $a > \max \delta$, and define the new operator valued measure

$$d\tilde{Q} = (1/C)dQ + (I + (Q(\mathbf{R})/C))\delta_a, \tag{6}$$

where δ_a is the Dirac measure. Of course $d\tilde{Q} \geq 0$, $\mathbf{supp}(d\tilde{Q}) \subset\subset \mathbf{R}$ and $\tilde{Q}(\mathbf{R}) = I$.

By Naimark's dilation theorem, see §I.6, there exists a larger Hilbert space K and a spectral measure dE on $\mathbf{R}$ with values in $L(K)$, so that

$$\tilde{Q}(\beta) = PE(\beta)|H, \quad \beta \in B(\mathbf{R}),$$

where P stands for the orthogonal projection of K onto H.

Relation (6) yields

$$(1/C)Q(\beta) = PE(\delta)E(\beta)E(\delta)P, \quad \beta \in B(\mathbf{R}). \tag{7}$$

If $A = \int_{\mathbf{R}} t dE(t)$ is the self-adjoint operator attached to the spectral measure E, then (7) gives for a fixed $z \in \mathbf{C} \setminus \mathbf{R}$:

$$(1/C)\int_{\mathbf{R}} dQ(t)/(t - z) = PE(\delta)(A - z)^{-1}E(\delta)P.$$

Thus, by denoting $K = \sqrt{C}E(\delta)P \in L(H,K)$, we finally obtain from (5) and the last identity:

$$\Theta(z) = K^*(A - z)^{-1}K, \quad z \in \mathbf{C} \setminus \mathbf{R}.$$

This completes the proof of Theorem 1.7.

Let us observe that the space H and the operators K and A in Theorem 1.7 above can be chosen in such a way that

$$H = \bigvee_{n=0}^{\infty} A^n K X.$$

In other words, the smallest invariant subspace of A containing the range of K is H

itself. This follows from the following proposition whose simple proof is left to the reader.

1.8. Proposition. *Let* $A \in L(H)$, $K \in L(X,H)$ *be operators such that* $A = A^*$ *and the space* H *is the smallest invariant subspace of* A *which contains the range of* K. *Then the phase operator is a complete unitary invariant for the pair* (A,K).

2. Determining functions

Some compressed resolvents arising from a pure hyponormal operator, which turned out to be complete unitary invariants, are studied in more detail. As an application of the phase operator theory one then defines the mosaics.

We shall freely interchange the notions of determining or determinental function. They designate the same object, cf. Pincus [2] and Carey - Pincus [6].

Let T be a pure hyponormal operator, with self-commutator $D = [T^*, T] = 2i[X, Y]$, $X = \mathbf{Re}\, T$, $Y = \mathbf{Im}\, T$. We denote throughout this section $\mathcal{X} = \overline{\mathbf{Ran}\, D} = (\mathbf{Ker}\, D^{\frac{1}{2}})^{\perp}$.

Recall that expressions like

$$E(z, w) = I + (1/2i)D^{\frac{1}{2}}(X - z)^{-1}(Y - w)^{-1}D^{\frac{1}{2}}, \quad \mathbf{Im}\, z, \mathbf{Im}\, w \neq 0$$

have appeared in the description of the unitary orbit of the pure hyponormal operator T, §II.4. The function $E(z,w) \in L(\mathcal{X})$ will be of extreme importance in what follows, and it will be called after Pincus [2], *the determining function* of the operator T. In fact there are at least two possibilities of choosing such a function (by interchanging X and Y), but its precise form is at our will.

Next we prove some algebraic properties of the function E.

2.1. Lemma. *For every pair* $(z,w) \in \mathbf{C}^2$ *with* $\mathbf{Im}\, z \cdot \mathbf{Im}\, w \neq 0$, *one has*

$$(Y - w)^{-1}(X - z)^{-1}D^{\frac{1}{2}}E(z,w) = (X - z)^{-1}(Y - w)^{-1}D^{\frac{1}{2}}.$$

Proof.

$$(Y - w)^{-1}(X - z)^{-1}D^{\frac{1}{2}}E(z,w) = (Y - w)^{-1}(X - z)^{-1}D^{\frac{1}{2}} +$$

$$+ (Y - w)^{-1}(X - z)^{-1}[X - z, Y - w](X - z)^{-1}(Y - w)^{-1}D^{\frac{1}{2}} = (X - z)^{-1}(Y - w)^{-1}D^{\frac{1}{2}}.$$

2.2. Lemma. *For every pair* $(z,w) \in \mathbf{C}^2$ *with* $\mathbf{Im}\, z \cdot \mathbf{Im}\, w \neq 0$, *one has*

$$E(\bar{z},\bar{w})^* = E(z,w)^{-1}.$$

Proof.

$$E(\bar{z},\bar{w})^* E(z,w) = (I - (1/2i)D^{\frac{1}{2}}(Y - w)^{-1}(X - z)^{-1}D^{\frac{1}{2}})\cdot$$

$$\cdot(I + (1/2i)D^{\frac{1}{2}}(X - z)^{-1}(Y - w)^{-1}D^{\frac{1}{2}}) = I - (1/2i)D^{\frac{1}{2}}(Y - w)^{-1}(X - z)^{-1}D^{\frac{1}{2}} +$$

$$+ (1/2i)D^{\frac{1}{2}}(X - z)^{-1}(Y - w)^{-1}D^{\frac{1}{2}} - (1/2i)D^{\frac{1}{2}}(Y - w)^{-1}(X - z)^{-1}[X - z\,,\, Y - w]\cdot$$

$$\cdot(X - z)^{-1}(Y - w)^{-1}D^{\frac{1}{2}} = I.$$

Similarly one proves the following identities.

2.3. Lemma. *For any non-real complex numbers* z, w_1, w_2 *the identity*

$$E(z,w_1)^{-1}E(z,w_2) = I + ((w_2 - w_1)/2i)D^{\frac{1}{2}}(Y - w_1)^{-1}(X - z)^{-1}(Y - w_2)^{-1}\sqrt{D}$$

holds.

In particular, for $w_2 = \bar{w}_1 = w$, we get

(8) $$E(z,\bar{w})^{-1}E(z,w) = I + (\mathbf{Im}\, w)D^{\frac{1}{2}}(Y - \bar{w})^{-1}(X - z)^{-1}(Y - w)^{-1}D^{\frac{1}{2}}.$$

2.4. Lemma. *For any non-real complex numbers* z_1, z_2, w_1, w_2, *with* $w_1 \neq \bar{w}_2$ *and* $z_1 \neq \bar{z}_2$, *one has*

$$2i(E^*(\bar{z}_1,w_2)E(z_1,w_1) - E^*(z_2,w_2)E(\bar{z}_2,w_1)/(w_1 - \bar{w}_2)(z_1 - \bar{z}_2)) =$$

$$= D^{\frac{1}{2}}(Y - \bar{w}_2)^{-1}(X - \bar{z}_2)^{-1}(X - z_1)^{-1}(Y - w_1)^{-1}D^{\frac{1}{2}}.$$

We conclude by putting together some of the properties of the determining function E, which turn out to be characteristic for it.

2.5. Proposition. *Let* T *be a pure hyponormal operator with determining function* E. *Then*

(i) $||I - E(z,w)|| = O((|\mathbf{Im}\, z\cdot\mathbf{Im}\, w|)^{-1})$ *for* $\mathbf{Im}\, z\cdot\mathbf{Im}\, w \to \infty$.

(ii) $\lim\limits_{|z|,|w|\to\infty} 2izw(I - E(z,w)) = D \geq 0,$

(iii) *The function* $E(z,w)$ *is analytic on* $\mathbf{C}^2 \setminus \mathbf{R}^2$ *and separately continuous for* $|z| \gg 0$ *and* w *fixed, or* $|w| \gg 0$ *and* z *fixed.*

(iv) *The kernel*

$$K(z_1,w_1\,;\,z_2,w_2) = 2i((E^*(\bar{z}_1,w_2)E(z_1,w_1) -$$

$$- E^*(z_2,w_2)E(\bar{z}_2,w_1))/(w_1 - \bar{w}_2)(z_1 - \bar{z}_2)$$

is positive definite.

Carey–Pincus [4] describe how a function E with properties (i) - (iv) above produces a hyponormal operator with E as a determining function. Since E is a complete unitary invariant for the pure hyponormal operator T, the correspondence $T \to E$ is faithful.

Next, we pay attention to the self-adjoint expression (8) and to its limits toward the real axis in the w-complex plane.

Let w be a complex number with positive imaginary part and denote $K(w) = (\mathbf{Im}\, w)^{\frac{1}{2}}(Y - w)^{-1}D^{\frac{1}{2}}$. Then by virtue of (8) we have

$$E(z,\bar{w})^{-1}E(z,w) = I + K(w)^*(X - z)^{-1}K(w).$$

By Carey's Theorem 1.4 we can write

$$E(z,\bar{w})^{-1}E(z,w) = \mathbf{exp} \int (B_w(t)dt)/(t - z), \tag{9}$$

where $B_w(\cdot)$ is the phase operator of the perturbation problem $X \mapsto X + K(w)K(w)^*$. It will be convenient to denote

$$B_\eta(t\ ,\ \lambda) = B_{\lambda+i\eta}(t) \quad \text{for } \lambda, \eta \in \mathbf{R},\ \eta > 0.$$

2.6. Lemma. *For a fixed* $\eta > 0$, *the function* $B_\eta : \mathbf{R} \times \mathbf{R} \to L(H)$ *is weakly***-measurable in both variables.*

Proof. The function $B_\eta(\cdot\ ,\ \lambda)$ is measurable for every λ by the definition of the phase operator.

Take a unit vector $\xi \in H$, and recall that

$$\langle B_\eta(t,\lambda)\xi\ ,\ \xi\rangle = \lim_{s \downarrow 0} (1/\pi)\, \mathbf{Im}\langle \mathbf{log}\, E(t + is\ ,\ \lambda - i\eta)^{-1}E(t + is\ ,\ \lambda + i\eta)\xi\ ,\ \xi\rangle,$$

hence $\langle B_\eta(\cdot\ ,\cdot)\xi\ ,\ \xi\rangle$ is a measurable function. Since $0 \le B_\eta \le I$ and H is a separable Hilbert space, the joint measurability follows.

The family $(B_\eta)_{\eta>0}$ lies in the unit ball of the von Neumann algebra $L^\infty(\mathbf{R}^2, L(X))$, hence it is relatively w^*-compact.

2.7. Definition. Any w^*-limit $B = \lim_j B_{\eta_j}$ with respect to a decreasing sequence $\eta_j \downarrow 0$ is called a *mosaic* of the operator T.

Notice that B is not necessarily unique. However, in the case when $[T^*, T]$ is trace class, more can be said about the limit functions B, and this will be the subject of the next sections of this chapter.

2.8. Lemma. *Let* B *be a mosaic of the operator* T *and let* $f \in C_0^\infty(\mathbf{R})$. *Then*

$$\pi D^{\frac{1}{2}} f(Y) D^{\frac{1}{2}} = \int_{\mathbf{R}^2} f(\lambda) B(t,\lambda)\, dt\, d\lambda.$$

Proof. By taking residues at infinity in (9) one finds

$$\eta D^{\frac{1}{2}}(Y - \lambda - i\eta)^{-1}(Y - \lambda + i\eta)^{-1} D^{\frac{1}{2}} = \int B_\eta(t,\lambda)\, dt.$$

Let F denote the spectral measure of the operator Y. Then

$$\int B_\eta(t,\lambda)\, dt = \eta D^{\frac{1}{2}} \int (dF(s)/((s - \lambda)^2 + \eta^2)) D^{\frac{1}{2}},$$

so that, for every test function $f \in C_0^\infty(\mathbf{R})$, one obtains

$$\int\int f(\lambda) B_\eta(t,\lambda)\, dt\, d\lambda = D^{\frac{1}{2}} \int \Big(\int (\eta f(\lambda)\, d\lambda/((s - \lambda)^2 + \eta^2)) \Big) dF(s) D^{\frac{1}{2}}.$$

By taking the weak limit on the sequence $\eta_j \downarrow 0$ which defines B, we get the desired identity.

2.9. Corollary. *Every mosaic* B *is a compactly supported function in both variables.*

Proof. The preceding lemma tells us that $\mathbf{supp}\, B \subset \mathbf{R} \times \sigma(Y)$ for every mosaic B.

On the other hand, every phase operator function $B_\eta(\cdot, \lambda)$ is supported by an interval $[-M, M]$, where

$$M = \| X \| + \eta \| (Y - \lambda - i\eta)^{-1} D (Y - \lambda + i\eta)^{-1} \|.$$

But $D = 2i[X, Y - \lambda - i\eta]$, so that

$$\eta \| (Y - \lambda - i\eta)^{-1} D (Y - \lambda + i\eta)^{-1} \| \le 2\eta \| Y - \lambda - i\eta \|^{-1} \| X \| +$$

$$+ (2\eta)^2 \| Y - \lambda - i\eta \|^{-1} \| Y - \lambda + i\eta \|^{-1} + 2\eta \| Y - \lambda + i\eta \|^{-1} \| X \|,$$

therefore

$$\sup_{\lambda,\eta \in \mathbf{R}} (\eta \| (Y - \lambda - i\eta)^{-1} D (Y - \lambda + i\eta)^{-1} \|) < \infty.$$

In fact, we shall prove later that $\mathbf{supp}\, B = \sigma(T)$ whenever T is pure and $[T^*, T]$ is trace-class.

3. The principal function

When the self-commutator of the hyponormal operator T is trace-class, then the function **Tr** $B(t,\lambda)$ is characterized by its complex moments, independently from the choice of an arbitrary (a priori) mosaic B. This function is central in the analysis of hyponormal operators and it will be studied in the next chapters.

Throughout this section, T is a pure hyponormal operator with trace-class self-commutator $D = [T^*, T]$.

3.1. Lemma. *If* $D \in C_1$, *then every mosaic* $B(t,\lambda)$ *is trace-class* a.e. *on* $(t,\lambda) \in \mathbf{R}^2$.

Proof. By taking residues at ∞ in (9) and by passing to the limit on a sequence $\eta_j \downarrow 0$, one finds

$$\pi D = \int_{\mathbf{R}^2} B(t,\lambda)dtd\lambda.$$

As $0 \leq B \leq I$ and $\mathbf{Tr}\, D < \infty$, we infer by Beppo–Levi's lemma that $\mathbf{Tr}\, B(t,\lambda)$ is an integrable function on $\mathbf{R}^2$, hence $\mathbf{Tr}\, B(t,\lambda) < \infty$ a.e.

3.2. Proposition. (Carey and Pincus). *Let* T *be a pure hyponormal operator with trace-class self-commutator and let* B *be a mosaic of it. Then, for large values of* $|z|$ *and* $|w|$ *one has*

$$\mathbf{Tr}\,\mathbf{log}\, E(z,w) = (1/2\pi i) \int_{\mathbf{R}^2} (\mathbf{Tr}\, B(t,\lambda)/(t-z)(\lambda - w))dtd\lambda. \tag{10}$$

Proof. Since $\|K(w)\|$ is uniformly bounded on w, and

$$E(z,\bar{w})^{-1}E(z,w) = I + K(w)^*(X - z)^{-1}K(w),$$

one has, for large values of $|z|$, a well defined log function with values in the ideal of trace-class operators. Moreover,

$$\mathbf{log}[E(z,\bar{w})^{-1}E(z,w)] = \int (B_w(t)/(t-z))dt.$$

For such values of $|z|$ one gets

$$\mathbf{Tr}(\mathbf{log}\, E(z,\bar{w})^{-1}E(z,w)) = \int (\mathbf{Tr}\, B_w(t)/(t-z))dt,$$

and, because the function **Tr log** is additive in a neighbourhood of the identity in C_1 (see the proof of Theorem VIII.1.8) one obtains

$$\mathbf{Tr}\,\mathbf{log}\, E(z,w) - \mathbf{Tr}\,\mathbf{log}\, E(z,\bar{w}) = \int(\mathbf{Tr}\, B_w(t)/(t - z))dt, \qquad |z| \gg 0.$$

Recall that $\mathbf{Im}\, w > 0$, so that, for every $\eta > 0$ we still have

$$\mathbf{Tr}\,\mathbf{log}\, E(z, w + i\eta) - \mathbf{Tr}\,\mathbf{log}\, E(z, \bar{w} - i\eta) = \int(\mathbf{Tr}\, B_{w+i\eta}(t)/(t - z))dt.$$

Let us consider the following functions:

$$f_\eta(z,\zeta) = \begin{cases} \mathbf{Tr}\,\mathbf{log}\, E(z, \zeta + i\eta) & \text{for } \mathbf{Im}\,\zeta > 0, \\ \mathbf{Tr}\,\mathbf{log}\, E(z, \zeta - i\eta) & \text{for } \mathbf{Im}\,\zeta < 0, \end{cases}$$

and

$$h_\eta(z,\zeta) = (1/2\pi i)\int_{\mathbf{R}}\int_{\mathbf{R}} (\mathbf{Tr}\, B_{\lambda+i\eta}(t)/(t - z)(\lambda - \zeta))dtd\lambda\,, \quad \eta > 0.$$

Notice that

$$h_\eta(z,\zeta) = (1/2\pi i)\int_{\mathbf{R}}(\mathbf{Tr}\,\mathbf{log}\, E(z, \lambda + i\eta) - \mathbf{Tr}\,\mathbf{log}\, E(z, \lambda - i\eta))(d\lambda/(g - \zeta))$$

and that the difference of the two traces is differentiable in λ for a fixed value of η.

Both functions f_η and h_η are analytic in ζ in a domain of the form $\mathbf{C} \setminus [-M , M]$. Moreover, by the (SPP) formulae the nontangential limits below exist and

$$f_\eta(z, \lambda + i0) - h_\eta(z, \lambda + i0) = f_\eta(z, \lambda - i0) - h_\eta(z, \lambda - i0),$$

almost everywhere on $\lambda \in \mathbf{R}$. Therefore $f_\eta(z,\cdot) - h_\eta(z,\cdot)$ is an entire function (for $|z| \gg 0$), vanishing at ∞. Accordingly, $f_\eta(z,\zeta) = h_\eta(z,\zeta)$.

Thus we see that, for a fixed $\eta > 0$, the function $f_\eta(\cdot\,, \zeta)$ has an analytic continuation in a domain like $\mathbf{C} \setminus [-M , M]$. By moving η into $\eta' > \eta$, remark that these functions are related by

$$f_{\eta'}(z,\zeta) = f_\eta(z, \zeta + i\,\mathrm{sgn}(\mathbf{Im}\,\zeta)(\eta' - \eta)).$$

Concluding this part of the proof, we have established the formula

$$\mathbf{Tr}\,\mathbf{log}\, E(z,\zeta) = (1/2\pi i)\int\int(\mathbf{Tr}\, B_\eta(t,\lambda)/(t - z)(\lambda - \zeta + i\eta))dtd\lambda,$$

whenever $0 < \eta < \mathbf{Im}\,\zeta$, and similarily for the lower half-plane.

Next, we prove that, for the decreasing sequence $\eta_j \downarrow 0$, for which $w^*\text{-}\mathbf{lim}_j\, B_{\eta_j} =$ $= B$, on has

$$\mathbf{lim}_{n\to\infty} \int\int(\mathbf{Tr}\, B_{\eta_j}(t,\lambda)/(t - z)(\lambda - \zeta))dtd\lambda = \int\int(\mathbf{Tr}\, B(t,\lambda)/(t - z)(\lambda - \zeta))dtd\lambda$$

for fixed values of z and ζ, $\mathbf{Im}\, z \neq 0$, $\mathbf{Im}\, \zeta > 0$.

Let $\varepsilon > 0$ be arbitrary and let $\{e_k\}_{k=0}^{\infty}$ be an orthogonal basis of H. We simply denote $B_j = B_{\eta_j}$, so that for every integer N we have

$$\lim_{j\to\infty} \left| \sum_{i=0}^{N} \iint (\langle (B_j(t,\lambda) - B(t,\lambda))e_i , e_i\rangle/(t - z)(\lambda - \zeta))dtd\lambda \right| = 0.$$

We may assume that, for an N large enough,

$$2\pi\, |\mathbf{Im}\, z|^{-1} \cdot |\mathbf{Im}\, \zeta|^{-1} \sum_{i=N}^{\infty} \langle De_i , e_i\rangle < \varepsilon/2.$$

Recall that $\iint B_j(t,\lambda)dtd\lambda = \pi D$, so that

$$\left| \sum_{i=N}^{\infty} \iint (\langle (B_j(t,\lambda) - B(t,\lambda))e_i , e_i\rangle/(t - z)(\lambda - \zeta))dtd\lambda \right| < \varepsilon/2.$$

By putting all together, we find that, for large values of N and j we have

$$\left| \iint ((\mathbf{Tr}\, B_j(t,\lambda) - \mathbf{Tr}\, B(t,\lambda))/(t - z)(\lambda - \zeta))dtd\lambda \right| < \varepsilon.$$

Since ε was arbitrary, the proposition is proved.

Let us remark that, because the left hand side of identity (10) is independent of the choice of the sequence η_j, the function $\mathbf{Tr}\, B(t,\lambda)$ inherits the same property. Thus we have found an intrinsic function associated to the hyponormal operator T.

3.3. Definition. The *principal function* of the (pure) hyponormal operator T with trace-class self-commutator is $g_T(t,\lambda) = \mathbf{Tr}\, B(t,\lambda)$, where B is a mosaic of T.

Notice that $g_T \in L^1(\mathbf{R}^2)$, $g_T \geq 0$ and $\mathbf{supp}\, g$ is compact.

The following lemma will be needed in the next chapter.

3.4. Lemma. *For large values of* $|z|$ *and* $|w|$ *the identity*

$$\mathbf{Tr}[(X - z)^{-1}(Y - w)^{-1}(X - z)^{-1}D] = (1/\pi)\int (g_T(t,\lambda)/(t - z)^2(\lambda - w))dtd\lambda$$

holds.

Proof. By taking **log** in an identity like that asserted by Lemma 2.3 we find

$$\mathbf{log}[E(z,w)E(\zeta,w)^{-1}] = ((\zeta - z)/2i)D^{\frac{1}{2}}(X - z)^{-1}(Y - w)^{-1}(X - \zeta)^{-1}D^{\frac{1}{2}} + O(|\zeta - z|^2).$$

By the above projection we infer

$$\mathbf{log}[E(z,w)E(\zeta,w)^{-1}] = (1/2\pi i)\int g_T(t,\lambda)(1/(t-z) - (1/(t-\zeta)))(dtd\lambda/(\lambda - w)),$$

hence

$$(1/\pi)\int(g_T(t,\lambda)/(t-z)(t-\zeta)(\lambda - w))dtd\lambda =$$

$$= \mathbf{Tr}\, D^{\frac{1}{2}}(X-z)^{-1}(Y-w)^{-1}(X-\zeta)^{-1}D^{\frac{1}{2}} + O(|z-\zeta|).$$

Finally, by taking $z = \zeta$ we find the expressions in the statement.

The principal function, like the phase shift, is much more computable when compared with a mosaic or with the phase operator. The next chapter will be entirely devoted to the principal function and its properties.

The rest of this chapter deals with another description of the mosaics, by means of which one proves some of their properties.

4. Symbol homomorphisms and mosaics

The most effective way of relating a mosaic with the initial hyponormal operator is provided by the singular integral model picture. In such a way one gains a remarkable identification between all mosaics and a certain expression in the symbols $S_Y^{\pm}(X)$. At this point the key steps in the proofs are the general theorems of perturbation theory presented in Chapter VIII.

The present section follows closely Carey and Pincus [5]. Though the results can be generalized to operators with trace-class self-commutator, we confine ourselves to deal only with the hyponormal case.

Let $T \in L(H)$ be a pure hyponormal operator with trace-class self-commutator and Cartesian decomposition $T = X + iY$. As usually one denotes $D = 2i[X, Y]$.

Recall that the symbols $S_Y^{\pm}(X)$ exist and

(11) $$X = S_Y^{\pm}(X) + \Gamma_Y^{\pm}([X, Y]),$$

see §VIII.3.

Our first aim is to prove the existence of the radial limits:

$$\lim_{\varepsilon \downarrow 0} E(z, \lambda - i\varepsilon)^{-1}E(z, \lambda + i\varepsilon) = \lim_{\varepsilon \downarrow 0} (I + D^{\frac{1}{2}}(Y - \lambda + i\varepsilon)^{-1}(X-z)^{-1}(Y - \lambda - i\varepsilon)^{-1}D^{\frac{1}{2}})$$

for $\mathbf{Im}\, z \neq 0$ and $\lambda \in \mathbf{R}$.

The main technical fact is contained in the next.

4.1. Lemma. *Let* S *be a trace-class operator lying in the domain of* Γ_Y^+. *Then*

$$\lim_{\varepsilon \downarrow 0} \| \varepsilon D^{\frac{1}{2}}(Y - \lambda + i\varepsilon)^{-1}(\Gamma_Y^+ S)(Y - \lambda - i\varepsilon)^{-1} D^{\frac{1}{2}} \|_2 = 0$$

for almost all $\lambda \in \mathbf{R}$.

Proof. Since $[\Gamma_Y^+ S, Y] = S$, one has

$$[\Gamma_Y^+ S, (Y - w)^{-1}] = -(Y - w)^{-1} S (Y - w)^{-1},$$

and

$$D^{\frac{1}{2}}(Y - \bar{w})^{-1}(\Gamma_Y^+ S)(Y - w)^{-1} D^{\frac{1}{2}} = D^{\frac{1}{2}}(Y - \bar{w})^{-1}(Y - w)^{-1}(\Gamma_Y^+ S) D^{\frac{1}{2}} - \tag{12}$$
$$-D^{\frac{1}{2}}(Y - \bar{w})^{-1}(Y - w)^{-1} S (Y - w)^{-1} D^{\frac{1}{2}}.$$

Next we can decompose $S = S_1 S_2$ with $S_1, S_2 \in C_2(H)$. By the resolvent equation, (12) yields

$$(\bar{w} - w) D^{\frac{1}{2}}(Y - \bar{w})^{-1}(\Gamma_Y^+ S)(Y - w)^{-1} D^{\frac{1}{2}} = D^{\frac{1}{2}}((Y - w)^{-1} - (Y - \bar{w})^{-1})(\Gamma_Y^+ S) D^{\frac{1}{2}} -$$
$$- (D^{\frac{1}{2}}((Y - w)^{-1} - (Y - \bar{w})^{-1}) S_1) S_2 (Y - w)^{-1} D^{\frac{1}{2}}.$$

At this moment we can apply Lemma VIII.4.3 which guarantees the existence of the limits

$$\lim_{\varepsilon \downarrow 0} 2i\varepsilon D^{\frac{1}{2}}(Y - \lambda + i\varepsilon)^{-1}(\Gamma_Y^+ S)(Y - \lambda - i\varepsilon)^{-1} D^{\frac{1}{2}} = \tag{13}$$
$$= (d/d\lambda)(D^{\frac{1}{2}} F_\lambda (\Gamma^+ S) D^{\frac{1}{2}}) - ((d/d\lambda)(D^{\frac{1}{2}} F_\lambda S_1)) S_2 (Y - \lambda - i0)^{-1} D^{\frac{1}{2}},$$

in the Hilbert–Schmidt norm, for almost all $\lambda \in \mathbf{R}$. The above F stands for the spectral measure of Y.

Let $(\xi_n)_{n=1}^N$ be an orthonormal system (finite or infinite) of vectors, which diagonalizes D. That is $D\xi_n = \lambda_n \xi_n$ for a suitable $\lambda_n > 0$, for every n. By Kato's inequality VI.4.3 every vector ξ_n is Y-smooth, see §VIII.3. Thus Proposition VIII.3.9 yields

$$\lim_{\varepsilon \downarrow 0} \varepsilon \langle D^{\frac{1}{2}}(Y - \lambda + i\varepsilon)^{-1}(\Gamma_Y^+ S)(Y - \lambda + i\varepsilon)^{-1} D^{\frac{1}{2}} \xi_m, \xi_n \rangle = 0 \tag{14}$$

for every m, n = 1, 2, . . . and almost all $\lambda \in \mathbf{R}$.

By deleting a Borel subset $\delta \subset \mathbf{R}$ of Lebesgue measure zero, we know that the limits (13) and (14) exist for every $\lambda \in \mathbf{R} \setminus \delta$ and m, n = 1, 2, ...

Fix a $\lambda \in \mathbf{R} \setminus \delta$ and denote

$$C_\lambda = \lim_{\varepsilon \downarrow 0} \varepsilon D^{\frac{1}{2}}(Y - \lambda + i\varepsilon)^{-1}(\Gamma_Y^+ S)(Y - \lambda - i\varepsilon)^{-1} D^{\frac{1}{2}}.$$

Since

$$|\langle C_\lambda \xi_m, \xi_n \rangle|^2 = \lim_{\varepsilon \downarrow 0} \varepsilon^2 \lambda_m \lambda_n |\langle (Y - \lambda + i\varepsilon)^{-1}(\Gamma_Y^+ S)(Y - \lambda - i\varepsilon)^{-1} \xi_m, \xi_n \rangle|^2$$

it follows that $\langle C_\lambda \xi_m, \xi_n \rangle = 0$ for any m, n = 1, 2... .

Therefore $C_\lambda = 0$ and Lemma 4.1 is proved.

By taking into account relation (13) and Lemma 2.3 we get the following.

4.2. Theorem. (Carey and Pincus). *Let* T *be a pure hyponormal operator with trace-class self-commutator* D *and Cartesian form* T = X + iY. *Let* F *denote the spectral measure of* Y.

Then for every $z \in \mathbf{C} \setminus \mathbf{R}$ *the limit*

$$\lim_{\varepsilon \downarrow 0} E(z, \lambda - i\varepsilon)^{-1} E(z, \lambda + i\varepsilon) = I + \pi(d/d\lambda) D^{\frac{1}{2}} F_\lambda (S_Y^+ X - z)^{-1} D^{\frac{1}{2}}$$

exists in the Hilbert–Schmidt norm, almost everywhere on $\lambda \in \mathbf{R}$.

For the proof it suffices to apply Lemma 4.1 and to remark that the operator $S_Y^+(X - z)^{-1} = (S_Y^+ X - z)^{-1}$ commutes with Y.

4.3. Corollary. *Let* T *be as in the theorem and consider* B *a mosaic of it. Then*

$$E(z, \lambda - i0)^{-1} E(z, \lambda + i0) = \mathbf{exp} \int (B(t,\lambda)/(t - z))dt. \tag{15}$$

Consequently, any two mosaics coincide.

Proof. Similar arguments to those presented above show that the limit $E(z, \lambda + i0)^{-1} E(z, \lambda - i0)$ exists in the Hilbert–Schmidt norm. Hence the operators $E(z, \lambda - i\varepsilon)^{-1} E(z, \lambda + i\varepsilon)$ as well as the limit $E(z, \lambda - i0)^{-1} E(z, \lambda + i0)$ are invertible, and, for $\mathbf{Im}\, z > 0$, all have non-negative imaginary part. Accordingly $\mathbf{log}\, E(z, \lambda - i0)^{-1} \cdot E(z, \lambda + i0)$ exists and

$$\lim_{\varepsilon \downarrow 0} \mathbf{log}\, E(z, \lambda - i\varepsilon)^{-1} E(z, \lambda + i\varepsilon) = \mathbf{log}\, E(z, \lambda - i0)^{-1} E(z, \lambda + i0),$$

in the uniform topology.

Let $\eta_j \downarrow 0$ be a decreasing sequence, such that $w^*\text{-}\lim_j B_{\eta_j} = B$. Then

$$\text{wo-}\lim_j \int (B_{\eta_j}(t,\lambda)/(t - z))dt = \int (B(t,\lambda)/(t - z))dt.$$

By comparing the above limits we find

$$\log E(z, \lambda - i0)^{-1}E(z, \lambda + i0) = \int (B(t,\lambda)/(t - z))dt.$$

This proves the corollary.

Thus, in the case of a hyponormal operator T with $\mathbf{Tr}\,[T^*, T] < \infty$, one can speak without ambiguity of the *mosaic* of T. It will be denoted B_T or simply B.

It is important to remark that the above arguments show the existence of the limits of the determining function $E(z, \lambda \pm i0)$ for all $z \in \mathbf{C} \setminus \mathbf{R}$ and for almost all λ, independently of z.

Our next aim is to relate the mosaic and the principal function of the hyponormal operator T to its singular integral model.

Suppose consequently $T = X + iY$ to be pure and take a direct integral decomposition

$$H = \int_{\sigma(Y)}^{\oplus} H(\lambda)d\lambda$$

which diagonalizes Y, that is $Y = M_\lambda$. Then the operator X can be represented by a singular integral operator

$$(X\xi)(\lambda) = \alpha(\lambda)\xi(\lambda) + (1/\pi i)\int(\beta^*(\lambda)\beta(s)/(s - \lambda))\xi(s)ds.$$

Recall that $\alpha(\lambda) = \alpha(\lambda)^*$ and $\beta(\lambda)$ are esentially bounded operator valued functions on the fibers $H(\lambda)$, respectively.

In Chapter VIII we found the symbols $S_Y^{\pm}(X)$ to be (see formula VIII (8)'):

$$S_Y^{\pm}(X)(\lambda) = \alpha(\lambda) \mp \beta^*(\lambda)\beta(\lambda), \quad \lambda \in \sigma(Y).$$

Let us consider an orthogonal system $(\xi_n)_n$ of vectors which diagonalize D, that is $D\xi_n = \lambda_n\xi_n$ for some $\lambda_n > 0$ and such that the space $X = \mathbf{Ran}\, D^-$ is spanned by (ξ_n). Since the operator D corresponds in the functional model to the integral operator

$$(D\xi)(\lambda) = (2/\pi)\beta^*(\lambda)\int\beta(t)\xi(t)dt,$$

in particular we obtain

$$(2/\pi)\beta^*(\lambda)\int\beta(t)\xi_n(t)dt = \lambda_n\xi_n(\lambda), \quad n \geq 1.$$

Fix the integers k and ℓ. Then

$$\langle (d/d\lambda)D^{\frac{1}{2}}F_\lambda(S_Y^+(X) - z)^{-1}D^{\frac{1}{2}}\xi_k, \xi_\ell\rangle =$$

$$= \lambda_k^{\frac{1}{2}}\lambda_\ell^{\frac{1}{2}}(d/d\lambda) < \int_{-\infty}^{\lambda} (S_Y^+(X)(t) - z)^{-1}\xi_k(t) , \xi_\ell(t)>dt =$$

$$= \lambda_k^{\frac{1}{2}}\lambda_\ell^{\frac{1}{2}} <(S_Y^+(X)(\lambda) - z)^{-1}\xi_k(\lambda) , \xi_\ell(\lambda)> =$$

$$= (2/\pi)<(S_Y^+(X)(\lambda) - z)^{-1}\beta^*(\lambda)\eta_k , \beta^*(\lambda)\eta_\ell>,$$

where we have denoted

$$\eta_k = (2/\pi\lambda_k)^{\frac{1}{2}}\int\beta(t)\xi_k(t)dt.$$

Notice that (η_k) provides another orthonormal basis of the space $X = \mathbf{Ran}\, D^-$. Indeed, for every k and ℓ we have:

$$<\eta_k , \eta_\ell> = (2/\pi)(1/\lambda_k\lambda_\ell)^{\frac{1}{2}}\int<\beta(t)\xi_k(t) , \beta(\lambda)\xi_\ell(\lambda)>dtd\lambda =$$

$$= (1/\lambda_k\lambda_\ell)^{\frac{1}{2}}\int(2/\pi)<\int\beta(\lambda)^*\beta(t)\xi_k(t)dt , \xi_\ell(\lambda)>_\lambda d\lambda = (1/\lambda_k\lambda_\ell)^{\frac{1}{2}}\int<(D\xi_k)(\lambda) , \xi_\ell(\lambda)>_\lambda d\lambda =$$

$$= (1/\lambda_k\lambda_\ell)^{\frac{1}{2}}<D\xi_k , \xi_\ell> = <\xi_k , \xi_\ell>.$$

By Theorem of Carey and Pincus we derive:

$$<E(z, \lambda - i0)^{-1}E(z, \lambda + i0)\xi_k , \xi_\ell> = <(I + \pi(d/d\lambda)D^{\frac{1}{2}}F_\lambda(S_Y^+(X) - z)^{-1}D^{\frac{1}{2}}\xi_k , \xi_\ell> =$$

$$= \delta_{k\ell} + \pi(2/\pi)<(S_Y^+(X)(\lambda) - z)^{-1}\beta^*(\lambda)\eta_k , \beta^*(\lambda)\eta_\ell> =$$

$$= \delta_{k\ell} + 2<\beta(\lambda)(\alpha(\lambda) - \beta^*(\lambda)\beta(\lambda) - z)^{-1}\beta^*(\lambda)\eta_k , \eta_\ell>,$$

for almost all $\lambda \in \mathbf{R}$.

In other terms, if $U \in L(X)$ denotes the unitary operator $U\xi_n = \eta_n$, we have proved the next identity

$$E(z, \lambda - i0)^{-1}E(z, \lambda + i0) = I_X + 2U^*\beta(\lambda)(S_Y^+(X)(\lambda) - z)^{-1}\beta^*(\lambda)U =$$

$$= I_X + 2U^*\beta(\lambda)(\alpha(\lambda) - \beta^*(\lambda)\beta(\lambda) - z)^{-1}\beta^*(\lambda)U.$$

By taking into account Definition 1.4 and relation (15) we have already proved the following.

4.4. Proposition. *The operator valued function* $B_T(\cdot,\lambda)$ *is the phase operator of the pair* $(S_Y^+(X)(\lambda) , \sqrt{2}\beta^*(\lambda)U)$, *for any* $\lambda \in \mathbf{R}$.

We may slightly modify the functional model of the operator T by replacing $\beta(\lambda)$

with $U^*\beta(\lambda)$ – this change is permitted because the kernel of the operator D is $(2/\pi)\beta^*(\lambda)\beta(t) = (2/\pi)\beta^*(\lambda)UU^*\beta(t)$. Consequently the factor U can be deleted in Proposition 4.4. On the new singular model we have

$$\int(B(t,\lambda)/(t-z))dt = I + 2\beta(\lambda)(\alpha(\lambda) - \beta^*(\lambda)\beta(\lambda) - z)^{-1}\beta^*(\lambda). \tag{16}$$

If g_T stands for the principal function of the operator T (that is $g_T = \mathbf{Tr}\, B_T$) then by taking traces in relation (16) we are led to the next.

4.5. Corollary. *The principal function* g_T *of the operator* T *coincides with the following phase shift:*

$$g_T(t,\lambda) = \phi(S_Y^+(X)(\lambda) \rightarrow S_Y^-(X)(\lambda))(t)$$

for every $t \in \mathbf{R}$ *and almost all* $\lambda \in \mathbf{R}$, *independently on* t.

Of course, a similar result holds for X replaced by Y. Such a change may, a priori, alter the mosaic and the principal function. In the next chapter we shall prove that the principal function has remarkable invariance properties, for instance

$$g_T(t,\lambda) = \phi(S_X^-(Y)(t) \rightarrow S_X^+(Y)(t))(\lambda).$$

Anyhow, Corollary 4.5 offers plenty of information about the principal function.

5. Properties of the mosaic

This final section collects some of the deepest results concerning hyponormal operators. Except for the last integrality theorem we present complete proofs to the properties of the mosaic. Though less computational than the results concerning the principal function, to which the next chapter is devoted, the next theorems (all due to Carey and Pincus) are of a high theoretical interest.

5.1. Theorem. *The mosaic* B_T *is a complete unitary invariant for a pure hyponormal operator with trace-class self-commutator.*

Proof. Let $T \in L(H)$, $T' \in L(H')$ be two hyponormal operators as in the statement and let us assume that there exists an unitary operator

$$U : X = \mathbf{Ran}\, D_T^- \rightarrow X' = \mathbf{Ran}\, D_{T'}^-,$$

with the property

$$UB_TU^* = B_{T'}.$$

Take the diagonalizations of the self-adjoint operators Y, Y' on the direct integrals of Hilbert spaces

$$H = \int^{\oplus} H(\lambda)d\lambda, \quad H' = \int^{\oplus} H'(\lambda)d\lambda,$$

respectively, and denote by a, b, a', b' the paramenters in the singular integral models of the operators X and X'. According to Propositions 4.4 and 1.8 we infer from the assumption of purity that there are unitary operators

$$V(\lambda) : H(\lambda) \rightarrow H'(\lambda),$$

such that

(17) $$V(\lambda)a(\lambda)V(\lambda)^* = a'(\lambda), \quad V(\lambda)b(\lambda)U^* = b'(\lambda),$$

for every $\lambda \in \mathbf{R}$. Therefore, for any $\xi \in X$ and $n \geq 0$, we get

(18) $$V(\lambda)a(\lambda)^n b(\lambda)\xi = a'(\lambda)^n b'(\lambda)U\xi.$$

Since the functions $a^n(\cdot)b(\cdot)\xi$ generate the Hilbert space H, relation (18) proves that $V(\cdot)$ is a measurable family of fibre preserving unitary operators (see § VI.5).

In conclusion, relations (17) imply that the operators T and T' are unitarily equivalent. This finishes the proof of the theorem.

Our next aim is to prove the central existence result concerning hyponormal operators. A few of its applications are included among the exercises.

5.2. Theorem. (Carey and Pincus). *Let X be a separable Hilbert space and let $B : \mathbf{R} \times \mathbf{R} \rightarrow C_1(X)$ be a measurable function with compact support and such that $0 \leq B \leq I$.*

Then there exists a Hilbert space H, a pure hyponormal operator $T \in L(H)$ with trace-class self-commutator, and a unitary operator

$$U : \mathbf{Ran}[T^*, T]^- \rightarrow (\mathbf{Ran} \int B(t,\lambda)dt\,d\lambda)^-,$$

such that $B_T = U^* B U$.

Proof. The operator T will be constructed as a singular integral operator acting on a direct integral of Hilbert spaces.

Throughout this proof we may assume that $X = (\mathbf{Ran} \int B(t,\lambda)dt\,d\lambda)^-$.

Fix a point $\lambda \in \mathbf{R}$. By Carey's Theorem 1.4 there exist an auxiliary Hilbert space

$H(\lambda)$ and bounded operators $K(\lambda) : X \rightarrow H(\lambda)$, $A(\lambda) : H(\lambda) \rightarrow H(\lambda)$, $A(\lambda)^* = A(\lambda)$, such that

(19) $$\mathbf{exp}\int (B(t,\lambda)/(t - z))dt = I_X + K(\lambda)^*(A(\lambda) - z)^{-1}K(\lambda), \quad \mathbf{Im}\, z \neq 0,$$

and

(20) $$H(\lambda) = \bigvee_{n=0}^{\infty} A(\lambda)^n K(\lambda)X.$$

We shall identify $H(\lambda)$ to a closed subspace $E_{j(\lambda)}$ belonging to an increasing sequence of subspaces $E_1 \subset E_2 \subset \ldots$ of a given Hilbert space E (see § VI.5).

By recalling the proof of Carey's theorem one finds

(21) $$\sup_{\lambda \in \mathbf{R}} \| K(\lambda) \| < \infty \quad \text{and} \quad \sup_{\lambda \in \mathbf{R}} \| A(\lambda) \| < \infty.$$

Moreover, the series expansion of both terms of equality (19) implies that $K^*(\lambda)A(\lambda)^n K(\lambda)$ are operator valued measurable functions on λ, for every natural n.

Let us consider the operator

$$C = \int B(t,\lambda)dtd\lambda = \int K^*(\lambda)K(\lambda)d\lambda.$$

By our assumption $X = (\mathbf{Ran}\, C)^-$ and $\mathbf{Tr}\, C < \infty$. Let (ξ_j) be an orthonormal basis of X that diagonalizes C, and denote

$$\xi_j^p(\lambda) = A(\lambda)^p K(\lambda)\xi_j, \quad \lambda \in \mathbf{R},\ j, p \in \mathbf{N}.$$

By taking into account the above remarks, the vectors $(\xi_j^p(\lambda))_{p,j}$ span $H(\lambda)$ and the scalar products

$$\langle \xi_j^p(\lambda) , \xi_k^q(\lambda) \rangle$$

are measurable functions.

The family of functions $(\xi_j^p)_{p,j}$ can be arranged into a single sequence. By orthogonalizing this sequence by the Gram - Schmidt procedure, one obtains a new sequence $(\psi_n)_{n=0}^{\infty}$. Put

$$\phi_n(\lambda) = \begin{cases} \psi_n(\lambda)/\| \psi_n(\lambda) \| , & \text{if } \psi_n(\lambda) \neq 0 \\ 0 & \text{, otherwise.} \end{cases}$$

By construction, it is plain to verify that the sequence (ϕ_n) fulfils the conditions needed in the second description of a direct integral of Hilbert spaces, cf. Lemma VI.5.3. Moreover, the functions

$$\langle A(\lambda)\phi_n(\lambda) , \phi_m(\lambda) \rangle \quad \text{and} \quad \langle K^*(\lambda)\phi_n(\lambda) , \xi_j \rangle$$

also depend measurably on λ; j, m, n $\in$ **N**.

In conclusion, we have a direct integral space $H = \int^{\oplus} H(\lambda)d\lambda$ spanned by the functions (ϕ_n), and some measurable families of operators on it:

$$\alpha(\lambda) : H(\lambda) \rightarrow H(\lambda), \quad \alpha(\lambda) = A(\lambda) + \beta^*(\lambda)\beta(\lambda),$$

$$\beta(\lambda) : H(\lambda) \rightarrow X, \quad \beta(\lambda) = (1/\sqrt{2})K(\lambda)^* \,|H(\lambda),$$

which are in addition essentially bounded in view of (21). Notice also that $\beta(\lambda)$ is a Hilbert–Schmidt operator for almost all $\lambda \in \mathbf{R}$.

At this point the singular integral model described in Chapter VII.2 is very close.

Namely, the following singular integral operators

$$(X\xi)(\lambda) = \alpha(\lambda)\xi(\lambda) + (1/\pi i)\int(\beta^*(\lambda)\beta(t)/(t - \lambda))\xi(t)dt,$$

$$(Y\xi)(\lambda) = \lambda\xi(\lambda)$$

are well defined. By Proposition VII.2.3 the operator $T = X + iY \in L(H)$ is pure hyponormal.

To finish the proof it suffices to remark that $V : X \rightarrow \mathbf{Ran}\, D_T^-$,

$$V\xi_j = \langle C\xi_j , \xi_j\rangle^{-\frac{1}{2}}\xi_j^0$$

is a unitary operator. Indeed,

$$\langle\xi_j^0 , \xi_k^0\rangle_H = \int\langle\xi_j^0(\lambda) , \xi_k^0(\lambda)\rangle d\lambda = \int\langle K(\lambda)^*K(\lambda)\xi_j , \xi_k\rangle d\lambda =$$

$$\langle C\xi_j , \xi_k\rangle = \langle C\xi_j , \xi_j\rangle^{\frac{1}{2}}\langle C\xi_k , \xi_k\rangle^{\frac{1}{2}}\langle\xi_j , \xi_k\rangle.$$

Then formula (16) applied to the mosaic of the operator T yields $B = V^*B_TV$.

The last part of this section is devoted to the integrality phenomenon of the mosaic. In other words we shall discuss conditions which imply that the mosaic is projection valued.

First, as a direct application of Proposition 1.6, we can state the following.

5.3. Proposition. *Let* $T = X + iY$ *be a pure hyponormal operator with trace-class self-commutator. If the symbol* $S_Y^+(X)(\lambda)$ *is purely singular for almost all* λ *belonging to an open interval* $I \subset \mathbf{R}$, *then* $B_T(t,\lambda)$ *is projection valued on* $\mathbf{R} \times I$, *almost everywhere.*

Of course, the same conclusion holds for $S_Y^-(X)$ or for $S_X^{\pm}(Y)$, with the natural modifications.

5.4. Corollary. *Assume that the imaginary part of the hyponormal operator* T *has finite multiplicity, almost everywhere. Then the mosaic* B_T *is projection valued and*

$$\operatorname{rank} B_T(t,\lambda) \leq n_Y(\lambda), \quad \text{a.e.}$$

By taking into account Corollary VII.2.4 we also derive from Proposition 5.3 the next.

5.5. Corollary. *The mosaic of a subnormal operator with finite rank self-commutator is projection valued.*

The last corollary is a part of a definitive and deep result, which we present without proof.

5.6. Theorem. (Carey and Pincus). *The mosaic of a subnormal operator is projection valued.*

For details see Carey and Pincus [9].

6. SUPPLEMENT : A spectral mapping theorem for the numerical range

It is the aim of this complement to present a spectral mapping theorem for the numerical range, due to Kato [2]. The reader is suposed to be acquainted with the basic properties of the numerical range, cf. for instance Halmos [4].

Let T be a bounded linear operator acting on the Hilbert space H. The *numerical range* of T is the set

$$W(T) = \{\langle T\xi , \xi\rangle \ ; \ \xi \in H, \ \|\xi\| = 1\}.$$

The following properties of the numerical range are known, see Halmos [4]:

(i) (Toeplitz - Hausdorff) the set W(T) is convex;

(ii) The closure of W(T) contains $\sigma(T)$;

(iii) (Toeplitz) If T is normal, then the closure of W(T) coincides with the convex hull of $\sigma(T)$.

We shall prove the next naturality theorem. Recall that $\mathbf{C}_+$ denotes the upper half-plane.

6.1. Theorem. (Kato). *Let* $T \in L(H)$ *be with* $W(T) \subset \bar{\mathbf{C}}_+$. *If* $f \in O(\bar{\mathbf{C}}_+)$, *then* $W(f(T)) \subset \mathbf{co}\, f(\mathbf{C}_+)$.

Proof. Take the Cayley transform

$$S = (T - i)(T + i)^{-1}.$$

By the spectral mapping theorem one obtains $1 \notin \sigma(S)$ and $\sigma(S) \subseteq \bar{\mathbf{D}}$, where $\mathbf{D}$ is the unit disk. We have

$$T = i(I + S)(I - S)^{-1},$$

and

$$f(T) = \lim_{r \uparrow 1} f_r(S),$$

where $f_r(z) = f(i(1 + rz)(1 - rz)^{-1})$, $z \in \mathbf{D}$.

Since $f_r(\bar{\mathbf{D}})$ is a closed subset of $f(\bar{\mathbf{C}}_+)$, it suffices to prove that $W(f_r(S))$ is contained in the convex hull of $f_r(\bar{\mathbf{D}})$.

To this end we have to know firstly that S is a contraction. Let $\xi \in H, \|\xi\| = 1$, and put $\eta = (T + i)^{-1}\xi$. Notice the estimate

$$|\langle S\xi, \xi\rangle| = |\langle T - i)\eta, (T + i)\eta\rangle| \leq \|(T - i)\eta\| \cdot \|(T + i)\eta\|.$$

But $1 = \|\xi\| = \|(T + i)\eta\|$, and

$$\|(T \pm i)\eta\|^2 = \|T\eta\|^2 + \|\eta\|^2 \pm 2i\,\mathbf{Im}\langle T\eta, \eta\rangle.$$

As $W(T) \subset \bar{\mathbf{C}}_+$, $\mathbf{Im}\,\langle T\eta, \eta\rangle \geq 0$, whence

$$\|(T - i)\eta\| \leq \|(T + i)\eta\| = 1.$$

This shows $|\langle S\xi, \xi\rangle| \leq 1$, as desired.

Next, let U denote the minimal unitary dilation of the contraction S, acting on the bigger Hilbert space K. We put $P = P^K_H$. By the Sz.-Nagy dilation theorem we infer

$$f_r(S) = Pf_r(U)|H, \quad r < 1,$$

see Sz.-Nagy and Foiaş [1]. Because U is a normal operator

$$W(f_r(U)) = \mathbf{co}\,\sigma(f_r(U)) = \mathbf{co}\,f_r\sigma(U)) \subset \mathbf{co}\,f_r(\bar{\mathbf{D}}).$$

In conclusion

$$W(f_r(S)) \subset W(f_r(U)) \subset \mathbf{co}\,f_r(\bar{\mathbf{D}}),$$

and the proof is complete.

6.2. Corollary. *Let* $T \in L(H)$ *with* $W(T)$ *contained in a compact convex set*

$K \subset \mathbf{C}$. *If* f *is an entire function, then* $W(f(T)) \subset \mathbf{co}\, f(K)$.

Actually the condition on f can be weakened; see Kato [2].

Notes. Section 1 is essentially reproduced after Carey [1]. In our opinion, the theory of the phase operator is less known and much less exploited. The determining functions appeared in Pincus [1], [2] and then they were intensively used in the work of Carey and Pincus, cf. almost all references. The proof of the basic Proposition 3.2 is reproduced after Carey and Pincus [6]. The contents of Section 4 is a part of the memoir Carey and Pincus [5]. We mention that the latter paper is essentially self--contained and it treats the theory of pairs of self-adjoint operators with trace-class commutator. Section 5 collects results from several papers: Carey and Pincus [4], [6], [9]. A mosaic constructed from the polar decomposition of a hyponormal operator was considered by Carey and Pincus [10]. Also, Pincus and Xia [1] have generalized the mosaic to some more general operators than the hyponormal operators.

EXERCISES

1*. (Carey). Let $A, K \in L(H)$, $A = A^*$. If 0 is an isolated eigenvalue of finite multiplicity of A, then there exists a deleted neighbourhood $U = (-\varepsilon, 0) \cup (0, \varepsilon)$ of ε, on which the phase operator B(t) is a harmonic projection valued function of bounded rank, commuting with v.p.$\int (B(x)/(x - t))dt$; and conversely.

2. Prove that there exists a pure hyponormal operator T, such that the operator $\mathbf{Re}\, T$ has infinite multiplicity and $\mathbf{rank}\, [T^*, T] < \infty$.

3. Prove that there exists a pure hyponormal operator T with the property $C(\sigma(T)) = R(\sigma(T))$.

4. Compute the mosaic of a cutting of a hyponormal operator along the real axis.

5. a) Show that there are hyponormal but not subnormal operators with projection valued mosaic.

b) The rectangular cuttings of a subnormal operator are not necessarily subnormal.

6. Prove that the mosaic is not weakly continuous in the $*$-strong operatorial topology.

Hint. Use Corollary 5.4.

7. Open problem. (Carey and Pincus [9]). Does the mosaic of a rationally cyclic hyponormal operator assume only projection values?

8*. (D. Xia [3]). **a)** Let T be an invertible semi-hyponormal operator represented by its singular integral model (cf. Exercise VII.9).

Prove that there exists a measurable function $B : \mathbf{T} \times \mathbf{R}_+ \to L(X)$ (X is the codomain of the operators a(z)) satisfying $0 \leq B \leq I$ and such that

$$I + a(z)(b(z) - \zeta)^{-1}a^*(z) = \mathbf{exp} \int_{\mathbf{R}} (B(z,t)/(t - \zeta))dt, \quad z \in \mathbf{T}.$$

b) Prove, by using angular cuttings, that **supp**(B) $\subset \sigma(T)$.

c) Prove the inequality

$$\|(T^*T)^{\frac{1}{2}} - (TT^*)^{\frac{1}{2}}\| \leq (1/2\pi)\int_{\sigma(T)} d\mu(\zeta)/|\zeta|.$$

9. Let T be a hyponormal operator with finite rank self-commutator. If there exists a non-trivial relation of the form

$$\sum_{k,\ell=0}^{N} a_{k\ell} X^k Y^b \xi = 0, \qquad \xi \in \mathbf{Ran}[T^*, T],$$

between the real and imaginary parts (X respectively Y) of T, then the mosaic B_T is projection valued, a.e.

Hint. Prove the finiteness a.e. of the multiplicity function of X and then use Proposition 4.4.

Chapter X

THE PRINCIPAL FUNCTION

By its construction, the mosaic of a hyponormal operator with trace-class self--commutator heavily depends on the Cartesian decomposition of the operator, whereas its trace – the principal function – has remarkable invariance properties. This fact is in close analogy with what happens in the case of a trace-class perturbation of self-adjoint operators, where the phase-shift is much more flexible than its operator analogue. The basic property of the principal function, whereby its invariant behaviour may be easily deduced, is a formula for the trace of the commutator between two smooth functions on the operator. By taking this formula, due to Carey - Pincus and Helton - Howe, as a definition of the principal function, the reader may ignore at a first glance the details of the construction of mosaics.

The aim of the present chapter is to discuss Helton and Howe's approach towards the trace formula, and then to present the main properties of the principal function. It is worth to mention that the principal function is one of the most important and refined tools in studying (hyponormal) operators with trace-class self-commutator.

1. Bilinear forms with the collapsing property

Let $T \in L(H)$ be a hyponormal operator with $[T^*, T] \in C_1(H)$ and principal function g_T. We already know that $g_T \in L^1_+(\mathbf{C})$ and $\mathbf{supp}(g_T) \subset \sigma(X) \times \sigma(Y)$, where $X = \mathbf{Re}\,T$ and $Y = \mathbf{Im}\,T$. In the previous chapter we have established the following identity

(1) $$\mathbf{Tr}[X^{n+1}Y^m, Y] = (1/2\pi i)\int_{\mathbf{C}} (n+1)x^n y^m g_T(x,y)dxdy,$$

which is valid for any natural numbers $n, m \in \mathbf{N}$. If one associates to a polynomial $p = \sum_{i,j=1}^{d} a_{ij}x^i y^j$ the ordered operator expression

$$p(X,Y) = \sum_{i,j=1}^{d} a_{ij} X^i Y^j,$$

then relation (1) yields

$$\mathbf{Tr}[p(X,Y), Y] = (1/2\pi i) \int_{\mathbf{C}} (\partial p/\partial x)(x,y) g_T(x,y) dxdy.$$

In fact a more symmetric formula holds, as follows:

1.1. Theorem. (Carey - Helton - Howe - Pincus). *Let* $T \in L(H)$ *be a hyponormal operator with trace-class self-commutator. For any pair of polynomials* $p, q \in \mathbf{C}[x,y]$, *one has*

(2) $$\mathbf{Tr}[p(X,Y), q(X,Y)] = (1/2\pi i) \int_{\mathbf{C}} ((\partial p/\partial x)(\partial q/\partial y) - (\partial p/\partial y)(\partial q/\partial x)) g_T dxdy,$$

where g_T *is the principal function of* $T = X + iY$.

This section is devoted to the proof of Theorem 1.1.

First of all we have to remark that the commutator $[p(X,Y), q(X,Y)]$ is trace-class. Further the reader will certainly recognize

$$J(p,q) = (\partial p/\partial x)(\partial q/\partial y) - (\partial p/\partial y)(\partial q/\partial x)$$

to be the Jacobian (or the Poisson bracket) of the pair of functions (p,q). Formula (2) is derived from (1) by a pure algebraic device which may be of an independent interest. Quite specifically, we shall investigate the properties of the bilinear form

$$b_T : \mathbf{C}[x,y] \times \mathbf{C}[x,y] \to \mathbf{C},$$

$$b_T(p,q) = \mathrm{tr}[p(X,Y), q(X,Y)], \quad p, q \in \mathbf{C}[x,y].$$

Obviously the form b_T is antisymmetric, and

(3) $$b_T(p \circ r, q \circ r) = 0,$$

for every $p, q \in \mathbf{C}[x]$ and $r \in \mathbf{C}[x,y]$. We call (3) (after Helton and Howe) the *collapsing property* of the form b_T. Our aim is to prove that an antisymmetric bilinear form b on $\mathbf{C}[x,y]$, which possesses the collapsing property is perfectly determined by its values $b(p,y)$, $p \in \mathbf{C}[x,y]$. This is done in the next.

1.2. Proposition. (N. Wallach). *Let* b *be an antisymmetric bilinear form on the algebra* $\mathbf{C}[x,y]$. *If* b *has the collapsing property, then there is a linear functional* ℓ *on* $\mathbf{C}[x,y]$, *such that*

(4) $$b(p,q) = \ell(J(p,q)); \quad p, q \in \mathbf{C}[x,y].$$

Proof. It is immediate that the class of antisymmetric bilinear forms on $\mathbf{C}[x,y]$ with the collapsing property is a vector space which contains every form $\ell \circ J$, where ℓ is a linear functional on $\mathbf{C}[x,y]$.

Fix a form b as in the statement, and define a linear functional $\ell : \mathbf{C}[x,y] \to \mathbf{C}$ by the formula:

$$\ell(p) = b(P,y), \quad p \in \mathbf{C}[x,y],$$

where P is any polynomial satisfying $(\partial P/\partial x) = p$. The definition is correct because a second polynomial P' with the property $(\partial P'/\partial x) = p$ will differ from P by a polynomial in y, $P' - P$, and, according to the collapsing property $b(P' - P, y) = 0$. Thus

$$b(p,y) = \ell(J(p,y))$$

for any $p \in \mathbf{C}[x,y]$. By replacing b with $b - \ell \circ J$, we consequently may assume that

(5) $$b(p,y) = 0, \quad p \in \mathbf{C}[x,y].$$

Then it remains to prove that $b = 0$.

Let us denote the radical of the form b by

$$R = \{q \in \mathbf{C}[x,y] \, ; \, b(p,q) = 0, \quad p \in \mathbf{C}[x,y]\}.$$

We claim that

(i) R is a subalgebra of $\mathbf{C}[x,y]$,

(ii) $x \in R$.

By (5), (i) and (ii) it would follow that $x, y \in R$, hence $R = \mathbf{C}[x,y]$ and the proof would be finished.

In order to prove (i), take $q_1, q_2 \in R$, $\alpha, \beta \in \mathbf{C}$ and $p \in \mathbf{C}[x,y]$. By virtue of the collapsing property

$$b((\alpha q_1 + \beta q_2 + p), (\alpha q_1 + \beta q_2 + p)^2) = 0.$$

Since $q_1, q_2 \in R$ and $b(p,p^2) = 0$, one finds

$$b(p, \alpha^2 q_1^2 + \beta^2 q_2^2 + 2\alpha\beta q_1 q_2 + 2\alpha q_1 p + 2\beta q_2 p) = 0.$$

But α, β are free parameters, whence the coefficient of $\alpha\beta$ must vanish:

$$b(p, q_1 q_2) = 0.$$

This proves (i).

Similarly, for any $\alpha \in \mathbf{C}$ and $n \in \mathbf{N}$ we infer from the collapsing property and (5):

$$b((x + \alpha y)^n , x + \alpha y) = 0,$$

$$b((x + \alpha y)^n, x) = 0,$$

therefore

$$b(x^j y^{n-j} , x) = 0, \quad 0 \leq j \leq n.$$

This proves the assertion (ii).

The proofs of Proposition 1.2 and Theorem 1.1 are complete.

2. Smooth functional calculus modulo trace-class operators and the trace formula

Helton and Howe's formula (2) naturally extends to more general functions than polynomials. Though further we do not make use of this strenghthened formula, it reveals the importance of the principal function in the context of smooth extensions of C^*-algebras. Below, we touch only tangentially the subject by restricting ourselves to generalizing Theorem 1.1. The reader may omit the present section without loss of continuity. For further developments consult Voiculescu [6] and Connes [1].

Throughout this paragraph X and Y are bounded self-adjoint operators on the Hilbert space H, subject to the commutator condition $[X, Y] \in C_1(H)$. Let $S(\mathbf{R}^2)$ denote the Schwarz space of rapidly decreasing functions at infinity.

There are at least two reasonable and quite classical ways of constructing a smooth functional calculus distribution for the pair X, Y,

$$\tau_{X,Y} : S(\mathbf{R}^2) \longrightarrow L(H),$$

with the following properties:

(i) $\tau_{X,Y}$ is linear, continuous and compactly supported on $\mathbf{R}^2$,

(ii) $\tau_{X,Y}(x) = X$, $\tau_{X,Y}(y) = Y$, $\tau_{X,Y}(\mathbf{1}) = I$, and

$$\mathbf{supp}\, \tau_{X,Y} \subset \sigma(X) \times \sigma(Y),$$

(iii) $\tau_{X,Y}(\phi)^* - \tau_{X,Y}(\bar{\phi}) \in C_1, \quad \phi \in S(\mathbf{R}^2)$,

(iv) $\tau_{X,Y}(\phi)\tau_{X,Y}(\psi) - \tau_{X,Y}(\phi\psi) \in C_1, \quad \phi, \psi \in S(\mathbf{R}^2)$.

Notice that condition (iv) implies $[\tau_{X,Y}(\phi) , \tau_{X,Y}(\psi)] \in C_1$. Our final aim is to compute $\mathrm{tr}[\tau_{X,Y}(\phi) , \tau_{X,Y}(\psi)]$ in terms of ϕ, ψ and the spectral data (read: principal function) of the pair of operators X, Y.

A first candidate for the distribution $\tau_{X,Y}$ is

$$\tau_1(\phi) = \int_{\mathbf{R}^2} \phi(x,y)dE(x)dF(y), \quad \phi \in S(\mathbf{R}^2),$$

where E and F are the spectral measures of the operators X and Y, respectively. Instead, we may take the uniform limit

(6) $$\tau_1(\phi) = \lim_{n\to\infty} \sum_{j,k\in\mathbf{Z}} \phi(j/n, k/n)E((j-1)/n, j/n]F((k-1)/n, k/n].$$

It is plain to check that τ is continuous in the uniform topology of $L(H)$. For details consult for instance Daleckii - Krein [1].

Condition (ii) is trivially satisfied. Moreover

$$\tau_1(p) = \sum_{i,j} a_{ij}X^iY^j,$$

for every polynomial $p(x,y) = \sum_{i,j} a_{ij}x^iy^j$. Thus τ_1 is an extension of the polynomial functional calculus introduced in the first section.

Before verifying properties (iii) and (iv) for the distribution τ_1, let us introduce a second candidate, namely

$$\tau_2(\phi) = (1/2\pi) \int_{\mathbf{R}^2} e^{itX}e^{isY}\hat{\phi}(t,s)dtds.$$

Here

$$\hat{\phi}(t,s) = (1/2\pi) \int_{\mathbf{R}^2} e^{-i(tx+sy)}\phi(x,y)dxdy$$

denotes the Fourier transform of the function $\phi \in S(\mathbf{R}^2)$. Since $e^{itX}e^{isY}$ is a unitary operator for every t, $s \in \mathbf{R}$, the definition of τ_2 makes sense. By a standard argument one proves that τ_2 is also continuous. Moreover $\tau_2(\phi)$ can be uniformly approximated by Riemann sums:

(7) $$\tau_2(\phi) = \lim_{n\to\infty} (1/2\pi n^2) \sum_{j,k\in\mathbf{Z}} e^{i(j/n)X}e^{i(k/n)Y}\hat{\phi}(j/n, k/n).$$

2.1. Lemma. *With the above notation,* $\tau_1 = \tau_2$.

Proof. Fix an element ϕ of $S(\mathbf{R}^2)$ and denote

$$v_{j,k}^{(n)}(x,y) = e^{i(jx/n + ky/n)}, \quad (x,y) \in \mathbf{R}^2,$$

for any j, $k \in \mathbf{Z}$ and $n \in \mathbf{N}$, $n \geq 1$. Then

$$e^{i(j/n)X}e^{i(k/n)Y} = \int_{\mathbf{R}^2} v_{j,k}^{(n)}(x,y)dE(x)dE(y) = \tau_1(v_{j,k}^{(n)}).$$

By virtue of (6) and by the continuity of τ_1 one gets:

$$2\pi\tau_2(\phi) = \lim_{n\to\infty} (1/n^2) \sum_{j,k} \tau_1(v_{j,k}^{(n)})\hat{\phi}(j/n\,,\,k/n) = \tau_1(\lim_{n\to\infty} (1/n^2) \sum_{j,k} v_{j,k}^{(n)}\hat{\phi}(j/n\,,\,k/n)) =$$

$$= \tau_1(\int e^{i(xs+yt)}\hat{\phi}(s,t)dsdt) = 2\pi\tau_1(\phi).$$

Let us denote $\tau_{X,Y} = \tau_1 = \tau_2$.

2.2. Proposition. *Let* X,Y *be a pair of self-adjoint operators with trace-class self commutator. The distribution* $\tau_{X,Y} : S(\mathbf{R}^2) \to L(H)$ *has properties* (i) - (iv).

Proof. We have already seen that $\tau_{X,Y}$ satisfies properties (i) and (ii).

Let $\rho = \max(\|X\|, \|Y\|)$. Then for any pair of natural numbers one obtains by a standard argument:

$$\|[X^n, Y^m]\|_1 \leq nm\rho^{n+m-2}\|[X, Y]\|_1.$$

Consequently, for any s, t $\in$ **R**,

$$\|[e^{itX}, e^{itY}]\|_1 \leq (\sum_{n,m=1}^{\infty} (|t|^n|s|^m\rho^{n+m-2}/(n-1)!(m-1)!))\|[X, Y]\|_1,$$

that is

$$(8) \qquad \|[e^{itX}, e^{itY}]\|_1 \leq |t||s|e^{\rho|t|}e^{\rho|s|}\|X, Y]\|_1.$$

If $M = e^{2\rho}|[X, Y]|_1$, then we assert that the next estimate

$$(9) \qquad \|[e^{itX}, e^{itY}]\|_1 \leq M(1+|t|)(1+|s|)$$

holds. Indeed, assume that $n - 1 < |t| \leq n$ and $m - 1 < |s| \leq m$, with n, m integers. Then the unitary operators $U = e^{i(t/n)X}$ and $V = e^{i(s/m)Y}$ satisfy by (8):

$$\|[U, V]\|_1 \leq M.$$

Accordingly

$$\|[e^{itX}, e^{isY}]\|_1 = \|[U^n, V^m]\|_1 \leq n\cdot m\|U\|\|V\|M,$$

and inequality (9) is proved.

In order to prove that the distribution $\tau_{X,Y}$ satisfies condition (iii), we pick $\phi \in S(\mathbf{R}^2)$ and compute $\tau_{X,Y}(\phi)^* - \tau_{X,Y}(\bar{\phi})$ as

$$\tau_2(\phi)^* - \tau_2(\bar{\phi}) = (1/2\pi)\int[e^{-isY}e^{-itX}\hat{\phi}(t,s) - e^{itX}e^{isY}\hat{\bar{\phi}}(t,s)]dtds =$$

$$=-(1/2\pi)\int[e^{itX}, e^{itY}]\hat{\phi}(t,s)dtds.$$

By (9) one finally obtains

$$\|\tau_{X,Y}(\phi)^* - \tau_{X,Y}(\bar{\phi})\|_1 \leq (M/2\pi)\int(|t|+1)(|s|+1)\|\hat{\bar{\phi}}(t,s)\|_1 dtds$$

and statement (iii) is proved.

Similarily one proceeds with property iv). Let ϕ, ψ be elements of $S(\mathbf{R}^2)$. Then, by Fubini's theorem

$$\tau_{X,Y}(\phi)\tau_{X,Y}(\psi) = (1/4\pi^2)\int_{\mathbf{R}^4} e^{itX}e^{isY}e^{it'X}e^{is'Y}\hat{\phi}(t,s)\hat{\psi}(t',s')dtdsdt'ds',$$

and

$$\tau_{X,Y}(\phi\psi) = (1/2\pi)\int_{\mathbf{R}^2} e^{itX}e^{isY}(\phi\psi)^{\wedge}(t,s)dtds = (1/4\pi^2)\int_{\mathbf{R}^2} e^{itX}e^{isY}(\hat{\phi} * \hat{\psi})(t,s)dtds$$

$$= (1/4\pi^2)\int_{\mathbf{R}^2} e^{i(t+t')X}e^{i(s+s')Y}\hat{\phi}(t,s)\hat{\psi}(t',s')dtdsdt'ds'.$$

Thus

$$\tau_{X,Y}(\phi)\tau_{X,Y}(\psi) - \tau_{X,Y}(\phi\psi) = (-1/4\pi^2)\int e^{itX}[e^{it'X}, e^{isY}]e^{is'Y}\hat{\phi}(t,s)\hat{\psi}(t',s')dtdsdt'ds'$$

and consequently

$$|\tau_{X,Y}(\phi)\tau_{X,Y}(\psi) - \tau_{X,Y}(\phi\psi)|_1 \leq (M/4\pi^2)\int(|s|+1)|\hat{\phi}(t,s)|dtds \cdot \int(|t|+1)|\hat{\psi}(t,s)|dtds$$

and the proof of Proposition 2.2 is complete.

The trace-norm estimates given above lead easily to the next.

2.3. Corollary. *With the above notations, the maps*

$$S(\mathbf{R}^2) \to C_1, \quad \phi \mapsto \tau_{X,Y}(\phi)^* - \tau_{X,Y}(\bar{\phi}),$$

$$S(\mathbf{R}^2)\times S(\mathbf{R}^2) \to C_1, \quad (\phi,\psi) \mapsto [\tau_{X,Y}(\phi), \tau_{X,Y}(\psi)]$$

are well defined and continuous.

Since $\mathbf{supp}\,\tau_{X,Y}$ is compact, the distribution $\tau_{X,Y}$ extends continuously to $E(\mathbf{R}^2)$ - the Fréchet space of smooth functions on $\mathbf{R}^2$.

Now our task is simple, because in the case of a hyponormal operator $T = X + iY \in L(H)$, with $[T^*, T] \in C_1(H)$, we can define a functional calculus modulo trace-class operators by

$$\tau_T(\phi) = \tau_{X,Y}(\phi), \quad \phi \in E(\mathbf{R}^2).$$

Moreover, for any polynomial $p \in \mathbf{C}[x,y]$,

$$\tau_T(p) = p(X,Y),$$

whence the two continuous bilinear forms on $E(\mathbf{R}^2) \times E(\mathbf{R}^2)$,

$$\mathbf{Tr}[\tau_T(\phi)\,,\,\tau_T(\psi)] \quad \text{and} \quad (1/2\pi i)\int J(\phi,\psi)g_T dxdy$$

coincide by Theorem 1.1 on $\mathbf{C}[x,y] \times \mathbf{C}[x,y]$. Since the last subspace is dense in $E(\mathbf{R}^2) \times E(\mathbf{R}^2)$ we have proved the following.

2.4. Theorem. *Let* $T \in L(H)$ *be a hyponormal operator with trace-class self-commutator. Then*

$$\mathbf{Tr}[\tau_T(\phi)\,,\,\tau_T(\psi)] = (1/2\pi i)\int_{\mathbf{R}^2} J(\phi,\psi)(x,y)g_T(x,y)dxdy, \tag{10}$$

for any $\phi, \psi \in E(\mathbf{R}^2)$.

It should be mentioned that the theorem is valid for any operator T with $[T^*, T] \in C_1$. Anyhow relation (10) characterizes the principal function g_T. Some consequences of this formula are presented in the next section. Henceforth (10) will be called *Helton and Howe's formula.*

Our next aim is to establish a more accurate evaluation of the support of the bilinear form

$$b_T : E(\mathbf{R}^2) \times E(\mathbf{R}^2) \to \mathbf{C}, \quad b_T(\phi,\psi) = \mathrm{tr}[\tau_T(\phi),\tau_T(\psi)],$$

attached to a hyponormal operator with trace-class self-commutator.

2.5. Proposition. *Assume that* $T \in L(H)$ *satisfies* $[T^*, T] \in C_1(H)$ *(and that it is hyponormal). Then*

$$\mathbf{supp}(b_T) \subset \sigma_{\mathbf{ess}}(T) \times \sigma_{\mathbf{ess}}(T).$$

Proof. In other terms the proposition asserts that

$$b_T(\phi,\psi) = \mathbf{Tr}[\tau_T(\phi)\,,\,\tau_T(\psi)] = 0,$$

whenever $\mathbf{supp}(\phi) \cap \sigma_{\mathbf{ess}}(T) = \emptyset$ or $\mathbf{supp}(\psi) \cap \sigma_{\mathbf{ess}}(T) = \emptyset$; $\phi, \psi \in E(\mathbf{R}^2)$.

To prove this assertion, consider the C^*-algebra $A(T)$ generated by T in $L(H)$, and its quotient $A_c(T) = A(T)/A(T) \cap K(H)$, where $K(H)$ is the ideal of compact operators. Due to the fact that $[T^*, T] \in C_1$, the algebra $A_c(T)$ is commutative. Moreover the Gelfand spectrum of $A_c(T)$ coincides with $\sigma_{ess}(T)$ - the essential spectrum of T.

Let $E(T)$ denote the involutive (non-closed) subalgebra of $A_c(T)$ generated by the operators $\tau_T(\phi)$, $\phi \in E(\mathbf{R}^2)$. It is plain to remark that $[E(T), E(T)] \subset C_1(H)$. Consider the restriction of the Gelfand transformation to $E(T)$:

$$\Gamma : E(T) \longrightarrow C(\sigma_{ess}(T)).$$

One easily computes $\Gamma(\tau_T(\phi)) = \phi | \sigma_{ess}(T)$, $\phi \in E(\mathbf{C})$.

We assert that the bilinear form $b_T(\phi, \psi)$ factorizes through $\Gamma \times \Gamma$, hence it depends only on $\phi | \sigma_{ess}(T)$ and $\psi | \sigma_{ess}(T)$. This follows from the next assertion.

Let $A, B \in E(T)$ and assume A or B to be compact. Then

$$\mathbf{Tr}[A , B] = 0. \qquad (11)$$

For the proof we may assume A compact and self-adjoint. Then there is a diagonalization

$$Ae_i = \lambda_i e_i , \quad \lambda_i \in \mathbf{R},$$

with respect to an orthogonal basis $(e_i)_{i=1}^{\infty}$ of H. Accordingly

$$\langle (AB - BA)e_i , e_i \rangle = \langle Be_i , Ae_i \rangle - \langle BAe_i , e_i \rangle = 0,$$

and relation (11) is proved.

The proof of Proposition 2.5 is complete.

The final part of this section is concerned with multiplicative commutators and with an exponential formula for their determinants, derived from the Helton and Howe formula.

The *multiplicative commutator* of the pair of invertible operators $A, B \in L(H)$ is the expression $ABA^{-1}B^{-1}$.

2.6. Lemma. *Let* $A, B \in L(H)$ *be such that* $[A, B] \in C_1(H)$. *Then* $e^A e^B e^{-A} e^{-B} - I \in C_1(H)$ *and*

$$\mathbf{det}(e^A e^B e^{-A} e^{-B}) = e^{\mathbf{Tr}[A,B]}.$$

Proof. By estimate (8) we infer $[e^B, e^{-A}] \in C_1$, whence the operator

$$e^A e^B e^{-A} e^{-B} - I = e^A [e^B, e^{-A}] e^{-B}$$

is trace-class.

Fix a parameter $t > 0$. Since the Hausdorff series converges in $C_1(H)$ for t small enough we have

$$e^{tA}e^{tB}e^{-tA}e^{-tB} = e^{t^2[A,B]+t^3C_3+\dots}$$

where the higher coefficients are iterated commutators, see Bourbaki [1], §II.7. Consequently $\mathbf{det}(e^{tA}e^{tB}e^{-tA}e^{-tB})$ is well defined and

$$\mathbf{det}(e^{tA}e^{tB}e^{-tA}e^{-tB}) = e^{t^2\mathbf{Tr}[A,B]}.$$

By virtue of the analyticity in t of both terms in the last expression, we finally obtain the desired identity.

2.7. Proposition. *Let* $T = X + iY$ *be a hyponormal operator with trace-class self-commutator and let* $\phi, \psi \in S(\mathbf{R}^2)$. *Then*

$$\mathbf{det}(e^{\phi(X,Y)}e^{\psi(X,Y)}e^{-\phi(X,Y)}e^{-\psi(X,Y)}) = \mathbf{exp}((1/2\pi i)\int_{\mathbf{R}^2} J(\phi,\psi)g_T d\mu).$$

For the proof simply apply Lemma 2.6 and Theorem 2.4.

3. The properties of the principal function

We collect below the main properties of the principal function. Throughout this section $T \in L(H)$ is a pure hyponormal operator with trace-class self-commutator $D = [T^*, T]$. As usually we denote by $X = \mathbf{Re}\, T$, $Y = \mathbf{Im}\, T$ and $g = g_T$.

We start with the following characterization of the principal function g:

(12) $$\mathbf{Tr}[p(T,T^*), q(T,T^*)] = (1/\pi)\int_{\mathbf{C}}(\bar\partial p\, \partial q - \partial p\, \bar\partial q) g d\mu; \quad p, q \in \mathbf{C}[z,\bar z],$$

which is a restatement of relation (2), and

(13) $$g(x + iy) = \mathbf{Tr}\, B_T(x,y) = \phi(S_Y^+(X)(y) \to S_Y^-(X)(y))(x), \text{ a.e.}$$

cf. Proposition IX.4.4. Here B_T stands for the mosaic of T and ϕ is the phase shift of the respective perturbation, see §IX.4.

It is quite clear from (12) that the principal function behaves as follows with respect to an affine transformation:

1) $g_{aT+b}(z) = g_T((z - b)/a)$, $a \neq 0$.

In particular $g_{iT}(z) = g(-iz)$, whence

2) $g(x+iy) = \phi(S_X^-(Y)(x) \to S_X^+(Y)(x))(y)$, a.e.

Let δ be a Borel subset of **R** and consider the cut-down T_δ of the operator T. That means (see § V.1) to take the compression of T onto the subspace $E(\delta)H$, denoting by E the spectral measure of X. Since the symbols $S_X^{\pm}(Y)$ lie in the commutant of X, relation 2) yields

3) $$g_{T_\delta}(x+iy) = \begin{cases} 0, & x \notin \delta, \\ g_T(x+iy), & x \in \delta. \end{cases}$$

Analogously

3') $$g_{T^\delta}(x+iy) = \begin{cases} 0, & y \notin \delta, \\ g_T(x+iy), & y \in \delta. \end{cases}$$

Our next aim is to compute the essential support of the function g. Let δ, δ' be two closed intervals of **R**, with the property $\sigma(T) \cap \mathrm{int}(\delta \times \delta') \neq \emptyset$. Since the operator T was supposed to be pure, Corollary VI.2.4 implies $\sigma((T_\delta)^{\delta'}) \neq \emptyset$. Therefore the principal function

$$\tilde{g} = \chi_{\delta\times\delta'} \cdot g$$

of the operator $(T_\delta)^{\delta'}$ does not vanish. The above $\chi_{\delta\times\delta'}$ denotes the characteristic function of the rectangle $\delta \times \delta'$.

By virtue of properties of the cut-downs of T, in particular Proposition VI.2.3, we have proved that every open rectangle which intersects $\sigma(T)$ intersects **supp**(g), too; and conversely. Thus we have established the remarkable property:

4) $\mathbf{supp}(g) = \sigma(T)$.

Immediatly one obtains by (13)

4') $\mathbf{supp}(B_T) = \sigma(T)$.

Now let us compute the phase shift of the non-negative perturbation $TT^* \to T^*T$. Fix an integer $n \geq 0$. Then we find by (12)

$$\mathbf{Tr}[(T^*T)^n - (TT^*)^n] = \mathbf{Tr}[T^*(TT^*)^{n-1}, T] = (1/\pi)\int_{\mathbf{C}} n|z|^{2n-2} g(z)d\mu(z) =$$

$$= (1/\pi)\int_0^\infty \int_0^{2\pi} nr^{2n-1} g(re^{i\theta})d\theta dr = (1/2\pi)\int_0^\infty n(r^2)^{n-1}\int_0^{2\pi} g(re^{i\theta})d\theta d(r^2) =$$

$$= (1/2\pi)\int_0^\infty nt^{n-1}\int_0^{2\pi} g(\sqrt{t}e^{i\theta})d\theta dt.$$

By applying the trace formula contained in Lemma VIII.1.4 we find:

5) $\phi_{(TT^* \to T^*T)}(t) = (1/2\pi)\int_0^{2\pi} g(\sqrt{t}e^{i\theta})d\theta$, for a.a. $t > 0$.

If $0 \in \sigma(T) \setminus \sigma_{\mathbf{ess}}(T)$, then T^*T is invertible and TT^* has 0 as an isolated eigenvalue of finite multiplicity. Moreover

$$\mathbf{dim\ Ker}\ TT^* = -\mathbf{ind}\ T.$$

Consequently Proposition VIII.1.6 shows that

$$\phi_{(TT^* \to T^*T)}(t) = -\mathbf{ind}\ T$$

for small values of $t > 0$. Therefore property 5) gives

$$\lim_{\varepsilon \downarrow 0} (1/\mu(D_\varepsilon)) \int_{D_\varepsilon} g d\mu = -\mathbf{ind}\ T,$$

where the limit is taken over all disks D_ε centered at 0, of radius ε. By property 1) the same relation holds for $T - z$, with small enough $|z|$:

$$\lim_{\varepsilon \downarrow 0} (1/\mu(D_\varepsilon)) \int_{D_\varepsilon} g(z + \zeta)d\mu(\zeta) = -\mathbf{ind}(T - z).$$

As the first limit coincides with g(z) almost everywhere, by Lebesgue theorem, we have proved the equality:

6) $g(z) = -\mathbf{ind}(T - z), \quad z \notin \sigma_{\mathbf{ess}}(T)$.

Returning to the description (13) of the principal function one gets,

7) $g(x + iy) \leq \mathbf{mult}_X(x); \quad x, y \in \mathbf{R}$,

where $\mathbf{mult}_X(x)$ is the von Neumann multiplicity function of the self-adjoint operator X, evaluated at the point $x \in \mathbf{R}$. Of course the complementary estimate also holds:

8) $g(x + iy) \leq \mathbf{mult}_Y(y); \quad x, y \in \mathbf{R}$.

The same relation (13) combined with Proposition VIII.1.6 yields:

9) If one of the self-adjoint operators X or Y has finite multiplicity, bounded by n, then the principal function g is integer valued and $g \leq n$.

Since $\phi_{(TT^* \to T^*T)} \leq \mathbf{rank}[T^*, T]$, property 5) implies the estimate:

10) $g \leq \mathbf{rank}[T^*, T]$.

In order to study the transformation law of the principal function under nonlinear maps, we need a slight generalization, see Clancey's book [5], pp. 106-107. Let α, β be real valued polynomials of two real variables x and y. We define the operators

$$A = \alpha(X,Y), \quad B = \beta(X,Y)$$

with an arbitrary choice of the order between X and Y in the monomials of α and β. Then A and B are self-adjoint modulo the ideal of trace-class operators.

Though the operator $A + iB$ is not necessarily hyponormal, it still has trace-class self-commutator, hence the expression

$$\mathbf{Tr}[p(A,B)\,,\,q(A,B)]$$

has a good meaning for all polynomials p, q.

By virtue of (12) one obtains

$$\mathbf{Tr}[p(A,B)\,,\,q(A,B)] = \mathbf{Tr}[p(\alpha,\beta)\,,\,q(\alpha,\beta)] = (1/2\pi i)\int_{\mathbf{R}^2} J(p(\alpha,\beta)\,,\,q(\alpha,\beta))g\,d\mu =$$

$$= (1/2\pi i)\int_{\mathbf{R}^2} J(p,q)(\alpha(x',y')\,,\,\beta(x',y'))\,J(\alpha,\beta)(x',y')g(x'+iy')dx'dy' =$$

$$= (1/2\pi i)\int_{\mathbf{R}^2} J(p,q)(x,y)g(x+iy)dxdy,$$

after an obvious change of coordinates.

In conclusion we have proved the following statement:

11) Let f be a polynomial of two complex variables. Though the operator $f(T)$ fails to be hyponormal in general, its tracial bilinear form can be represented by

$$\mathbf{Tr}[p(f(T),f(T)^*)\,,\,q(f(T),f(T)^*)] = (1/2\pi i)\int_{\mathbf{R}^2} J(p,q)g_{f(T)}d\mu,$$

$p, q \in \mathbf{C}[z,\bar{z}]$, where

$$g_{f(T)}(z) = \sum_{f(\zeta)=z} \mathrm{sgn}(J(\mathbf{Re}\,f,\mathbf{Im}\,f)(\zeta))g(\zeta).$$

For instance, if the operator $f(T)$ is hyponormal, then its principal function is precisely $g_{f(T)}$ above.

As a corollary we derive the next property.

12) Let n be a positive integer. Then

$$g_{T^n}(z) = \sum_{k=1}^{n} g(\zeta^k),$$

where $\zeta^n = z$.

Above of course, as well as elsewhere, only the class of the function g in $L^1(\mathbf{R}^2)$ is computed.

The last property of the principal function to be discussed in this section is a formula due to Clancey [8]. It relates the unitary invariant Γ_T to the principal function.

13) Let T be a hyponormal operator whose self-commutator D satisfies $\mathbf{Tr}\, D^{\frac{1}{2}} < \infty$. Then

$$\mathbf{Tr}\, D^{\frac{1}{2}} C(\lambda)^* = (1/\pi)\int g_T(z)(z-\lambda)^{-1} d\mu(z),$$

for almost every $\lambda \in \mathbf{C}$.

Recall that $C(\lambda)$ is the unique contraction which satisfies the relations:

$$(T-\lambda)^* C(\lambda) = D^{\frac{1}{2}}, \quad C(\lambda)^* \,|\, \mathbf{Ker}(T-\lambda)^* = 0$$

see § II.3.

In order to prove property 13) we may assume $\lambda = 0$ and $\int g_T(z)|z|^{-1} d\mu(z) < \infty$; this is due to the fact that the integral $\int g_T(z)|z-\lambda|^{-1} d\mu(z)$ is finite for almost all $\lambda \in \mathbf{C}$. We shall denote for simplicity $C = C(0)$.

Let K be the unique contraction which satisfies

$$T^* = KT, \quad K\,|\,\mathbf{Ker}\, T^* = 0.$$

Then

$$[K,T]T = (KT - TK)T = T^*T - TT^* = D = D^{\frac{1}{2}} C^* T,$$

whence we infer the identity

$$D^{\frac{1}{2}} C^* = [K,T].$$

Notice that $K = \text{so-}\mathbf{lim}_{\varepsilon \downarrow 0}\, T^*(T^*T+\varepsilon)^{-1} T^*$.

Since by hypothesis the operator $[K,T]$ is trace-class, one gets by Helton and Howe's formula:

$$\mathbf{Tr}\, D^{\frac{1}{2}} C^* = \mathbf{lim}_{\varepsilon \downarrow 0}\, \mathbf{Tr}[T^*(T^*T+\varepsilon)^{-1}T^*, T] = \mathbf{lim}_{\varepsilon \downarrow 0}\, (1/\pi)\int ((\bar{z}^2 z + 2\varepsilon\bar{z})/(\bar{z}z+\varepsilon)^2) g_T(z) d\mu(z) =$$

$$= (1/\pi)\int g_T(z) z^{-1} d\mu(z).$$

The last equality was possible by the Lebesgue dominated convergence theorem.

We remark finally that property 13) is equivalent, by taking derivatives, to the relation

$$g_T = -\bar{\partial}\, \mathbf{Tr}(D^{\frac{1}{2}} C^*),$$

which in turn implies by obvious modifications the next formula

$$\mathbf{Tr}\, \Gamma_T = g_T.$$

4. Berger's estimates

In a couple of papers, C.A. Berger [1], [2] developed an original method to obtain bounds from above for the principal function by certain invariants like the rational multiplicity. His estimates have interesting consequences to some positive results concerning the existence of invariant subspaces for hyponormal operators. The present section deals exclusively with Berger's techniques.

4.1. Lemma. *Let* $\tilde{T}$ *be a trace-class perturbation of a hyponormal operator* $T \in L(H)$ *with* $[T^*, T] \in C_1(H)$. *Then*

$$\mathbf{Tr}[p(\tilde{T},\tilde{T}^*), q(\tilde{T},\tilde{T}^*)] = \mathbf{Tr}[p(T,T^*), q(T,T^*)]$$

for every pair of polynomials p, q.

For the proof it sufficies to recall that $\mathbf{Tr}[A, B] = 0$, whenever $A \in C_1(H)$ and $B \in L(H)$.

Next we need some smooth cuttings of a hyponormal operator. Accordingly, let ϕ, ψ be real valued polynomials of a real variable and denote for every $T = X + iY$:

$$T_{\phi,\psi} = \phi(Y)X\phi(Y) + i\psi(\phi(Y)X\phi(Y))Y\psi(\phi(Y)X\phi(Y))$$

The reader will easily recognize $T_{\phi,\psi}$ as a double cutting of the operator T, see §V.1. Moreover a computation performed in §V.1 yields

$$[T^*_{\phi,\psi}, T_{\phi,\psi}] = \psi(\phi(Y)X\phi(Y))\phi(Y)[T^*, T]\phi(Y)\psi(\phi(Y)X\phi(Y)).$$

Thus $T_{\phi,\psi}$ is hyponormal whenever the operator T is hyponormal.

4.2. Lemma. *Let* $\tilde{T}$ *be a trace-class perturbation of a hyponormal operator* T *with* $\mathbf{Tr}[T^*, T] < \infty$. *For any real polynomials* ϕ *and* ψ *one has*

$$\mathbf{Tr}[\tilde{T}^*_{\phi,\psi}, \tilde{T}_{\phi,\psi}] = (1/\pi) \int_{\mathbf{R}^2} \phi(y)^2\psi(\phi(y)^2x)^2 g_T(x + iy)dxdy.$$

Proof. By Lemma 4.1 we may assume $\tilde{T} = T$. Then apply Helton and Howe's formula (12).

Now we are able to prove a first of Berger's estimates.

4.3. Theorem. (Berger). *Let* T *and* S *be hyponormal operators with trace-class self-commutator. Assume* A *is a trace-class operator with* **Ker** $A =$ **Ker** $A^* = 0$ *and such* *that*

$$SA = AT.$$

Then

$$g_S \leq g_T \quad \text{a.e.}$$

The theorem is still true for A Hilbert–Schmidt, cf. Berger [2].

Proof. For any real $t \in \mathbf{R}$, let us consider the graph G_{tA} of the operator $tA \in L(H)$ and the orthogonal projection onto $G_{tA} \subset H + H$:

$$P_t = \begin{bmatrix} (I + t^2A^*A)^{-1} & (I + t^2A^*A)^{-1}tA^* \\ tA(I + t^2A^*A)^{-1} & tA(I + t^2A^*A)tA^* \end{bmatrix}.$$

It is quite clear that $P_t - P_0 \in C_1(H + H)$ for every $t \in \mathbf{R}$. Thus the operator

$$B_t = \begin{bmatrix} T & 0 \\ 0 & S \end{bmatrix} P_t = \begin{bmatrix} T & 0 \\ 0 & 0 \end{bmatrix} + \begin{bmatrix} T & 0 \\ 0 & S \end{bmatrix} (P_t - P_0)$$

is a trace-class perturbation of $T \oplus 0$. Moreover, by relation $SA = AT$ we learn that G_{tA} is an invariant subspace for the operator $T \oplus S$. Whence

(14) $$[B_t^*, B_t] = P_t(T\oplus S)^*(T\oplus S)P_t - (T\oplus S)P_t(T\oplus S)^* \geq P_t[(T\oplus S)^*, (T\oplus S)]P_t.$$

Let ϕ, ψ be two polynomials with real coeficients depending on a single variable, and define for $B_t = X_t + iY_t$:

$$R_t = \phi(Y_t)X_t\phi(Y_t),$$

$$I_t = \psi(R_t)Y_t\psi(R_t),$$

so that $(B_t)_{\phi,\psi} = R_t + iI_t$ is the Cartesian decomposition. In view of (14) note that

(15) $$\mathbf{Tr}[(B_t)^*_{\phi,\psi}, (B_t)_{\phi,\psi}] \geq \mathbf{Tr}\, \psi(R_t)\phi(Y_t)P_t[(T\oplus S)^*, (T\oplus S)]P_t\phi(Y_t)\psi(R_t).$$

On the other hand, Lemma 4.1 gives

$$\mathbf{Tr}[(B_t)^*_{\phi,\psi}, (B_t)_{\phi,\psi}] = \mathbf{Tr}[T^*_{\phi,\psi}, T_{\phi,\psi}].$$

Since $\text{so-}\lim\limits_{t\to\infty} P_t = \begin{bmatrix} 0 & 0 \\ 0 & I \end{bmatrix}$ we also have

$$\text{so-}\lim_{t\to\infty} \phi(Y_t) = \phi(\mathbf{Im}\, S),$$

$$\text{so-}\lim_{t\to\infty} \psi(R_t) = \psi(\phi(\mathbf{Im}\, S)(\mathbf{Re}\, S)\phi(\mathbf{Im}\, S)).$$

By passing to the limit in (15) and by using Fatou's theorem we find:

$$\mathbf{Tr}[T^*_{\phi,\psi}, T_{\phi,\psi}] \geq \mathbf{Tr}[S^*_{\phi,\psi}, S_{\phi,\psi}].$$

The explicit computation of the two terms offered by Lemma 4.2 finally gives, by varying ϕ and ψ:

$$g_T(x + iy) \geq g_S(x + iy), \quad \text{a.e.,}$$

and the proof is complete.

4.4. Corollary. *Let* T *be a cyclic hyponormal operator. Then* $g_T \leq 1$.

Proof. Of course we may take T pure and after a homothety we may assume $\|T\| < 1$.

Let us define an operator $A : l^2(\mathbf{N}) \to H$ by

$$Ae_n = T^n\xi, \quad n \geq 0,$$

where ξ is a cyclic vector for T, and (e_n) denotes the standard basis of l^2. Then the operator A intertwines the unilateral shift U_+ and T, that is $AU_+ = TA$. It is plain to check that $\mathbf{Ker}\, A^* = 0$. Also $\mathbf{Ker}\, A \neq \emptyset$ if and only if there exists an analytic function f in the unit disk, so that $f(T) = 0$. As the operator T was supposed to be pure, this is not possible and hence $\mathbf{Ker}\, A = 0$.

It remains to prove that the operator A is trace-class. Indeed, let $\|T\| < r < 1$, and define $A_2 : l^2 \to l^2$ by

$$A_2e_n = r^n e_n, \quad n \geq 0,$$

and $A_1 : l^2 \to H$ by

$$A_1e_n = (T/r)^n\xi.$$

Therefore $A = A_1A_2$ and one can easily verify that A_1 and A_2 are Hilbert–Schmidt operators.

By applying Berger's Theorem,

$$g_T \leq g_{U_+} = \chi_D$$

and the corollary is proved.

This corollary has in its turn an interesting application, as follows:

4.5. Theorem. (Berger). *Let* T *by a hyponormal operator. For a sufficiently high* n, *the operator* T^n *has a non-trivial invariant subspace.*

Proof. Of course we may assume T pure and with cyclic vector. Then the self-commutator $[T^*, T]$ is trace-class by Berger and Shaw's inequality, Corollary VI.1.4.

Using property 12) of the principal function one easily remarks that there is an $n \geq 0$ such that the set

$$\{z \in \mathbf{C} \,;\, g_{T^n}(z) = \sum_{k=1}^{n} g_T(z^{k/n}) > 1\}$$

has non-trivial Lebesgue measure.

But the above corollary still holds for the not necessarily hyponormal operator T^n, as one can deduce from an inspection of the proof of Theorem 4.3. Consequently

$$\| g_{T^n} \|_\infty > 1$$

and the proof is over.

Actually, the estimate given in Corollary 4.4 can be generalized to the following statement.

4.6. Proposition. *Let* T *be a hyponormal operator with finite rational cyclic multiplicity* m(T). *Then*

$$g_T \leq m(T).$$

Proof. Berger and Shaw's inequality enables us to speak of the principal function of T.

First, let us assume that the operator T is m-cyclic. Then, by repeating the argument given in the proof of Corollary 4.4 one constructs an intertwining operator A:

$$A(U_+ \oplus \ldots \oplus U_+) = (1 + \|T\|)^{-1} TA,$$

such that **Ker** A = **Ker** $A^* = 0$. Then the estimate $g_T \leq m$ follows by Berger's Theorem 4.3.

In the general case let $\xi_1, \ldots, \xi_m$ be a system of vectors, such that

$$\bigvee\{f(T)\xi_j\ ;\ 1 \le j \le m,\ f \in \mathbf{Rat}(\sigma(T))\} = H.$$

Fix a countable dense subset $(\lambda_k)_{k=1}^{\infty}$ of $\rho(T)$.

For every natural n we define the space

$$H_n = \bigvee\{f(T)\xi_j\ ;\ 1 \le j \le m,\ f \in \mathbf{Rat}(\sigma(T)) \text{ with poles among } \lambda_1, \dots, \lambda_n\}.$$

It is immediate that H_n are invariant subspaces for the operator T, and that $H_n \subset H_{n+1}$, $H_n \uparrow H$. Moreover each H_n is an m-cyclic space, with the cyclic vectors

$$\prod_{k=1}^{n} (T - \lambda_k)^{-1}\xi_j, \quad 1 \le j \le m.$$

Thus the proof of the proposition will be finished by the next result.

4.7. Lemma. *Let* $T \in L(H)$ *be hyponormal with trace-class self-commutator and let* $H_n \subset H$ *be an increasing sequence of closed* T*-invariant subspaces, such that* $H_n \uparrow H$ *and* $g_{T|H_n} \le M < \infty$, $n \ge 1$. *Then*

$$g_T = \text{w-}\lim_{n\to\infty} (g_{T|H_n}).$$

Proof. Let us denote by P_n the orthogonal projection onto H_n, and put $T_n = T|H_n$. The operators T_n are still hyponormal and:

$$[T_n^*, T_n] \ge P_n[T^*, T]P_n.$$

Again by using the device with the double smooth cutting $T_{\phi,\psi}$, we recall that

$$[T_{\phi,\psi}^*, T_{\phi,\psi}] = \psi(\phi(Y)X\phi(Y))\phi(Y)[T^*, T]\phi(Y)\psi(\phi(Y)X\phi(Y)),$$

cf. Lemma 4.2 and its preliminaries.

If one denotes by $T_n = X_n + iY_n$ the corresponding cartesian decompositions, then

$$\text{so-}\lim_{n\to\infty} \psi(\phi(Y_n)X_n\phi(Y_n))\phi(Y_n)P_n = \psi(\phi(Y)X\phi(Y))\phi(Y)$$

for any polynomials $\phi, \psi \in \mathbf{R}[t]$. Thus

$$\lim_{n\to\infty} \mathbf{Tr}[(T_n)_{\phi,\psi}^*, (T_n)_{\phi,\psi}] = \mathbf{Tr}[T_{\phi,\psi}^*, T_{\phi,\psi}].$$

Then Lemma 4.2 yields

$$\lim_{n\to\infty} \int_{\mathbf{R}^2} \phi(y)^2\psi(\phi(y)^2x)^2(g_{T_n} - g_T)(x + iy)dxdy = 0$$

for any $\phi, \psi \in \mathbf{R}[t]$. Since the sequence $(g_{T_n})_n$ is uniformly bounded in the norm of $L^1(\mathbf{R}^2)$, the proof is over. This finishes also the proof of Proposition 4.6.

Notes. The principal function was introduced by Pincus [1] in the case of operators with one dimensional self-commutator. Section 1 is reproduced after Helton and Howe [1], [2]. Theorems 1.1 and 2.4 crown the efforts of Helton and Howe [1] and Carey and Pincus [5]. Quite specifically, Helton and Howe proved formula (10) for a positive measure instead of $g_T d\mu$, and Carey and Pincus proved that this measure is absolutely continuous with respect to the Lebesgue measure, by identifying the weight with the principal function, as it has been defined in Chapter IX. The properties of the principal function are collected from literature, cf. Helton and Howe [1], Carey and Pincus [2], [6], [7] and Clancey [5], [8]. Section 4 is reproduced with minor modification after Berger [1], [2].

EXERCISES

1. Let $T = X + iY$ be a hyponormal operator with trace-class self-commutator. Prove the formula

$$\mathbf{det}[(X - z)^{-1}(Y - w)^{-1}(X - z)(Y - z)] = \mathbf{exp}\Big((1/2\pi i) \int_{\mathbf{R}^2} (g_T(x + iy)/(x - z)(y - w))dxdy\Big),$$

for any $z, w \in \mathbf{C} \setminus \mathbf{R}$.

2. Let $N \in L(H)$ be a normal operator and $P \in L(H)$ an orthogonal projection, such that $NP = PNP$.

a) The subnormal operator $S = N|PH$ has trace-class self-commutator iff $[N, P] \in C_2$.

b) If $[N, P] \in C_2$, then for any analytic function $f \in O(\sigma(S) \cup \sigma(N))$ one has

$$\|[f(N), P]\|_2^2 = (1/\pi) \int_{\mathbf{C}} |f'|^2 g_S d\mu.$$

3. a) (Putnam). If T is a hyponormal operator with the property that T^n is normal for some positive integer n, then T is normal.

b) Generalize the above statement to f(T), where f is an analytic function defined in a neighbourhood of the spectrum of T.

4. Let T be a pure hyponormal operator and consider the unique contraction $K(\lambda)$ subject to the conditions

$$K(\lambda)(T - \lambda) = (T - \lambda)^*, \quad K(\lambda)|\,\mathbf{Ker}(T - \lambda)^* = 0, \quad \lambda \in \mathbf{C}.$$

Prove that, if f is a rational function with poles off $\sigma(T)$ and the operator $[K(\lambda), f(T)]$ is trace-class for almost all $\lambda \in \mathbf{C}$, one has

$$\mathbf{Tr}[K(\lambda), f(T)] = (1/\pi)\int \partial f(z)(z - \lambda)^{-1} g_T(z)d\mu(z),$$

μ-almost everywhere.

5. (Putinar [4]). Let K be a compact subset of $\mathbf{C}$ with positive planar density at every-one of its points and denote by R_K the unique pure hyponormal operator with one dimensional self-commutator which satisfies $\sigma(R_K) = K$.

Prove that a pure hyponormal operator T satisfies $g_T(z) = \mathbf{mult}_{\mathbf{Re}\,T}(\mathbf{Re}\, z)$ a.e. if and only if $\sigma(T) \cap (\{\lambda\} \times \mathbf{R})$ is connected for every $\lambda \in \mathbf{R}$ and T is unitarily equivalent to $P(R_K \otimes I_L)P$, where L is a Hilbert space and P is a projection belonging to the commutant of $\mathbf{Re}(R_K)\otimes I$, such that $\mathbf{mult}_{\mathbf{Re}\,T} = \mathbf{mult}_{P(\mathbf{Re}\,R_K \otimes I)}$.

6. Open problem. (Voiculescu [6]). Let $T, S \in L(H)$ be pure hyponormal operators with trace-class self-commutator. If $g_T = g_S$, prove that there is a normal operator $N \in L(H)$ and a unitary $U \in L(H \oplus H)$, such that

$$U(T \oplus N)U^* - S \oplus N \in C_2.$$

Chapter XI

OPERATORS WITH ONE DIMENSIONAL SELF-COMMUTATOR

The class of operators with one dimensional self-commutator is the best understood among all classes of hyponormal operators. Beginning with the first memoir of Pincus [1] and continuing with a large number of papers up to the recent report of Clancey [9], these operators have constituted the test field for any new idea related to hyponormal operators.

The present chapter is mainly concerned with the recent unifying developments due to Clancey. Thanks to his efforts a natural relationship between operators with one dimensional self-commutator and function theory of two real variables has become relevant.

Although the principal methods and facts which have been already presented in the preceding chapters can be shortened or improved in the case of operators with one dimensional self-commutator, we confine ourselves to presenting a selection of the most recent results in the field. The reader will easily fill this gap by following a few guiding marks in the problems section.

1. The global local resolvent

Let $T \in L(H)$ be an operator with one dimensional self-commutator: $[T^*, T] = \pm \xi \otimes \xi$. Throughout this chapter we deal exclusively with hyponormal operators like T above, that means

$$D = [T^*, T] = \xi \otimes \xi.$$

Since the space $X = [T^*, T]H = \mathbf{C} \cdot \xi$ is one dimensional, the contractive operator valued function $C : \mathbf{C} \to L(X)$ is determined by the formula

$$C(z) = T_z^{*-1}\xi \otimes \xi / \|\xi\|. \tag{1}$$

Indeed, $D^{\frac{1}{2}} = \xi \otimes \xi / \|\xi\|$ and $T_z^{*-1}\xi$ denotes the unique solution of $T_z^* x = \xi$ which is orthogonal to **Ker** T_z^*, see §II.4.

The vector valued function $z \mapsto T_z^{*-1}\xi$ was introduced by Putnam [10] and we call it the *global resolvent* of T^*, *localized* at the vector ξ. We shall organize our approach around this object. First we present a few properties of the global local resolvent.

Because C(z) is a contraction, (1) yields

(2) $$\|T_z^{*-1}\xi\| \leq 1.$$

Another property inherited from those of C, is

(3) $$T_z^{*-1}\xi \in H_p(T),$$

where $H_p(T)$ denotes the pure part of the Hilbert space H, with respect to the operator T. Note that T is irreducible if and only if it is pure.

Let us assume, for a fixed point $z \in \mathbf{C}$, that $\mathbf{Ker}\, T_z^* \cap H_p \neq 0$, and let η denote the unique (!) element in $\mathbf{Ker}\, T_z^* \cap H_p(T)$ which satisfies $\langle\xi, \eta\rangle = 1$. Then

$$(T_z^*T_z - T_zT_z^*)\eta = T_z^*T_z\eta = \langle\eta, \xi\rangle\xi = \xi,$$

so that $T_z^{*-1}\xi = T_z\eta$. Then

$$\|T_z^{*-1}\xi\|^2 = \|T_z\eta\|^2 = \langle T_z^*T_z\eta, \eta\rangle = \langle\xi, \eta\rangle = 1,$$

therefore we have proved that

(4) $$\|T_z^{*-1}\xi\| = 1, \quad \bar{z} \in \sigma_p(T^* | H_p(T)).$$

The main specific result concerning the function $T_z^{*-1}\xi$ is contained in the next.

1.1. Proposition. *Let* T *be a hyponormal operator with one dimensional self--commutator. The global local resolvent* $T_z^{*-1}\xi$ *is weakly continuous on* $\mathbf{C}$, *and strongly continuous off the set* $\{z \in \sigma(T) \mid \|T_z^{*-1}\xi\| < 1\}$.

Proof. Let $\{z_n\}$ be a sequence of complex numbers, with the limit z_0. Any weak limit x_0 of a subsequence of $T_{z_n}^{*-1}\xi$ satisfies $T_{z_0}^*x = \xi$. Hence the function $T_z^{*-1}\xi$ is weakly continuous off the set $\delta = \overline{\sigma_p(T^* | H_p(T))}$. Since, by (4), $\|T_{z_0}^{*-1}\xi\| = 1$ whenever $z_0 \in \delta$ and $T_{z_0}^{*-1}\xi$ is the solution of $T_{z_0}^*x = \xi$ of minimum norm, and because $\|x_0\| \leq 1$, we infer $T_{z_0}^{*-1}\xi = x_0$. This proves the weak continuity on the whole $\mathbf{C}$.

By the weak continuity property one obtains $\|T_{z_o}^{*-1}\xi\| \leq \underline{\lim}_k \|T_{z_k}^{*}\xi\| \leq 1$. Therefore, if $\|T_{z_0}^{*-1}\xi\| = 1$, then necessarily $\lim_k \|T_{z_k}^{*-1}\xi\| = \|T_{z_0}^{*-1}\xi\|$. This proves the strong continuity part of the statement, and finishes the proof of the proposition.

The exceptional set $\{z \in \sigma(T) \mid \|T_z^{*-1}\xi\| < 1\}$ may be non-empty as we shall see later. However it will be proved that this set is rather thin.

2. The kernel function

Throughout this section T is an irreducible hyponormal operator with one dimensional self-commutator, $[T^*, T] = \xi \otimes \xi$. One denotes by g its principal function. Recall that $0 \leq g \leq 1$ and $\mathbf{supp}(g) = \sigma(T)$, § X.4.

The complete unitary invariants $C^*(z)C(w)$ and Γ described in Chapter II have simpler forms, namely

$$C^*(z)C(w) = \langle T_z^{*-1}\xi, T_w^{*-1}\xi\rangle \text{ and } \Gamma = g.$$

We denote for simplicity

$$k_T(z,w) = \langle T_w^{*-1}\xi, T_z^{*-1}\xi\rangle; \quad z, w \in \mathbf{C}.$$

The equation (II.9) which relates the two invariants becomes, still at the level of distributions:

$$(\bar{w} - \bar{z})\bar{\partial}_z k_T(z,w) = g(z)(1 - k_T(z,w)). \tag{5}$$

The purpose of this section is to solve explicitly this equation, with the kernel k_T as unknown and g as a given data. We already know by Theorem VII.3.3 that the condition $\lim_{|z|+|w|\to\infty} |k_T(z,w)| = 0$ and the positivity of k_T insures the existence and uniqueness of the solution to this problem. More specifically we shall prove the following.

2.1. Theorem. (Clancey). *Let* T *be a hyponormal operator with one dimensional self-commutator and principal function* g. *Then*

$$k_T(z,w) = 1 - \mathbf{exp}((-1/\pi)\int g(\zeta)d\mu(\zeta)/(\zeta - z)(\bar{\zeta} - \bar{w})). \tag{6}$$

Notice that the function under the integral is summable for $z \neq w$. We take by convention the exponential to be zero whenever $\int g(\zeta)d\mu(\zeta)/(|\zeta - z|^2) = \infty$.

Proof. Let w be a fixed complex number and let $z \in \mathbf{C} \setminus \{w\}$. Then equation (5) implies

$$\bar{\partial}_z[(1 - k_T(z,w))\mathbf{exp}((1/\pi)\int g(\zeta)d\mu(\zeta)/(\zeta - z)(\bar{\zeta} - \bar{w}))] = 0.$$

Therefore

(7) $$(1 - k_T(z,w)) = f_w(z)\mathbf{exp}((-1/\pi)\int g(\zeta)d\mu(\zeta)/(\zeta - z)(\bar{\zeta} - \bar{z})), \quad z \neq w,$$

where $f_w \in \mathcal{O}(\mathbf{C} \setminus \{w\})$. By taking limits when $z \longrightarrow \infty$, one gets $f_w(\infty) = 1$.

Since the left hand side of (7) is separately continuous by Proposition 1.1, it remains to prove that the function

$$F(z,w) = \mathbf{exp}(-1/\pi)\int g(\zeta)d\mu(\zeta)/(\zeta - z)(\bar{\zeta} - \bar{w}))$$

is separately continuous on $\mathbf{C}^2$.

Indeed, in that case Liouville's theorem would imply $f_w \equiv 1$.

Because the function F(z,w) is separately continuous off the diagonal z = w, a property inherited from the convolution product in its expression, and because

$$\lim_{z\to w} (1 - k_T(z,w)) = \lim_{z\to w} |1 - k_T(z,w)|,$$

the proof of the theorem reduces to the next lemma reproduced after Clancey [7], pp. 451-453.

2.2. Lemma. *For any* $w \in \mathbf{C}$, $\lim_{z\to w} |F(z,w)| = |F(w,w)|$.

Proof. We may assume without loss of generality that w = 0. Then

$$|F(z,0)| = \mathbf{exp}((-1/\pi)\int \mathbf{Re}(\zeta/(\zeta - z))(g(\zeta)/|\zeta|^2)d\mu(\zeta)).$$

It is sufficient to prove that

(8) $$\lim_{z\to 0} \int \mathbf{Re}(\zeta/(\zeta - z))(g(\zeta)/|\xi|^2)d\mu(\zeta) = \int (g(\zeta)/|z|^2)d\mu(\zeta).$$

Let us denote

$$h(\zeta,z) = \mathbf{Re}(\zeta/(\zeta - z)) = \tfrac{1}{2}(1 + (|\zeta|^2 - |z|^2)/|\zeta - z|^2),$$

and

$$\ell = \int (g(\zeta)/|\zeta|^2)d\mu(\zeta).$$

For a fixed $\lambda \in \mathbf{R}$, the level set $\{\zeta \mid h(\zeta,z) = \lambda\}$ is the circle

$$\Gamma_\lambda = \{\zeta \mid |\zeta - (1 + \mu/2)z| = |(1 - \mu/2)z|\},$$

where $\mu = \lambda/(\lambda - 1)$, $\lambda, \mu \neq 1$. In the limiting case $\lambda = 1$, Γ_1 is the line through w perpendicular to w.

Let $\lambda \neq 1$ and map the unit disc $\mathbf{D} = \{u \mid |u| \leq 1\}$ onto the interior of Γ_λ by a linear transformation. By changing variables $\zeta \to u$, one obtains for $\lambda < 1$:

$$(9) \quad \int_{h(\cdot,z)\leq\lambda} h(\zeta,z)(d\mu(\zeta)/|\zeta|^2) = \int_{\mathbf{D}} (1 - (\lambda - 1/(u - 1) - (\lambda - 1)/(\bar{u} - 1))d\mu(u)/|u - (2\lambda - 1)|^2.$$

Similarly for $\lambda > 1$ one finds

$$\int_{h(\cdot,z)\geq\lambda} h(\zeta,z)(d\mu(\zeta)/|\zeta|^2) =$$

$$= \int_{\mathbf{D}} (1 - ((\lambda - 1)/(u - 1)) - ((\lambda - 1)/(\bar{u} - 1)))d\mu(u)/|u - (2\lambda - 1)|^2.$$

In particular one deduces from the last expressions that for every $\varepsilon > 0$ there is a constant $M > 1$ so that

$$(10) \quad \int_{|h(\cdot,z)|\geq M} |k(\zeta,z)|\,d\mu(\zeta)/|\zeta|^2 < \varepsilon.$$

Case 1. Assume $\ell < \infty$. Let $\varepsilon > 0$ and chose $M > 2$ and $\delta > 0$ such that (10) holds and

$$\int_{|h(\cdot,w)|>M} (g(\zeta)/|\zeta|^2)d\mu(\zeta) < \varepsilon.$$

This is possible because the set $|h(\cdot, w)| > M > 2$ is contained in the disc $|\zeta| < 2|z| < 2\delta$. Then

$$\left|\int h(\zeta,z)(g(\zeta)/|\zeta|^2)d\mu(\zeta) - \ell\right| \leq \int_{|h(\cdot,z)|\leq M} |h(\zeta,z) - 1|(g(\zeta)/|\zeta|^2)d\mu(\zeta) + 2\varepsilon.$$

Because $\lim_{z\to 0} |h(\zeta,z) - 1|$ for a fixed ζ, the proof of Case 1 is over.

Case 2. Assume $\ell = \infty$. By Fatou's lemma

$$\int (g(\zeta)/|\zeta|^2)d\mu(\zeta) < \varliminf_{z\to 0} \int_{h(\cdot,z)\geq 0} h(\zeta,z)(g(\zeta)/|\zeta|^2)d\mu(\zeta),$$

by remarking that $\{\zeta \mid k(\zeta,z) \geq 0\} \subset \{\zeta \mid |\zeta| \leq |z|\}$. Since

$$\sup_{0<|z|<\delta} \int_{h(\cdot,z)\leq 0} h(\zeta,z)(g(\zeta)/|\zeta|^2)d\mu(\zeta) < \infty$$

by (9), the lemma is proved.

This establishes the proof of Clancey's theorem.

We must remark that we do not know an elementary proof of the fact that, for an arbitrary function $0 \leq g \leq 1$ with compact support, the kernel

$$1 - \mathbf{exp}((-1/\pi)\int(g(\zeta)d\mu(\zeta)/(\zeta - z)(\overline{\zeta} - \overline{w}))$$

is separately continuous and positive definite.

Let us assume $|z|$ and $|w|$ large. Then

$$1 - \langle T_z^{*-1}\xi , T_w^{*-1}\xi\rangle = \mathbf{det}(I - T_z^{*-1}(\xi\otimes\xi)T_w^{-1}) =$$

$$= \mathbf{det}(I - T_z^{*-1}[T_z^* , T_w]T_w^{-1}) = \mathbf{det}(T_z^{*-1}T_w T_z^* T_w^{-1}).$$

Thus we can state the next.

2.3. Lemma. *For large values of* $|z|$ *and* $|w|$, *the identity*

$$k_T(z,w) = 1 - \mathbf{det}(T_w^{*-1}T_z T_w^* T_z^{-1})$$

holds.

A couple of other similar kernels appear in Clancey [9] and Pincus - Xia - Xia [1]. Since any of these kernels is a complete unitary invariant, they can be deduced, at least theoretically, from k_T, and finally from g_T. We make a choice by which only the kernel k_T will be discussed in the sequel.

By taking residues at infinity in (6), one finds

(11) $$\langle\xi , T_z^{*-1}\xi\rangle = (1/\pi)\int(g(\zeta)/(\zeta - z))d\mu(\zeta).$$

This formula has interesting applications to cyclic vector problems, see Clancey [10].

We summarize below for the convenience of the reader the characteristic properties of the kernel k_T, as they follow from Theorem VII.3.3.

2.4. Proposition. *Let* g *be a compactly supported measurable function on* $\mathbf{C}$, *so that* $0 \leq g \leq 1$. *There is a unique kernel function* $k \in L^\infty(\mathbf{C}^2)$, *so that*

i) $\lim_{|z|+|w|\to\infty} |k(z,w)| = 0,$

ii) k *is positive definite,*

iii) *equation* (5) *is satisfied.*

We note that, due to the same Theorem VII.3.3, for a function $g \in L^1(\mathbf{C})$, $g \geq 0$, the above problem has a solution k if and only if $\mathbf{ess\,sup}_{\lambda\in\mathbf{C}} |g(\lambda)| \leq 1$.

3. A functional model

It was proved by Pincus in [1] that the irreducible hyponormal operators with one dimensional self-commutator are parametrized by their principal function, and moreover, that any compactly supported measurable function between 0 and 1 is the principal function of such an operator. The present section contains a cons-tructive way of producing the operator with prescribed principal function. In fact we repeat Theorem VII.3.3 in the above particular case. However, due to Clancey's theorem proved in the preceding section we have an explicit formula for the corresponding functional Hilbert space.

Let $g \in L^{\infty}(\mathbf{C})$, $0 \leq g \leq 1$, and denote

$$k(z,w) = 1 - \mathbf{exp}((1/\pi)\int(g(\zeta)/(\zeta - z)(\overline{\zeta} - \overline{z}))d\mu(\zeta)); \quad z, w \in \mathbf{C}.$$

We know indirectly that k is a positive definite kernel. Let us consider the following scalar product on the space $D(\mathbf{C})$ of test functions:

$$<\phi , \psi>_g = \int k(z,w)\,\overline{\partial}\phi(z)\,\partial\,\overline{\psi}(w)\,d\mu(z)d\mu(w),$$

$\phi, \psi \in D(\mathbf{C})$.

Let $\mathbf{H}_g$ be the Hilbert space separate completion of $D(\mathbf{C})$ in the norm $\|\cdot\|_g$. Then the following expressions

$$(T\phi)(z) = z\phi(z) - (1/\pi)\int(g(\zeta)\phi(\zeta)/(\overline{\zeta} - \overline{z}))d\mu(\zeta),$$

$$(T^*\phi)(z) = \overline{z}\phi(z), \quad \phi \in D(\mathbf{C}),$$

extend to bounded operators on H_g. Moreover, T^* is the adjoint of T, and

$$([T^* , T]\phi)(z) = (1/\pi)\int g(\zeta)\phi(\zeta)d\mu(\zeta),$$

$$<[T^* , T]\phi , \phi> = \left|\int g(\zeta)\phi(\zeta)d\mu(\zeta)\right|^2.$$

The above functions which do not belong to $D(\mathbf{C})$ are considered as representatives of classes in H_g.

In view of Theorem IX.5.1 we can state the next.

3.1. Proposition. *The operator* T *is the unique irreducible hyponormal operator with one dimensional self-commutator whose principal function coincides with* g.

A similar model, in which the operator T is diagonalized instead of T^*, is described in Pincus - Xia - Xia [1].

4. The spectrum and the principal function

The perturbation theoretical methods led Carey and Pincus to a characterization of the essential, continuous and residual parts of the spectrum of a hyponormal operator with one dimensional self-commutator in terms of its principal function. Our aim is to state these results and only to sketch their proofs, see Carey and Pincus [2].

The main fact is contained in the next.

4.1. Theorem. *Let* T *be a pure hyponormal operator with one dimensional self--commutator. A point* $\lambda \in \mathbf{C}$ *belongs to the residual spectrum of* T *if and only if, for a compact ball* B *centered at* λ,

$$\int_B ((1 - g_T(\zeta))/|\zeta - \lambda|^2) d\mu(\zeta) < \infty.$$

Proof. We may assume $\lambda = 0$. Recall that $0 \in \sigma_r(T) \Leftrightarrow 0 \in \sigma_p(T^*)$.

Let us denote $A = TT^*$ and $[T^*, T] = \xi \otimes \xi$. We consider the one dimensional perturbation problem $A \rightsquigarrow A + \xi \otimes \xi = T^*T$. The interaction space is $K = \overline{\mathbf{span}}_{n \geq 0}\{A^n\xi\}$.

It is plain to remark, in view of the irreducibility of T, that $0 \in \sigma_p(T^*)$ if and only if $0 \in \sigma_p(A|K)$. But $0 \in \sigma_p(A|K)$ if and only if the scalar measure $d\mu_\xi = d\langle E\xi, \xi\rangle$ has a point mass at 0, where E denotes the spectral measure of A.

Let δ denote the phase shift of the perturbation problem $TT^* \longrightarrow T^*T$. Then

$$1 + \int_{\mathbf{R}} d\mu_\xi(t)/(t - z) = \mathbf{det}(I + (A - z)^{-1}(\xi \otimes \xi)) = \mathbf{exp}(\int_{\mathbf{R}} (\delta(t)/(t - z))dt),$$

for every nonreal z, see § VIII.1.

By assertion VIII.5.11 the measure μ_ξ has a point mass at 0 if and only if

$$\int_0^1 ((1 - \delta(t))/t)dt < \infty.$$

In virtue of property X. 3) of the principal function one gets:

$$\delta(t) = (1/2\pi)\int_0^{2\pi} g_T(\sqrt{t}e^{i\theta})d\theta, \tag{12}$$

so that, after some immediate computations

$$\int_0^1 ((1 - \delta(t))/t)dt = (1/2\pi)\int_0^1\int_0^{2\pi} ((1 - g_T(\sqrt{t}e^{i\theta}))/t)d\theta dt = (1/2\pi)\int_{B(0,1)} ((1 - g_T(\zeta))/|\zeta|^2)d\mu(\zeta).$$

This establishes the theorem.

Similarly, the question whether the range of the operator T_z is closed or not turns out to have a complete answer in terms of the behaviour of the principal function.

4.2. Theorem. *Let* T *be a pure hyponormal operator with one dimensional self--commutator. The point* $\lambda \in \mathbf{C}$ *belongs to the essential spectrum of* T *if and only if the principal function is not almost everywhere equal to* 0 *or* 1 *in a neighbourhood of* λ.

Proof. We may assume $\lambda = 0$. Let us denote as above $A = TT^*$ and $B = T^*T$, so that $B - A = \xi\otimes\xi$. Again δ will denote the phase shift of the perturbation problem $A \longrightarrow B$.

The proof relies on the following equivalence

$$0 \in \sigma_{ess}(T) \Longleftrightarrow \xi \in H_{ac}(A) \cap H_{ac}(B)$$

derived from § VIII.5.

But we have already remarked that $\xi \notin H_{ac}(A) \cap H_{ac}(B)$ if and only if the function δ is not a.e. a characteristic function near 0, cf. § VIII.5.

Since $0 \le g_T \le 1$ and δ is related to g_T by relation (12), the proof of the theorem is over.

Notes. The global resolvent localized at a vector has been successfully used and investigated by Putnam [10], Radjabalipour [3], Clancey and Wadhwa [1], Clancey [9]. Sections 2 and 3 rely essentially on Clancey's recent work [7], [9]. Section 4 is part of the paper of Carey and Pincus [2]. The first parametrization of the unitary orbits of operators with rank one self-commutator was obtained by Pincus in his fundamental work [1].

EXERCISES

1. (Howe [1]). A sequence of operators $T_n \in L(H)$ is said to be *$*$-strongly convergent* to the operator T if so-$\lim_n T_n = T$ and so-$\lim_n T_n^* = T^*$.

If $T_n \in L(H)$ is a sequence of hyponormal operators $*$-strongly convergent to T, then

$$\sigma(T) \subset \overline{\bigcup_n \sigma(T_n)}.$$

2. (Clancey [7]). Let T_n be a $*$-strongly convergent sequence to T, and suppose $[T_n^*, T_n] = \xi_n \otimes \xi_n$ and $[T^*, T] = \xi \otimes \xi$.

If $\lim_n \xi_n = \xi$, then for any $z \in \mathbf{C}$,

$$\text{w-}\lim_n (T_n^* - \bar{z})^{-1} \xi_n = T_z^{*-1} \xi,$$

and

$$\lim_n (T_n^* - \bar{z})^{-1} \xi_n = T_z^{*-1} \xi$$

for z off the set $\{\lambda \in \overline{\bigcup_n \sigma(T_n)} \mid \| T_z^{*-1} \xi \| < 1\}$.

Hint. The proof is similar to the proof of Proposition 1.1.

3. Compute $(U_+^* - \bar{z})^{-1} \xi$, where U_+ is the unilateral shift and $[U_+^*, U_+] = \xi \otimes \xi$.

4. (Clancey [8]). Let T be a hyponormal operator with one dimensional self-commutator $[T^*, T] = \xi \otimes \xi$ and let $f \in \mathcal{O}(\sigma(T))$. Then

$$\langle f(T)\xi, T_z^{*-1} \xi \rangle = (1/\pi) \int (f(\zeta) g_T(\zeta)/(\zeta - z)) d\mu(\zeta).$$

Hint. Prove first the identity for the function $f(\zeta) = (\zeta - \lambda)^{-1}$, $\lambda \notin \sigma(T)$, by using (11) and the resolvent equation.

5. Prove by using (11) the following generalization of Ahlfors and Beurling's inequality: for every function $f \in L^1(\mathbf{C}) \cap L^\infty(\mathbf{C})$, $f \geq 0$, one has

$$\sup_{z \in \mathbf{C}} \left| \int f(\zeta) d\mu(\zeta)/(\zeta - z) \right| \leq \sqrt{\pi} \, \| f \|_1^{\frac{1}{2}} \| f \|_\infty^{\frac{1}{2}}.$$

6. Prove the next estimate

$$\mathbf{dist}_{C(K)}(f, R(K)) \leq (2/\sqrt{\pi}) \| \bar{\partial} f \|_{1,K}^{\frac{1}{2}} \| \bar{\partial} f \|_{\infty,K}^{\frac{1}{2}},$$

where K is a compact subset of $\mathbf{C}$ and $f \in C^1(K)$.

7*. Find a direct proof of the positivity of the kernel

$$1 - \exp((-1/\pi) \int g(\zeta) d\mu(\zeta)/(\zeta - z)(\bar{\zeta} - \bar{w})),$$

where g is a measurable function with compact support such that $0 \leq g \leq 1$.

8. a) Prove that $\| T_z^{*-1} \xi \| < 1$ if and only if $\int g(\zeta)/|\zeta - z|^2 d\mu(\zeta) < \infty$.

b) Let $A = \{z \in \sigma(T) \mid \|T_z^{*-1}\xi\| < 1\}$. Prove that

$$\int_A g(\zeta)d\mu(\zeta) = 0.$$

9. a) Show that an irreducible hyponormal operator T with one dimensional self-commutator and uniform multiplicity 1 for **Re**(T) is unitarily equivalent with the singular integral operator:

$$(Sf)(x) = xf(x) + i[a(x)f(x) - (1/\pi i)\int_\delta b(x)\overline{b(t)}f(t)/(t - x)dt],$$

where $f \in L^2(\delta)$; $a, b \in L^\infty(\delta)$, $a \in \mathbf{R}$ and $b \neq 0$ a.e. on the bounded Borel set δ.

b) Prove that for every $x \in \delta$ and $z \in \mathbf{C} \setminus \mathbf{R}$, the principal function g of S satisfies:

$$\mathbf{exp}(\int_{\mathbf{R}} g(x + iy)/(y - z)dy) = (a(x) + |b(x)|^2 - z)/(a(x) - |b(x)|^2 - z).$$

c) The spectrum of S is the set

$$\sigma = \{x + iy \mid a(x) - |b(x)|^2 \leq y \leq a(x) + |b(x)|^2, \quad x \in \delta\}.$$

and $g = \chi_\sigma$.

10. Prove that formula

$$\mathbf{exp}(-\pi^{-1} \int_{D(a,r)} |\zeta - z|^{-2}d\mu(\zeta)) = 1 - r^2|z - a|^2$$

for $a \in \mathbf{C}$, $r > 0$ and $|z - a| \geq r$.

Chapter XII

APPLICATIONS

In this last chapter we present a few applications of the theory of hyponormal operators to other areas of mathematical analysis. To be more specific we are dealing with the following problems:

– the parametrization of commutative pairs of unbounded self-adjoint operators which do not strongly commute,

– some asymptotic expansions of Toeplitz determinants, and

– a moment problem in two dimensions.

These examples are far from exhausting the possible applications of hyponormal operators. On the other hand their variety and depth illustrate the strength of the methods developed in the previous chapters.

1. Pairs of unbounded self-adjoint operators

In a celebrated paper, Nelson [1] pointed out a pathology concerning the commutation properties of unbounded self-adjoint operators. Namely, he proved that there are unbounded self-adjoint operators which commute on a common dense domain but whose spectral measures fail to commute. A systematic study of this phenomenon was carried out in a series papers by Schmüdgen [1] and Schmüdgen and Friedrich [1].

In the present section we classify after Schmüdgen [1] the simplest class of pairs of unbounded self-adjoint operators which commute on a dense domain but whose spectral measures do not commute. A main tool in this classification is the parametrization of bounded operators with one dimensional self-commutators by their principal function.

Throughout this section $D(\mathrm{A}) \subset H$ denotes the domain of a (self-adjoint) operator A acting on the separable Hilbert space H. We introduce after Schmüdgen [1] the following class of operators.

1.1. Definition. A pair (A,B) of self-adjoint operators on H belongs to the class $\mathbf{N}_1$ if there exists a linear subspace D of H such that:

1) $D \subset D(AB) \cap D(BA)$ and $AB\xi = BA\xi$ for $\xi \in D$,

2) D is dense in H,

3) $A|D$ and $B|D$ are essentially self-adjoint.

We shall be concerned with pairs $(A,B) \in \mathbf{N}_1$ which do not *strongly commute*. This means that the spectral measures of A and B fail to commute. Notice that in this case neither of the operators A or B can be bounded.

A *pair* $(A,B) \in \mathbf{N}_1$ is said to be *irreducible* if there is no non-trivial decompositions $A = A_1 \oplus A_2$ and $B = B_1 \oplus B_2$ corresponding to $H = H_1 \oplus H_2$.

Next we assume $\sigma(A) \neq \mathbf{R}$ and $\sigma(B) \neq \mathbf{R}$, for a fixed pair $(A,B) \in \mathbf{N}_1$. Thus for some $\alpha \in \mathbf{R} \setminus \sigma(A)$ and $\beta \in \mathbf{R} \setminus \sigma(B)$, $X = (A - \alpha)^{-1}$ and $Y = (B - \beta)^{-1}$ are bounded self--adjoint operators on H. Let P denote the orthogonal projection onto the closed subspace $\overline{\mathbf{Ran}[X, Y]}$.

A first quantitative information about the deviation of the pair (A,B) from being strongly commutative is given by the integer

$$d(A,B) = \mathbf{dim}\, PH,$$

called the *defect number* of the pair (A,B).

Notice that A and B strongly commute if and only if X and Y commute, or equivalently, if and only if $d(A,B) = 0$.

First we establish a dictionary which relates the commutation properties of the pair (A,B) to those of the pair of bounded self-adjoint operators (X,Y).

1.2. Proposition. *With the above notations,*

a) $D(A,B) := XY(I - P)H$ *is the largest linear space which fulfils condition* 1) *in Definition* 1.1,

b) $D(A,B) = D(B,A)$,

c) $D(A,B)$ *is dense in* H *if and only if* $\overline{\mathbf{Ran}[X, Y]} \cap \mathbf{Ran}(XY) = \{0\}$,

d) $A|D(A,B)$ *is essentially self-adjoint if and only if* $\overline{\mathbf{Ran}[X, Y]} \cap \mathbf{Ran}\, Y = \{0\}$.

Proof. a) Assume $\xi \in D(AB) \cap D(BA)$ has the property $AB\xi = BA\xi$. Then the vector $\eta = (A - \alpha)(B - \beta)\xi = (B - \beta)(A - \alpha)\xi$ lies in the kernel of $[X, Y]$. Accordingly $\eta = (I - P)\eta$ and $\xi = XY(I - P)\eta = YX(I - P)\eta$ belongs to $D(A,B) = XY(I - P)H$.

Conversely, if $\xi = XY(I - P)\eta$, then $\xi = YX(I - P)\eta$, since $[X, Y](I - P) = 0$. Therefore $\xi \in D(AB) \cap D(BA)$ and $AB\xi = BA\xi$ as desired.

b) follows from the identity definition

c) The subspace $D(A,B) = \mathbf{Ran}(YX(I - P))$ is dense in H if and only if $\mathbf{Ker}((I - P)XY) = \{0\}$. But $(I - P)XY\xi = 0$ is equivalent to $XY\xi \in \overline{\mathbf{Ran}[X, Y]}$.

d) Because $X = (A - \alpha)^{-1}$ is bounded, the operator $A \mid D(A,B)$ is essentially self--adjoint if and only if $(A - \alpha)D(A,B) - Y(I - P)H$ is dense in H. This in turn is equivalent to $\mathbf{Ker}((I - P)Y) = 0$. Since $\mathbf{Ker}\, Y = 0$ one finally obtains $\mathbf{Ker}((I - P)Y) = 0$ if and only if $\mathbf{Ran}\, Y \cap PH = \{0\}$. This completes the proof of the proposition.

Let us remark from the above proof that the space $D(A,B)$ as well as the conditions on X and Y do not depend on the choice of the points $\alpha \in \mathbf{R} \setminus \sigma(A)$ and $\beta \in \mathbf{R} \setminus \sigma(B)$.

The main result of this section refers to pairs $(A,B) \in \boldsymbol{N}_1$ with defect number one. In this case the membership in $\boldsymbol{N}_1$ has a simple intrinsic characterization as follows.

1.3. Lemma. *Let* A *and* B *be self-adjoint operators with* $\sigma(A) \neq \mathbf{R}$ *and* $\sigma(B) \neq \mathbf{R}$. *If* $(A,B) \in \boldsymbol{N}_1$, *then*

$$D(A,B) = D(AB) \cap D(A) = D(BA) \cap D(B) = D(AB) \cap D(BA).$$

Conversely, if $\mathbf{Ran}[X, Y]$ *is closed,* $D(A,B)$ *is dense in* H, *and* $D(AB) \cap D(A) = D(BA) \cap D(B) = D(AB) \cap D(BA)$, *then* $(A,B) \in \boldsymbol{N}_1$.

Proof. Assume $(A,B) \in \boldsymbol{N}_1$. In view of Proposition 1.2,

$$D(AB) \cap D(A) \supseteq D(AB) \cap D(BA) = YXH \cap XYH \supset D(A,B).$$

On the other hand, if $\xi = YX\eta = X\zeta \in D(AB) \cap D(A)$ for some $\eta, \zeta \in H$, then $X(\zeta - Y\eta) = [Y, X]\eta = 0$ by condition d) in Proposition 1.2. Since $\mathbf{Ker}\, X = 0$ we get $\zeta = Y\eta$, whence $\eta \in \mathbf{Ker}[X, Y]$. Accordingly $\xi = YX(I - P)\eta \in D(A,B)$, and the converse inclusion is proved.

In order to prove the second assertion we use Proposition 1.2. Because the domain $D = D(A,B)$ is supposed to be dense in H and 1) holds, it remains to be proved that $\mathbf{Ran}[X, Y] \cap \mathbf{Ran}\, X = \{0\}$ and similarly for Y. Assume $\xi = [X, Y]\eta = X\zeta$ for some $\eta, \zeta \in H$. Then the vector $X\zeta - XY\eta = -YX\eta$ lies in $D(AB) \cap D(A)$ hence, by our assumption, it lies also in $D(BA)$. Take a $\theta \in H$ such that

$$X\zeta - XY\eta = -YX\eta = XY\theta.$$

Consequently $X(\zeta - Y\eta - Y\theta) = 0$ and $\zeta = Y\eta + Y\theta$, that is $\zeta \in \mathbf{Ran}(Y)$ and $\xi \in \mathbf{Ran}(XY)$. By Proposition 1.2. c) we infer $\xi = 0$. This concludes the proof of Lemma 1.3.

By putting together the above results we have already obtained the following correspondence:

A pair $(A,B) \in \mathbf{N}_1$ with $\alpha \in \mathbf{R} \setminus \sigma(A)$, $\beta \in \mathbf{R} \setminus \sigma(B)$ and $d(A,B) = 1$ is completely determined by the bounded self-adjoint operators $X = (A - \alpha)^{-1}$, $Y = (B - \beta)^{-1}$ subject to the conditions:

i) $\mathbf{Ker}\, X = \mathbf{Ker}\, Y = 0$,

ii) $[X, Y] = \pm i\xi \otimes \xi$, $\xi \neq 0$,

iii) $\xi \notin (\mathbf{Ran}\, X) \cup (\mathbf{Ran}\, Y)$.

And conversely, any self-adjoint operators $X, Y \in L(H)$ which satisfy i), ii), iii) give rise to a pair $(A,B) \in \mathbf{N}_1$, where $A = X^{-1} + \alpha$ and $B = Y^{-1} + \beta$.

In view of the obvious symmetry between A and B in $(A,B) \in \mathbf{N}_1$ we can assume that $i[X, Y] \geq 0$. Thus to a pair $(A,B) \in \mathbf{N}_1$ with $d(A,B) = 1$ and $\alpha \notin \mathbf{R} \setminus \sigma(A)$, $\beta \notin \mathbf{R} \setminus \sigma(B)$ we associate the hyponormal operator with one dimensional self-commutator $T = X + iY$.

Of course two pairs in $\mathbf{N}_1$ as above are unitarily equivalent if and only if so are the corresponding hyponormal operators; and similarly for irreducibility.

In order to complete this picture we need the following.

1.4. Lemma. *Let* $T \in L(H)$ *be a pure hyponormal operator with one dimensional self-commutator and principal function* $g = g_T$. *Then* $\mathbf{Ran}[T^*, T] \cap \mathbf{Ran}(\mathbf{Re}\, T) \neq \{0\}$ *if and only if*

$$\int_{\mathbf{R}^2} (g(x + iy)/x^2)dxdy < \infty. \tag{1}$$

Proof. Let us denote $[T^*, T] = \xi \otimes \xi$ and $X = \mathbf{Re}\, T$. It is plain that the condition $\mathbf{Ran}[T^*, T] \cap \mathbf{Ran}(\mathbf{Re}\, T) \neq \{0\}$ is equivalent to $\xi \in \mathbf{Ran}\, X$.

Assume $\xi = X\eta$ for some $\eta \in H$. Then the limit $\lim_{\varepsilon \downarrow 0} (X + i\varepsilon)^{-1}\xi$ exists and equals η, because the operator T was supposed to be pure and consequently X has no point spectrum. Conversely, if $\eta = \lim_{\varepsilon \downarrow 0} (X + i\varepsilon)^{-1}\xi$ exists, then obviously $\xi = X\eta$.

Thus we are led to proving the equivalence between the existence of $\lim_{\varepsilon \downarrow 0} (X + i\varepsilon)^{-1}\xi$ and condition (1).

Assume $\lim_{\varepsilon \downarrow 0} (X + i\varepsilon)^{-1}\xi$ exists. Then Helton and Howe's formula, see §X.2, yields for $\varepsilon > 0$:

$$\|(X + i\varepsilon)^{-1}\xi\|^2 = \langle (X^2 + \varepsilon^2)^{-1}\xi, \xi\rangle = 2i\mathbf{Tr}((X^2 + \varepsilon^2)^{-1}[X, Y]) =$$

$$= 2i\mathbf{Tr}[X, (X^2 + \varepsilon^2)^{-1}Y] = (1/\pi) \int_{\mathbf{R}^2} (g(x + iy)/(x^2 + \varepsilon^2))dxdy.$$

By using Fatou's Lemma one obtains (1).

Conversely, if condition (1) holds, then a similar computation gives for $\varepsilon, \delta > 0$:

$$||(X + i\varepsilon)^{-1}\xi - (X + i\delta)^{-1}\xi||^2 = (1/\pi) \int_{\mathbf{R}^2} g(x + iy)\,|(1/(x + i\varepsilon)) - (1/(x + i\delta))|^2\, dxdy.$$

As

$$|(1/(x + i\varepsilon)) - (1/(x + i\delta))| \leq 2/x^2.$$

and the integral (1) is finite, Lebesgue's dominated convergence theorem shows that $\lim_{\varepsilon \downarrow 0} (X + i\varepsilon)^{-1}\xi$ exists. This completes the proof of the lemma.

Similarly one proves under the same assumptions that $\mathbf{Ran}[T^*, T] \cap \mathbf{Ran}(\mathbf{Im}\, T) \neq \{0\}$ if and only if

$$\int_{\mathbf{R}^2} (g(x + iy)/y^2)\,dxdy < \infty.$$

In conclusion we have proved the following.

1.5. Theorem. (Schmüdgen). *Let* A, B *be an irreducible pair of unbounded self-adjoint operators with* $0 \notin \sigma(A) \cap \sigma(B)$. *Denote* $X = A^{-1}$, $Y = B^{-1}$ *and assume* $\mathbf{rank}[X, Y] = 1$.

Then $(A,B) \in \mathbf{N}_1$ *if and only if the principal function* g *of the operator* $T = X + iY$ *satisfies*

$$\int_{\mathbf{R}^2} x^{-2}\,|g(x + iy)|\,dxdy = \int_{\mathbf{R}^2} y^{-2}\,|g(x + iy)|\,dxdy = \infty.$$

In the above statement $g = -g_{T^*}$ if the operator T^* is hyponormal, that is when $i[X, Y] \leq 0$.

1.6. Corollary. *There is a one-to-one correspondence between the following sets:*

$$\left\{\begin{array}{c}\textit{Unitary orbits of irreducible pairs } (A,B) \in \mathbf{N}_1 \\ \textit{with defect number one and } 0 \notin \sigma(A) \cap \sigma(B)\end{array}\right\}$$

and

$$\left\{\begin{array}{c} g \in L^1(\mathbf{R}^2),\ \mathbf{supp}(g)\ \textit{compact, either } 0 \leq g \leq 1 \textit{ or } -1 \leq g \leq 0 \\ \textit{and } \int_{\mathbf{R}^2} x^{-2}\,|g(x + iy)|\,dxdy = \int_{\mathbf{R}^2} y^{-2}\,|g(x + iy)|\,dxdy = \infty \end{array}\right\}.$$

In particular this corollary offers plenty of examples of pairs of operators belonging to $\mathbf{N}_1$ and with positive defect number.

2. The Szegö limit theorem

A classical theorem of Szegö computes the asymptotic behaviour of the determinants of the truncations of an infinite Toeplitz matrix. Recently, Basor and Helton [1] have remarked that Szegö's theorem and some related asymp-totic expansions are available by the theory of the principal function. Although this new method applies to a broad class of operators, in the present section we confine ourselves to discussing the simplest case of Toeplitz operators with smooth symbols.

In what follows $\mathbf{T}$ denotes the unit circle, $P : L^2(\mathbf{T}) \to H^2(\mathbf{T})$ is the orthogonal projection onto the Hardy space, $\mathbf{D}_+ = \{z \in \mathbf{C} \mid |z| < 1\}$ and $\mathbf{D}_- = \{z \in \mathbf{C} \mid |z| > 1\} \cup \{\infty\}$.

Recall that the Toeplitz operator T_ϕ with smooth symbol $\phi \in E(\mathbf{T})$ acts by the formula

$$T_\phi : H^2(\mathbf{T}) \longrightarrow H^2(\mathbf{T}), \quad T_\phi(f) = P(\phi f), \quad f \in H^2(\mathbf{T}).$$

Among Toeplitz operators, a distinguished element is the unilateral shift U_+. Its symbol is given by the position function $z = e^{i\theta}$, $\theta \in [0,2\pi]$, on $\mathbf{T}$. Moreover, if $p \in \mathbf{C}[z]$ is a polynomial, then

$$p(U_+) = T_p. \tag{2}$$

If $\phi \in E(\mathbf{T})$ has the Fourier series

$$\phi(e^{i\theta}) = \sum_{k=-\infty}^{\infty} a_k e^{ik\theta},$$

then the smoothness of ϕ is equivalent to:

$$\sum_{k=-\infty}^{\infty} |k|^p |a_k| < \infty, \qquad \text{for all } p > 0. \tag{3}$$

The matrix of T_ϕ corresponding to the natural basis $\{e^{ik\theta}\}_{k=0}^{\infty}$ of $H^2(\mathbf{T})$ is a *Toeplitz matrix:*

$$T_\phi \sim \begin{bmatrix} a_0 & a_{-1} & a_{-2} & a_{-3} & \cdots \\ a_1 & a_0 & a_{-1} & a_{-2} & \cdots \\ a_2 & a_1 & a_0 & a_{-1} & \cdots \\ a_3 & a_2 & a_1 & a_0 & \\ \cdot & \cdot & \cdot & \cdot & \cdot \\ \cdot & \cdot & \cdot & \cdot & \cdot \\ \cdot & \cdot & \cdot & \cdot & \cdot \end{bmatrix}. \tag{4}$$

Form this representation and (3) one easily finds $T_{\phi\psi} - T_\phi T_\psi \in C_1(H)$ for every pair $\phi, \psi \in E(\mathbf{T})$. Similarly one gets $T_\phi^* = T_{\overline{\phi}}$; also $\|T_\phi\| \leq \|\phi\|_{\infty,\mathbf{T}}$, for any $\phi \in E(\mathbf{T})$.

Thus the map

$$E(\mathbf{T}) \rightarrow L(H^2(\mathbf{T}))/C_1(H^2(\mathbf{T})), \quad \phi \mapsto [T_\phi]$$

is a one-to-one $*$-homomorphism, i.e. a smooth functional calculus for the operator U_+, modulo the trace-class ideal.

2.1. Lemma. *For any functions* $\phi, \psi \in E(\overline{\mathbf{D}}_+)$, *the following formula holds*

(5)
$$\mathbf{Tr}[T_\phi, T_\psi] = (1/\pi) \int_{\mathbf{D}_+} (\overline{\partial}\phi\, \partial\psi - \partial\phi\, \overline{\partial}\psi) d\mu.$$

Proof. First take ϕ and ψ to be real analytic polynomials. By observation (2) and Theorem X.2.5 we infer

$$\mathbf{Tr}[T_\phi, T_\psi] = \mathbf{Tr}[\phi(U_+, U_+^*), \psi(U_+, U_+^*)] =$$

$$= (1/2\pi i) \int_{\mathbf{D}_+} ((\partial\phi/\partial\overline{z})(\partial\psi/\partial z) - (\partial\psi/\partial\overline{z})(\partial\phi/\partial z) d\overline{z} \wedge dz.$$

In the general case an approximation argument concludes the proof.

In fact, only the values of the function ϕ and ψ on $\mathbf{T}$ are needed in order to compute $\mathbf{Tr}[T_\phi, T_\psi]$. For this let us decompose a function $\phi \in E(\mathbf{T})$ into

(6)
$$\phi = \phi_+ + \phi_-,$$

where $\phi_+ = P_\phi$. Thus the functions $\phi_\pm$ extend analytically to $\mathbf{D}_\pm$, respectively; or equivalently, the functions ϕ_+ and $\overline{\phi}_-$ extend analytically to $\mathbf{D}_+$.

2.2. Corollary. *For any smooth functions* $\phi = \phi_+ + \phi_-$ *and* $\psi = \psi_+ + \psi_-$, *one has*

(7)
$$\mathbf{Tr}[T_\phi, T_\psi] = (1/2\pi i) \int_0^{2\pi} (\phi_-(\partial\psi_+/\partial\theta) - \psi_-(\partial\psi_+/\partial\theta))(e^{i\theta}) d\theta.$$

Proof. Observe by relation (2) that

(8)
$$[T_{\phi_+}, T_{\psi_+}] = [T_{\phi_-}, T_{\psi_-}] = 0.$$

Therefore only $\mathbf{Tr}[T_{\phi_-}, T_{\psi_+}]$ has to be computed.

Since $(\partial f/\partial z) dz|_{\mathbf{T}} = (\partial f(e^{i\theta})/\partial\theta) d\theta$ for any analytic function f in a neighbourhood of **T**, Stokes' theorem gives

$$\int_{D_+} (\bar\partial\phi_-)(\partial\psi_+)d\mu = (1/2i)\int_{\mathbf{T}} \phi_-(\partial\psi_+/\partial\theta)d\theta.$$

Then relation (7) follows from (5).

Any invertible function $\phi \in E(\mathbf{T})$ has, analogously to (6), a multiplicative factorization

(9) $$\phi(e^{i\theta}) = e^{in\theta}\phi_+(e^{i\theta})\phi_-(e^{i\theta}), \quad n \in \mathbf{Z},$$

where the functions $\phi_\pm$ extend analytically to invertible functions on $D_\pm$. Of course, the integer n is the degree of the continuous map $\phi : \mathbf{T} \longrightarrow \mathbf{C}^*$.

Indeed, the function $\phi(\theta)e^{-in\theta}$ has degree zero, whence it admits a smooth logarithm ψ. Then decompose additively $\psi = \psi_+ + \psi_-$ and define

$$\phi_\pm = e^{\psi_\pm}.$$

The variant of Szegő's theorem presented below refers to Toeplitz operators T_ϕ with invertible smooth symbol ϕ with degree zero. For this class of functions one introduces the *geometrical mean* as the complex number

$$G(\phi) = \mathbf{exp}((1/2\pi)\int_{-\pi}^{\pi} \mathbf{log}\, \phi(\theta)d\theta).$$

Here **log** is the branch of the logarithm defined on $\mathbf{C} \setminus (-\infty,0]$ and normalized so that $\mathbf{log}(1) = 0$. As we have already remarked, $\mathbf{log}\,\phi$ exists and moreover it extends to a smooth function on **T**.

If $\phi = \phi_+\phi_-$ is the factorization (9) of an invertible function $\phi \in E(\mathbf{T})$ of degree zero, then a straightforward computation shows that

$$G(\phi) = ((1/2\pi)\int_0^{2\pi} \phi_+(\theta)d\theta)((1/2\pi)\int_0^{2\pi} \phi_-(\theta)d\theta).$$

In other terms, the geometrical mean of the function $\phi = \phi_+\phi_-$ coincides with the product of the zero'th Fourier coefficients of the factors $\phi_\pm$.

In order to state Szegő's theorem we need the following notation. For every natural number n denote by P_n the orthogonal projection of $H^2(\mathbf{T})$ onto the subspace spanned by the vectors $e_o, e_1, \ldots, e_n$. Then define

$$T_n(\phi) = P_n T_\phi P_n, \quad \phi \in E(\mathbf{T}).$$

If the function ϕ factors as above into $\phi = \phi_+\phi_-$, then for every $n \in \mathbf{N}$

(10) $$P_nT_{\phi_+} = P_nT_{\phi_+}P_n, \quad T_{\phi_-}P_n = P_nT_{\phi_-}P_n.$$

Also, due to the fact that ϕ_+ and $\overline{\phi}_-$ are boundary values of analytic functions on $\mathbf{D}_+$, one easily obtains

(11) $$T_\phi = T_{\phi_-}T_{\phi_+}.$$

This decomposition can be visualized on the matrix picture. Namely, if a_k, $a_k^\pm$ denote the Fourier coefficients of the functions ϕ, $\phi_\pm$, respectively, then:

$$\begin{bmatrix} a_0 & a_{-1} & a_{-2} \cdots \\ a_1 & a_0 & a_{-1} \cdots \\ a_2 & a_1 & a_0 \;\cdots \\ \cdot & \cdot & \cdot \;\; \cdot \\ \cdot & \cdot & \cdot \;\; \cdot \\ \cdot & \cdot & \cdot \;\; \cdot \end{bmatrix} = \begin{bmatrix} a_0^- & a_1^- & \cdots \\ & a_0^- & * \\ & & \ddots \\ & 0 & \end{bmatrix} \begin{bmatrix} a_0^+ & & 0 \\ a_1^+ & a_0^+ & \\ & & \ddots \\ & * & \end{bmatrix}.$$

The main result of this section compares asymptotically the determinants of the truncations of the above three matrices. Quite specifically we can state the following:

2.3. Theorem. (Szegö, Widom). *Let ϕ be an invertible smooth function on* $\mathbf{T}$ *which admits the factorization $\phi = \phi_+\phi_-$, where $\phi_+, \overline{\phi}_- \in H^2(\mathbf{T})$ and they are smooth.*

Then

(12) $$\lim_{n\to\infty} (\det T_n(\phi)/G(\phi)^{n+1}) = \det (T_\phi T_{\phi^{-1}}) = \exp((1/2\pi i)\int_0^{2\pi} ((1/\phi_+)(\partial\phi_+/\partial\theta)\log\theta_-)d\theta).$$

Proof. Remark that the expression $T_\phi T_{\phi^{-1}} = T_{\phi_-}T_{\phi_+}T_{\phi_-^{-1}}T_{\phi_+^{-1}} = T_{\phi_-}T_{\phi_+}(T_{\phi_-})^{-1}\cdot(T_{\phi_+})^{-1}$ is a multiplicative commutator, whence it belongs to the determinant class $I + C_1$, see Lemma X.2.6. Moreover, Proposition X.2.7 and Corollary 2.2 imply the second equality in (12).

In order to prove the first equality, let $n > 0$ be fixed. The above matrix representations give:

$$\det T_n(\phi_+) = (a_0^+)^{n+1}, \quad \det T_n(\phi_-) = (a_0^-)^{n+1}.$$

On the other hand

$$G(\phi) = a_0^+ \cdot a_0^-.$$

By taking relations (10) into account one finds

$$T_n(\phi) = P_n T_\phi P_n = P_n T_{\phi_-} T_{\phi_+} P_n = P_n T_{\phi_+} T_{\phi_+^{-1}} T_{\phi_-} T_{\phi_+} T_{\phi_-^{-1}} T_{\phi_-} P_n =$$

$$= T_n(\phi_+) P_n T_{\phi_+^{-1}} T_{\phi_-} T_{\phi_+} T_{\phi_-^{-1}} P_n T_n(\phi_-).$$

Therefore

$$\mathbf{det}\, T_n(\phi)/G(\phi)^{n+1} = \mathbf{det}\, T_n(\phi)/\mathbf{det}\, T_n(\phi_+)\mathbf{det}\, T_n(\phi_-) = \mathbf{det}\, P_n T_{\phi^{-1}} T_{\phi_-} T_{\phi_+} T_{\phi_-^{-1}} P_n.$$

Since $\mathbf{det}\, T_{\phi_+^{-1}} T_{\phi_-} T_{\phi_+} T_{\phi_-^{-1}} = \mathbf{det}\, T_{\phi_-} T_{\phi_+} T_{\phi_-^{-1}} T_{\phi_+^{-1}}$ exists,

$$\lim_{n\to\infty} \mathbf{det}\, P_n T_{\phi_+^{-1}} T_{\phi_-} T_{\phi_+} T_{\phi_-^{-1}} P_n = \mathbf{det}(T_\phi T_{\phi^{-1}}),$$

and the proof is complete.

The above scheme of proof works for more general classes of Toeplitz operators, cf. Widom [1] and Basor - Helton [1].

An immediate application of the Szegö's limit theorem is contained in the next.

2.4. Corollary. *Let ϕ, ψ be invertible smooth functions on $\mathbf{T}$, which admit factorizations $\phi = \phi_+\phi_-$, $\psi = \psi_+\psi_-$ with $\phi_+, \psi_+, \overline{\phi}_-, \overline{\psi}_- \in H^2(\mathbf{T}) \cap E(\mathbf{T})$.*

Then

(13) $$\lim_{n\to\infty} (\mathbf{det}\, T_n(\phi\psi)/\mathbf{det}\, T_n(\phi)\mathbf{det}\, T_n(\psi)) = \mathbf{det}(T_{\phi_-} T_{\psi_+} T_{\phi_-}^{-1} T_{\psi_+}^{-1})\, \mathbf{det}(T_{\psi_-} T_{\phi_+} T_{\psi_-}^{-1} T_{\phi_+}^{-1}).$$

Proof. It is plain to observe that $G(\phi\psi) = G(\phi)G(\psi)$. According to Theorem 2.3 one finds that the limit (13) exists and equals

$$\mathbf{exp}((1/2\pi i) \int_0^{2\pi} ((\partial(\mathbf{log}\, \phi_+\psi_+)/\partial\theta)\mathbf{log}\, (\phi_-\psi_-) -$$

$$- (\partial(\mathbf{log}\, \phi_+)/\partial\theta)\mathbf{log}\, \phi_- - (\partial(\mathbf{log}\, \psi_+)/\partial\theta)\mathbf{log}\, \psi_-)d\theta) =$$

$$= \mathbf{exp}((1/2\pi i) \int_0^{2\pi} ((\partial(\mathbf{log}\, \phi_+)/\partial\theta)\mathbf{log}\, \psi_- + (\partial(\mathbf{log}\, \psi_+)/\partial\theta)\mathbf{log}\, \phi_-)d\theta).$$

Again an application of Proposition X.2.7 and Corollary 2.2 ends the proof.

A non-trivial generalization of Corollary 2.4 to the case of Toeplitz operators with discontinuous symbol appears in Basor and Helton [1].

3. A two dimensional moment problem

In Theorem I.2.4 a characterization was given of the moments of a positive Borel measure, compactly supported on $\mathbf{R}^2$, by a positivity and boundedness condition. In this section we refine this result, by determining the structure of the moments of such a measure, which in addition is absolutely continuous, with a bounded weight, with respect to the planar Lebesgue measure. The final result is quite similar to Theorem I.2.4, the only difference consisting in the form of the respective positive definite kernel.

The present section closely follows Putinar [5]. The main tools to deal with in the sequel are two distinct parametrizations of pure hyponormal operators with one dimensional self-commutator, cf. Chapter XI.

We begin with some formal transformations of a virtual moments sequence.

Let $(a_{mn})_{m,n=0}^{\infty}$ be a double sequence of complex numbers with the property:

(14) $$a_{mn} = a_{nm}, \quad m, n \in \mathbf{N},$$

and let δ be a positive real. We shall associate to these data a kernel

$$K_\delta : \mathbf{N}^2 \times \mathbf{N}^2 \longrightarrow \mathbf{C}$$

whose properties will reflect the fact that a_{mn} are the moments of a measure like $f d\mu$, $f \geq 0$, $f \in L^{\infty}_{comp}(\mathbf{R}^2)$.

First take two commuting indeterminates X, Y and consider the formal series

(15) $$\sum_{m,n=0}^{\infty} b_{mn} X^{m+1} Y^{n+1} = 1 - \mathbf{exp}((-1/\delta) \sum_{k,\ell=0}^{\infty} a_{k\ell} X^{k+1} Y^{\ell+1}).$$

Notice that the superposition of the exponential series and its argument is possible because the last series has no term of zero degree. Of course b_{mn} are polynomials in $a_{k\ell}$, with $k \leq m$ and $\ell \leq n$.

In order to define the kernel K_δ, recall that $\iota = (1,0)$, $\kappa = (0,1)$ denote the generators of the additive semigroup $\mathbf{N}^2$, and $\theta = (0,0)$ stands for its neutral element, see Chapter XI .

Let us denote $b(\alpha) = b_{mn}$ if $(m,n) = \alpha \in \mathbf{N}^2$ and take $b(\alpha)$ to be zero by convention if at least one of the entries of α is negative.

The kernel K_δ will be recursively defined according to the following rules:

(i) $K_\delta(\theta,\alpha) = K_\delta(m\kappa, n\kappa) = b(\alpha), \quad \alpha = (m,n) \in \mathbf{N}^2,$

(ii) $K_\delta(\alpha,\beta) = \overline{K_\delta(\beta,\alpha)}$, and

$$\text{(iii)}\quad K_\delta(\alpha+\iota, \beta) - K_\delta(\alpha, \beta+\kappa) = \sum_{r=0}^{\infty} K_\delta(\alpha, r\iota) b(\alpha - (r+1)\iota)$$

for any $\alpha, \beta \in \mathbf{N}^2$.

Let us remark that assumption (14) guarantees the consistency and sufficiency of the generating rules (i), (ii), (iii), for the definition of any value $K_\delta(\alpha,\beta)$, $\alpha, \beta \in \mathbf{N}^2$.

At this moment we can state the main result of this section.

3.1. Theorem. *Let* $(a_{mn})_{m,n=0}^{\infty}$ *be a sequence of complex numbers satisfying* $a_{mn} = \overline{a}_{nm}$, $n, m \in \mathbf{N}$.

There exists a positive measure $d\nu$, *compactly supported on* $\mathbf{C}$, *absolutely continuous with bounded weight with respect to the Lebesgue measure, and such that*

$$a_{mn} = \int_{\mathbf{C}} z^m \bar{z}^n d\nu(z), \quad m, n \in \mathbf{N},$$

if and only if there exists a $\delta > 0$ *such that:*

a) *The associated kernel* K_δ *to* (a_{mn}) *is positive definite, and*

b) *The shift* $S_{(1,0)}$ *is bounded with respect to* K_δ.

Moreover, in this case $d\nu/d\mu \leq \delta/\pi$ *and* **supp** $(d\nu)$ *is contained in a ball centered at* 0 *and of radius* $\|S_{(1,0)}\|_{K_\delta}$.

For terminology see §I.6.

Proof. The body of our proof is the parametrization of pure hyponormal operators with one dimensional self-commutator by one of the discrete unitary invariants introduced in Chapter II §5.

Let $T \in L(H)$ be a pure hyponormal operator with $[T^*, T] = \xi \otimes \xi$, for some vector $\xi \in H$, $\xi \neq 0$. In Theorem II.4.3 we have established that the double sequence

$$G_T(\alpha) = \langle T^m T^{*n} \xi, \xi \rangle, \quad \alpha = (m,n) \in \mathbf{N}^2$$

determines the unitary orbit of the operator T. The characteristic property of the unitary invariant G_T can be derived from Propositions II.4.4 and II.4.1, as follows.

3.2. Lemma. *Let* $b : \mathbf{N}^2 \to \mathbf{C}$. *Assume* $b(m,n) = \overline{b(n,m)}$ *for any* $m, n \in \mathbf{N}$, *and consider the kernel* $N : \mathbf{N}^2 \times \mathbf{N}^2 \to \mathbf{C}$ *attached to* b *by the rules* (i), (ii) *and* (iii).

Then there exists a pure hyponormal operator T *with one dimensional self-commutator, satisfying* $G_T = b$, *if and only if*

a) *The kernel* K *is positive definite, and*

b) *The shift* S_ι *is bounded with respect to* N.

On the other hand, the operator T above is determined up to unitary equivalence by its principal function g_T. Moreover, Clancey's theorem XI.2.1 gives a simple relation between the two invariants G_T and g_T, namely:

(16) $$\sum_{m,n=0}^{\infty} G_T(m,n)z^{-m-1}w^{-n-1} = 1 - \mathbf{exp}((-1/\pi)\int(g_T(\zeta)/(\zeta - z)(\bar{\zeta} - w))d\mu(\zeta)).$$

We need relation (16) only for large values of $|z|$ and $|w|$, when the series in the left-hand side is absolutely convergent and the function under the integral is summable.

Now the proof of Theorem 3.1 becomes straightforward.

To begin with, consider an integrable positive function f on $\mathbf{C}$, such that **supp**(f) is compact and $f \leq M$ a.e., $M < \infty$.

Let $g = f/M$ and take by Carey and Pincus' theorem IX.5.2 a pure hyponormal operator $T \in L(H)$ with $[T^*, T] = \xi \otimes \xi$ and $g = g_T$ in $L^1(\mathbf{C})$.

By comparing (15) and (16) in the formal series ring, one finds with the above notations that $b(\alpha) = G_T(\alpha)$. Here b is associated by (15) to the moments of the function f, with $\delta = M\pi$. Thus Lemma 3.2 shows that the corresponding kernel $K_{M\pi}$ fulfils conditions a) and b) in Theorem 3.1.

Conversely, assume that the sequence $(a_{mn})_{m,n=0}^{\infty}$ satisfies $a_{mn} = \bar{a}_{nm}$ for $n, m \in \mathbf{N}$ and assertions a) and b) hold for a $\delta > 0$.

Since the kernel K_δ associated to (a_{mn}) has properties a) and b) of Lemma 3.2, there exists a pure hyponormal operator T with $[T^*, T] = \xi \otimes \xi$ and $G_T(\alpha) = K_\delta(\theta, \alpha)$, $\alpha \in \mathbf{N}^2$. Then relations (15) and (16) yield

(17) $$\mathbf{exp}((1/\delta) \sum_{k,\ell=0}^{\infty} a_{k\ell} z^{-k-1}\bar{w}^{-\ell-1}) = \mathbf{exp}((1/\pi)\int(g_T(\zeta)/(\zeta - z)(\bar{\zeta} - \bar{w}))d\mu(\zeta)),$$

for large values of $|z|$ and $|w|$.

But the matrix (a_{mn}) is positive definite as condition a) shows in particular, whence by taking $z = w$, both expressions under the exponential in (17) become positive. Therefore

$$a_{mn} = (\delta/\pi)\int \zeta^m \bar{\zeta}^n g_T(\zeta)d\mu(\zeta), \quad m, n \in \mathbf{N},$$

and the sufficiency part of Theorem 3.1 is also proved.

The last statement follows from the observation that $\mathbf{supp}(g_T) = \sigma(T)$ and $\|T\|$ coincides with the norm of the shift S_1 on the Hilbert space with reproducing kernel N_δ, cf. § XI.3.

This completes the proof of Theorem 3.1.

Notes. For Nelson's example see also Reed and Simon [2]. Continuous phenomena which extend the Szegő limit theorem have been investigated by using different methods by Kac [1], Dym [1], [2] and more recently by Widom [2]. For an account to the multidimensional moment problem see Ahiezer [1], Fuglede [1] and Berg [1].

EXERCISES

1. Set $A = (\mathbf{Re}\, U_+)^{-1}$ and $B = (\mathbf{Im}\, U_+)^{-1}$ as unbounded operators, where U_+ denotes the unilateral shift. Prove that the self-adjoint operators A and B commute on a dense domain but they do not strongly commute.

Hint. Use Schmüdgen's theorem.

2. With the notations of Ex.1, let $D = \{\phi \in D(A^2B) \cap D(BA^2) \cap D(BA);\ AB\phi = BA\phi$ and $A^2B\phi = BA^2\phi\}$.

Show that the operator $A^2|D$ is not essentially self-adjoint.

Hint. $D = XY(\mathbf{Ran}[X^2,Y])^{\perp}$, where $A^2D \perp \mathbf{Ran}[X,Y]$.

3. a). Prove Theorem 2.3 under the weaker condition $\phi \in C^3(\mathbf{T})$.

b). Generalize Theorem 2.3 to block Toeplitz matrices, i.e. with symbol in $GL(d, E(\mathbf{T}))$.

4. For any invertible smooth function $\phi \in E(\mathbf{T})$ of zero degree, compute the limit

$$\lim_{n\to\infty} (\log\det T_n(\phi) - \mathbf{Tr}\, T_n(\log\phi)).$$

5. a). With the notations in Theorem 3.1, prove that, if the kernel K_δ satisfies conditions a) and b) then so does K_γ, for any $\gamma \geq \delta$.

b) Prove the equality

$$\operatorname*{ess\,sup}_{z\in\mathbf{C}}(d\nu/d\mu(z)) = \inf\{\delta/\pi\ ;\ K_\delta \text{ fulfils a) and b)}\}.$$

6*. Let $(a_{mn})_{m,n=0}^{\infty}$ denote the moments of a measurable function $g : \mathbf{C} \to [0,1]$. If

$$\sigma = \{z \in \mathbf{C} \mid |\alpha_i(z)| \leq 1,\ |\beta_j(z)| \geq 1,\quad i,j \in \mathbf{N}\}$$

is a compact set defined by the linear functions α_i, β_j, then $\mathbf{supp}(g) \subset \sigma$ if and only if the kernel K_π is positive definite and the operators $\alpha_i(S_{(1,0)})$, $\beta_j(S_{(1,0)})$ are contractive, respectively expansive, $i, j \in \mathbf{N}$, with respect to K_π.

REFERENCES

Adamjan, V.M.; Pavlov, B.S.

1. The trace formula for dissipative operators (Russian), *Vestnik Leningrad. Univ. Mat. Mekh. Astronom.*, **7**(1979), 5-8.

Agler, J.

1. Hypercontractions and subnormality, *J. Operator Theory*, **13**(1985), 203-217.

Ahlfors, L.; Beurling, A.

1. Conformal invariants and function theoretic null sets, *Acta Math.*, **83**(1950), 101-129.

Ahiezer, N.I.

1. *The classical moment problem and some related questions of analysis* (Russian), Fizmatgiz, Moscow, 1961.

Ahiezer, N.I.; Krein, M.G.

1. Über Fouriersche Reihen beschrankter summierbarer Functionen und ein neues Extremumproblem. I, *Zap. Harkov Mat. Obshch.*, **9**(1934), 9-23; II, ibid. **10** (1934), 3-32.
2. *Some questions in the theory of moments*, Transl. Math. Monographs, vol.**2**, Amer. Math. Soc., Providence, R.I., 1962.

Albrecht, E.

1. On decomposable operators, *Integral Equations Operator Theory*, **2**(1979), 1-10.

Albrecht, E.; Chevreau, B.

1. Invariant subspaces for l^p-operators having Bishop's property (β) on a large part of their spectrum, *J. Operator Theory*, **18**(1987), 339-372.

Alexander, H.

1. Projections of polynomial hulls, *J. Funct. Anal.*, **13**(1973), 13-19.

Ando, T.

1. On hyponormal operators, *Proc. Amer. Math. Soc.*, **14**(1963), 290-291.

Apostol, C.

1. The spectral flavour of Scott Brown's techniques, *J. Operator Theory*, **6**(1981), 3-12.

Apostol, C.; Clancey, K.F.

1. Local resolvents of operators with one-dimensional self-commutators, *Proc. Amer. Math. Soc.*, **58**(1976), 158-162.

2. Generalized inverses and spectral theory, *Trans. Amer. Math. Soc.*, **215**(1976), 293-300.

Aronszajn, N.; Donoghue Jr., W.F.

1. On exponential representations of analytic functions in the upper half-plane with positive imaginary part, *J. Analyse Math.*, **5**(1956-1957), 321-388.

Asano, K.

1. Notes on Hilbert transforms of vector-valued functions in the complex plane and their boundary values, *Proc. Japan Acad. Ser. A Math. Sci.*, **43**(1967), 572--577.

Atzmon, A.

1. A moment problem for positive measures on the unit disc, *Pacific J. Math.*, **59**(1975), 317-325.

Axler, S.

1. The Bergman space, the Bloch space, and commutators of multiplication operators, *Duke Math. J.*, **53**(1986), 305-332.

Axler, S.; Shapiro, J.H.

1. Putnam's theorem, Alexander's spectral area estimate, and VMO, *Math. Ann.*, **271**(1985), 161-183.

Basor, E.; Helton, J.W.

1. A new proof of the Szegö limit theorem and new results for Toeplitz operators with discontinuous symbol, *J. Operator Theory*, **3**(1980), 23-39.

Baumgärtel, H.; Wollenberg, M.

1. *Mathematical scattering theory*, Birkhäuser-Verlag, Basel-Boston-Stuttgart, 1984.

Ben-Jacob, M.G.

1. Hyponormal operators compact to a W^*-algebra, *Bull. London Math. Soc.*, **3** (1981), 229-230.

Berberian, S.

1. A note on hyponormal operators, *Pacific J. Math.*, **12**(1962), 1171-1175.

Berg, C.

1. The multidimensional moment problem and semi-groups, in *Proc. Symp. Appl. Math.*, vol.**37**(1987), pp.110-124.

Berger, C.A.

1. Sufficiently high powers of hyponormal operators have rationally invariant subspaces, *Integral Equations Operator Theory*, **1**(1978), 444-447.

2. Intertwined operators and the Pincus principal function, *Integral Equations Operator Theory*, **4**(1981), 1-9.

Berger, C.A.; Ben-Jacob, M.G.

1. Trace-class self-commutators, *Trans. Amer. Math. Soc.*, **1**(1983), 75-91.

Berger, C.A.; Shaw, B.L.

1. Self-commutators of multicyclic hyponormal operators are always trace class, *Bull. Amer. Math. Soc.*, **79**(1973), 1193-1199.

2. Intertwining, analytic structure, and the trace norm estimate, in *Proceedings of a conference on operator theory*, Lecture Notes in Math., vol.**345**, Springer--Verlag, Berlin-Heidelberg-New York, 1973, pp. 1-6.

Bergman, S.

1. *The kernel function and conformal mapping*, Math. Surveys, vol.**5**, Amer. Math. Soc., Providence, R.I., 1950.

Birman, M.S.

1. Existence criteria of wave operators, *Izv. Akad. Nauk SSSR Ser. Mat.*, **27**(1963), 883-906.

2. Local existence criteria of wave operators, *Izv. Akad. Nauk SSSR Ser. Mat.*, **32** (1968), 914-942.

Birman, M.S.; Entina, S.B.

1. Stationary methods in abstract scattering theory (Russian), *Izv. Akad. Nauk SSSR Ser. Math.*, **31**(1967), 401-430.

Birman, M.S.; Krein, M.G.

1. On the theory of wave operators and of scattering operators (Russian), *Dokl. Akad. Nauk SSSR*, **144**(1962), 457-478.

Birman, M.S.; Solomjak, M.Z.

1. Remarks on the phase shift function (Russian), *Zap, Nauchn. Sem. Leningrad Otdel. Mat. Inst. Steklow (LOMI)*, **27**(1972), 33-46.

Bishop, E.

1. Spectral theory for operators on a Banach space, *Trans. Amer. Math. Soc.*, **86** (1957), 414-445.

2. A duality theorem for an arbitrary operator, *Pacific J. Math.*, **9**(1959), 379-397.

Bourbaki, N.

1. *Groupes et algèbres de Lie*, Chapitre II, Hermann, Paris, 1972.

Bram, J.

1. Subnormal operators, *Duke Math. J.*, **22**(1955), 75-94.

de Branges, L.

1. Perturbation of self-adjoint transformations, *Amer. J. Math.*, **84**(1962), 543--580.

Brennan, J.E.

1. Invariant subspaces and rational approximation, *J. Funct. Anal.*, **7**(1971), 285-310.

2. Point evaluations, invariant subspaces, and approximation in the mean by polynomials, *J. Funct. Anal.*, **34**(1979), 407-480.

Brown, L.

1. The determinant invariant for operators with trace class self-commutator, in *Proceedings of a conference on operator theory*, Lecture Notes in Math., vol. **345**, Springer-Verlag, Berlin-Heidelberg-New York, 1973.

Brown, S.

1. Some invariant subspaces for subnormal operators, *Integral Equations Operator Theory*, **1**(1978), 310-333.

2. Hyponormal operators with thick spectra have invariant subspaces, *Ann. of Math.*, **125**(1987), 93-103.

Bunce, J.W.; Deddens, J.A.

1. On the normal spectrum of a subnormal operator, *Proc. Amer. Math. Soc.*, **69** (1978), 103-108.

Carey, R.W.

1. A unitary invariant for pairs of self-adjoint operators, *J. Reine Angew. Math.*, **283**(1976), 294-312.

Carey, R.W.; Pepe, W.

1. The phase shift and singular measures, *Indiana Univ. Math. J.*, **22**(1973), 1049-1064.

Carey, R.W.; Pincus, J.D.

1. The structure of intertwining isometries, *Indiana Univ. Math. J.*, **22**(1973), 679-703.

2. An invariant for certain operator algebras, *Proc. Nat. Acad. Sci. USA*, **71**(1974), 1952-1956.

3. Eigenvalues of seminormal operators, examples, *Pacific J. Math.*, **52**(1974), 347-357.

4. An exponential formula for determining functions, *Indiana Univ. Math. J.*, **23**(1974), 1031-1042.

5. Construction of seminormal operators with prescribed mosaic, *Indiana Univ. Math. J.*, **23**(1974), 1155-1165.

6. Commutators, symbols, and determining functions, *J. Funct. Anal.*, **19**(1975), 50-80.

7. Mosaics, principal functions, and mean motion in von Neumann algebras, *Acta Math.*, **138**(1977), 153-218.

8. Principal functions, index theory, geometric measure theory, and function algebras, *Integral Equations Operator Theory*, **2**(1979), 441-483.

9. Mean motion, principal functions, and the zeros of Dirichlet series, *Integral*

Equations Operator Theory, **2**(1979), 484-502.

10. An integrality theorem for subnormal operators, *Integral Equations Operator Theory*, **4**(1981), 10-44.

11. Almost commuting pairs of unitary operators and flat currents, *Integral Equations Operator Theory*, **4**(1981), 45-122.

12. Principal currents, *Intergal Equations Operator Theory*, **8**(1985), 614-640.

Clancey, K.F.

1. Examples of non-normal seminormal operators whose spectra are not spectral sets, *Proc. Amer. Math. Soc.*, **24**(1970), 797-800.

2. Seminormal operators with compact self-commutators, *Proc. Amer. Math. Soc.*, **26**(1970), 447-454.

3. Completeness of eigenfunctions of seminormal operators, *Acta Sci. Math. (Szeged)*, **39**(1977), 31-37.

4. On the local spectra of seminormal operators, *Proc. Amer. Math. Soc.*, **72**(1978), 473-479.

5. *Seminormal operators*, Lecture Notes in Math., vol.**742**, Sringer-Verlag, Berlin--Heidelberg-New York, 1979.

6. Toeplitz models for operators with one-dimensional self-commutator, in *Dilation theory, Toeplitz operators, and other topics*, Birkhäuser-Verlag, Basel--Boston-Stuttgart, 1981, pp. 81-107.

7. A kernel for operators with one-dimensional self-commutator, *Integral Equation Operator Theory*, **7**(1984), 441-458.

8. The Cauchy transform of the principal function associated with a non-normal operator, *Indiana Univ. Math. J.*, **34**(1985), 21-32.

9. Hilbert space operators with one-dimensional self-commutators, *J. Operator Theory*, **13**(1985), 265-289.

Clancey, K.F.; Conway, J.B.; Raphael, M.

1. On a conjecture of Carey and Pincus, *Integral Equations Operator Theory*, **6**(1983), 158-159.

Clancey, K.F.; Putnam, C.R.

1. The spectra of hyponormal integral operators, *Comment. Math. Helv.*, **46**(1971), 451-456.

Clancey, K.F.; Rogers, D.D.

1. Cyclic vectors and seminormal operators, *Indiana Univ. Math. J.*, **27**(1978), 689--696.

Clancey, K.F.; Wadhwa, B.L.

1. Local spectra of seminormal operators, *Trans. Amer. Math. Soc.*, **280**(1983), 415-428.

Clary, S.

1. Equality of spectra of quasi-similar hyponormal operators, *Proc. Amer. Math.*

Soc., **53**(1975), 88-90.

Colojoară, I.; Foiaş, C.

1. *Theory of generalized spectral operators*, Gordon and Breach, New York, 1968.

Connes, A.

1. Non-commutative differential geometry, *Inst. Hautes Études Sci. Publ. Math.*, **62**(1986), 257-359.

Conway, J.B.

1. The dual of a subnormal operator, *J. Operator Theory*, **5**(1981), 195-211.
2. *Subnormal operators*, Pitman, Boston, 1981.
3. Spectral properties of certain operators on Hardy spaces of planar regions, *Integral Equations Operator Theory*, **10**(1987), 659-706.

Conway, J.B.; Olin, R.F.

1. A functional calculus for subnormal operators. I, *Bull. Amer. Math. Soc.*, **82**(1976), 259-261; II, *Mem. Amer. Math. Soc.*, **184**(1977).

Conway, J.B.; Putnam, C.R.

1. An irreducible subnormal operator with infinite multiplicities, *J. Operator Theory*, **13**(1985), 291-297.

Conway, J.B.; Szymanski, W.

1. Linear combinations of hyponormal operators, *Rocky Mountain J. Math.*, to appear.

Cowen, C.C.; Long, J.J.

1. Some subnormal Toeplitz operators, *J. Reine Angew. Math.*, **351**(1984), 216-220.

Cowen, M.J.; Douglas, R.G.

1. Complex geometry and operator theory, *Acta Math.*, **141**(1978), 187-261.

Curto, R.E.; Muhly, P.S.; Xia, J.

1. Hyponormal pairs of commuting operators, preprint, 1987.
2. Toeplitz operators on flows, preprint, 1987.

Curto, R.E.; Muhly, P.S.; Xia, D.

1. A trace estimate for p-hyponormal operators, *Integral Equations Operator Theory*, **6**(1983), 507-514.

Curto, R.E.; Salinas, N.

1. Generalized Bergman kernels and the Cowen-Douglas theory, *Amer. J. Math.*, **106**(1984), 447-488.

Daleckii, Ju.L.; Krein, S.G.

1. *Integration and differentiation of functions of Hermitian operators and applications to the theory of perturbations*, AMS Translations, ser.2, vol.**47**, Providence, R.I.

Dibrell, P.; Campbell, J.T.

1. Hyponormal powers of composition operators, *Proc. Amer. Math. Soc.*, **4**(1988), 914-918.

Douglas, R.G.

1. On majorization, factorization and range inclusion of operators on Hilbert space, *Proc. Amer. Math. Soc.*, **17**(1966), 413-415.
2. *Banach algebra techniques in operator theory*, Academic Press, New York, 1972.

Dunford, N.; Schwartz, J.

1. *Linear operators. I*, Interscience, New York, 1958.
2. *Linear operators. II*, Interscience, New York, 1964.
3. *Linear operators. III*, Interscience, New York, 1971.

Dym, K.

1. Trace formulas for a class of Toeplitz-like operators, *Israel J. Math.*, **27**(1977), 21-48.
2. Trace formulas for pair operators, *Integral Equations Operator Theory*, **2**(1978), 152-175.

Dynkin, E.M.

1. Functional calculus based on Cauchy-Green's formula (Russian), in *Researches in linear operators and function theory. III*, Leningrad. Otdel. Mat. Inst. Steklov. (LOMI), 1972, pp. 33-39.
2. Pseudoanalytic extension of smooth functions. The uniform scale (Russian), in *Theory of functions and functional analysis*, Central Econom. Mat. Inst. Acad. Sci. SSSR, Moscow, 1976, pp. 40-73.
3. Methods of the theory of singular integrals (the Hilbert transform and Calderon-Zygmund theory) (Russian), in *Encyclopedia of Mathematics*, vol.15, Itogy Nauky i Techn., Sovrem. Problemy Math., Moscow, 1987, pp. 197-292.

Embry, M.R.

1. A generalization of the Halmos-Bram criterion for subnormality, *Acta Sci. Math. (Szeged)*, **35**(1973), 61-64.

Eschmeier, J.

1. *Analytische Dualität und Tensorprodukte in der Mehrdimensionalen Spektraltheorie*, Habilitationsschrift, Münster, 1986.
2. Invariant subspaces and Bishop's condition (β), to appear.

Eschmeier, J.; Putinar, M.

1. Bishop's condition (β) and rich extensions of operators, *Indiana Univ. Math. J.*, to appear.

Fan, P.

1. A note on hyponormal weighted shifts, *Proc. Amer. Math. Soc.*, **92**(1984), 271--272.

Foiaş, C.

1. Spectral maximal spaces and decomposable operators in Banach spaces, *Arch. Math.*, **14**(1963), 341-349.

2. A dilation theorem for hyponormal operators, *Acta Sci. Math. (Szeged)*, **37**(1975), 155-159.

Fong, C.K.

1. On M-hyponormal operators, *Studia Math.*, **65**(1979), 1-5.

Friedrichs, K.O.

1. On the perturbation of continuous spectra, *Comm. Pure Appl. Math.*, **1**(1948), 361-406.

2. *Perturbation of spectra in Hilbert space*, Lecture in Applied Mathematics, vol.III, Proceedings of the Summer Seminar Boulder, Colorado, 1960, Amer. Math. Soc., Providence, R.I., 1965.

Fuglede, B.

1. The multidimensional moment problem, *Expo. Math.*, **1**(1983), 47-65.

Gahov, F.D.

1. *Boundary problems* (Russian), Fizmatgiz, Moscow, 1958.

Gelfand, I.M.; Shilov, G.E.

1. *Generalized functions* (Russian), vol.III, Fizmatgiz, Moscow, 1958.

Gohberg, I.C.

1. An application of the theory of normed rings to integral singular equations (Russian), *Uspekhi Mat. Nauk.*, **2**(1952), 149-156.

Gohberg, I.C.; Krein, M.G.

1. *Introduction to the theory of linear nonselfadjoint operators*, Transl. Math. Monographs, vol.**18**, Amer. Math. Soc., Providence, R.I., 1969.

Gohberg, I.C.; Krupnik, N.Ja.

1. *Introduction to the theory of one-dimensional singular integral operators* (German), Birkhäuser-Verlag, Basel-Boston-Stuttgart, 1979.

Grothendieck, A.

1. *Produits tensoriels topologiques et espaces nucleaires*, Mem. Amer. Math. Soc., **16**(1955).

Halmos, P.R.

1. Normal dilations and extensions of operators, *Summa Brasil. Math.*, **2**(1950), 125-134.

2. *Introduction to Hilbert space and the theory of spectral multiplicity*, Chelsea, New York, 1951.

3. Commutators of operators, *Amer. J. Math.*, **74**(1952), 237-240.

4. *A Hilbert space problem book*, van Nostrand, Princeton, N.J., 1967.

Halmos, P.R.; Lumer, G.; Schäffer, J.J.

1. Square roots of operators, *Proc. Amer. Math. Soc.*, **4**(1953), 142-149.

Harrington, D.J.; Whitley, R.

1. Seminormal composition operators, *J. Operator Theory*, **11**(1984), 125-135.

Hartman, J.

1. A hyponormal weighted shift whose spectrum is not a spectral set, *J. Operator Theory*, 8(1982), 401-403.

Hastings, W.H.

1. The approximate point spectrum of a subnormal operator, *J. Operator Theory*, 5(1981), 119-126.

Helson, H.

1. *The spectral theorem*, Springer-Verlag, Berlin-Heidelberg-New York, 1986.

Helton, J.W.; Howe, R.

1. Integral operators, commutator traces, index and homology, in *Proceedings of a conference on operator theory*, Lecture Notes in Math., vol.**345**, Springer--Verlag, Berlin-Heidelberg-New York, 1973.
2. Trace of commutators of integral operators, *Acta Math.*, **135**(1975), 271-305.

Herlogtz, G.

1. *Über Potenzreihen mit positiven reelem Teil im Einheitskreis*, Berl. Verhandl. sächs. Acad. Wiss. Leipzig Math.-Phys. Klasse, **63**(1911).

Howe, R.

1. A functional calculus for hyponormal operators, *Indiana Univ. Math. J.*, **23**(1974), 631-644.

Hunt, R.A.; Muckenhoupt, B.; Wheeden, R.L.

1. Weighted norm inequalities for the conjugate function and the Hilbert transform, *Trans. Amer. Math. Soc.*, **176**(1973), 227-251.

Iavrian, V.A.; Krein, M.G.

1. On the spectral shift functions for perturbations of positive operators (Russian), *J. Operator Theory*, **6**(1981), 155-191.

Istrățescu, V.

1. Hyponormal operators (Romanian), *Stud. Cerc. Mat.*, I: **19**(1967), 423-437; II: **19**(1967), 439-449; III: **20**(1968), 1159-1168.

Itô, T.; Wong, T.K.

1. Subnormality and quasinormality of Toeplitz operators, *Proc. Amer. Math. Soc.*, **34**(1972), 157-164.

Janas, J.

1. Spectral properties of doubly commuting hyponormal operators, *Ann. Polon.*

Math., **XLIV**(1984), 185-195.

2. On unbounded hyponormal operators, preprint, 1987.

Kac, M.

1. Toeplitz matrices, translation kernels, and a related problem in probability theory, *Duke Math. J.*, **21**(1954), 501-509.

Kato, T.

1. Perturbation of continuous spectra by trace class operators, *Proc. Japan Acad. Ser. A Math. Sci.*, **33**(1957), 260-264.
2. Some mapping theorems for numerical range, *Proc. Japan Acad. Ser. A Math. Sci.*, **41**(1965), 652-655.
3. Wave operators and similarity for some non-selfadjoint operators, *Math. Ann.*, **162**(1966), 258-279.
4. *Perturbation theory for linear operators*, Springer-Verlag, Berlin-Heidelberg--New York, 1966.
5. Smooth operators and commutators, *Studia Math.*, **31**(1968), 535-546.

Kato, T.; Kuroda, S.T.

1. The abstract theory of scattering, *Rocky Mountain J. Math.*, **1**(1971), 127-171.

Kolmogorov, A.N.

1. Stationary sequences in Hilbert spaces (Russian), *Vestnik Moskov. Univ. Ser. I. Mat. Mech.*, **6**(1941), 1-40.

Koppelman, W.; Pincus, J.D.

1. Spectral representation for finite Hilbert transform, *Math. Z.*, **71**(1959), 399--407.

Köthe, G.

1. *Toplogical vector spaces. II*, Springer-Verlag, Berlin-Heidelberg-New York, 1979.

Krein, M.G.

1. On trace formula in perturbations theory (Russian), *Mat. Sb.*, **33**(1953), 597-626.
2. Perturbation determinants and a formula for the traces of unitary and self--adjoint operators (Russian), *Dokl. Akad. Nauk. SSSR*, **144**(1962), 268-271.
3. On perturbation determinants and trace formulae for some classes of pairs of operators (Russian), *J. Operator Theory*, **17**(1981), 123-188.
4. *Topics in differential and integral equations and operator theory*, Birkhäuser--Verlag, Basel-Boston-Stuttgart, 1983.

Krein, M.G.; Nudelman, A.A.

1. *Markov moment problem and extremal problems*, Transl. Math. Monographs, Amer. Math. Soc., Providence, R.I., 1977.

Lambert, A.

1. Hyponormal composition operators, *Bull. London Math. Soc.*, **18**(1986), 395-400.

Langer, H.

1. Eine Erweiterung der Spurformel der Störungstheorie, *Math. Nachr.*, **30**(1965), 123-135.

Martin, M.

1. Hermitian geometry and involutive algebras, *Math. Z.*, **188**(1985), 359-382.
2. An operator theoretic approach to analytic functions into a Grassmann manifold, *Math. Balkanica N.S.*, **1**(1987), 45-58.

Martin, M.; Putinar, M.

1. A unitary invariant for hyponormal operators, *J. Funct. Anal.*, **72**(1987), 297-232.

Mattila, K.

1. Complex strict and uniform convexity and hyponormal operators, *Math. Proc. Cambridge Philos. Soc.*, **96**(1984), 483-493.
2. A class of hyponormal operators and weak*-continuity of hermitian operators, *Ark. Mat.*, **25**(1987), 265-274.

Mikhlin, S.G.

1. *Linear partial differential equations* (Russian), Vys. Skola, Moskow, 1979.

Mikhlin, S.G.; Prössdorf, S.

1. *Singular integral operators*, Springer-Verlag, Berlin-Heidelberg-New York, 1988.

Miller, T.L.; Olin, R.F.; Thomson, J.E.

1. *Subnormal operators and representations of algebras of bounded analytic functions and other uniform algebras*, Mem. Amer. Math. Soc., **354**(1986), Providence, R.I.

Morrel, B.B.

1. A decomposition for some operators, *Indiana Univ. Math. J.*, **23**(1973), 497-511.

Mourre, E.

1. Absence of singular continuous spectrum for certain self-adjoint operators, *Comm. Math. Phys.*, **78**(1981), 391-408.

Muhly, P.S.

1. A note on commutators and singular integrals, *Proc. Amer. Math. Soc.*, **54** (1976), 117-121.

Muskhelishvili, N.I.

1. *Singular integral equations* (Russian), Fizmatlit, Moscow, 1968.

Nelson, E.

1. Analytic vectors, *Ann. of Math.*, **70**(1959), 572-615.

Nielsen, O.A.

1. *Direct integral theory*, Lecture Notes in Pure and Applied Math., vol.**61**, Marcel Dekker, New York-Basel, 1980.

Noether, F.

1. Über eine Klasse singulärer Integralgleichungen, *Math. Ann.*, **82**(1921), 42-63.

Parthasarathy, K.R.; Schmidt, K.

1. *Positive definite kernels, continuous tensor products and central limit theorems of probability theory*, Lecture Notes in Math., vol.**272**, Springer-Verlag, Berlin-Heidelberg-New York, 1972.

Pearson, D.B.

1. A generalization of Birman's trace theorem, *J. Funct. Anal.*, **28**(1978), 182-186.

Pincus, J.D.

1. Commutators and systems of singular integral equations. I, *Acta Math.*, **121** (1968), 219-249.
2. The determining function method in the treatment of commutator systems, in *Colloq. Math. Soc. János Bolyai 5. Hilbert space operators*, Tihany, Hungary, 1970.
3. The spectrum of seminormal operators, *Proc. Nat. Sci. USA*, **68**(1971), 1684--1685.
4. Symmetric singular integral operators, *Indiana Univ. Math. J.*, **23**(1973), 537--556.
5. Unitary equivalence modulo the trace-class for self-adjoint operators, *Amer. J. Math.*, **98**(1976), 481-514.

Pincus, J.D.; Xia, D.

1. Mosaic and principal function of hyponormal and semi-hyponormal operators, *Integral Equations Operator Theory*, **4**(1981), 134-150.

Pincus, J.D.; Xia, D.; Xia, J.

1. The analytic model of a hyponormal operator with rank one self-commutator, *Integral Equations Operator Theory*, **7**(1984), 516-535.

Privalov, I.J.

1. *Boundary properties of analytic functions* (Russian), Moscow, 1950.

Prössdorf, S.

1. *Einige Klassen singulärer Gleichungen*, Akademie-Verlag, Berlin, 1974.

Putinar, M.

1. Hyponormal operators are subscalar, *J. Operator Theory*, **12**(1984), 385-395.
2. Extensions scalaires et noyaux distribution des opérateurs hyponormaux, *C. R. Acad. Sci. Paris Ser. I. Math.*, **301**:15(1985), 739-741.
3. Hyponormal operators and eigendistributions, in *Advances in invariant sub-*

spaces and other results of operator theory, Birkhäuser-Verlag, Basel-Boston--Stuttgart, 1968, pp. 249-273.

4. Extreme hyponormal operators, in *Special classes of linear operators and other topics*, Birkhäuser-Verlag, Basel-Boston-Stuttgart, 1988, pp. 249-265.

5. A two-dimensional moment problem, *J. Funct. Anal.*, to appear.

Putnam, C.R.

1. On the continuous spectra of singular boundary value problems, *Canad. J. Math.*, **6**(1954), 420-426.

2. Commutators and absolutely continuous operators, *Trans. Amer. Math. Soc.*, **87**(1958), 513-525.

3. Commutators, perturbations and unitary spectra, *Acta Math.*, **106**(1961), 215--232.

4. Commutators, absolutely continuous spectra and singular integral operators, *Amer. J. Math.*, **86**(1964), 310-316.

5. *Commutation properties of Hilbert space operators and related topics*, Springer-Verlag, Berlin-Heidelberg-New York, 1967.

6. An inequality for the area of hyponormal spectra, *Math. Z.*, **116**(1970), 323-330.

7. The spectra of semi-normal singular integral operators, *Canad. J. Math.*, **22**(1970), 134-150.

8. Ranges of normal and subnormal operators, *Michigan Math. J.*, **18**(1971), 33-36.

9. Trace norm inequalities for the measure of hyponormal spectra, *Indiana Univ. Math. J.*, **21**(1972), 775-780.

10. Resolvent vectors, invariant subspaces and sets of zero capacity, *Math. Ann.*, **205**(1973), 165-171.

11. Hyponormal operators and spectral multiplicity, *Indiana Univ. Math. J.*, **28**(1979), 701-709.

12. A polar area inequality for hyponormal spectra, *J. Operator Theory*, **4**(1980), 191-200.

13. Hyponormal operators with minimal spectral area, *Houston J. Math.*, **6**(1980), 135-139.

14. Singly generated hyponormal C^*-algebras, *J. Operator Theory*, **11**(1984), 243--254.

15. Real parts of normal extensions of subnormal operators, *Illinois J. Math.*, **31**(1987), 240-247.

Radjabalipour, M.

1. On subnormal operators, *Trans. Amer. Math. Soc.*, **211**(1975), 377-389.

2. On majorization and normality of operators, *Proc. Amer. Math. Soc.*, **62**(1977), 105-110.

3. Ranges of hyponormal operators, *Illinois J. Math.*, **21**(1977), 70-75.

4. Decomposable operators, *Bull. Iranian Math. Soc.*, **9**(1978), 1-49.

5. Hyponormal operators and Dunford's condition (β), *Math. Ann.*, **272**(1985), 567--575.

Reed, M.; Simon, B.

1. *Methods of modern mathematical physics. I: Functional Analysis*, Academic Press, New York-San Francisco-London, 1972.

2. *Methods of modern mathematical physics. II: Fourier analysis*, Ibid., 1975.

3. *Methods of modern mathematical physics. III: Scattering theory*, Ibid., 1979.

4. *Methods of modern mathematical physics. IV: Analysis of operators*, Ibid., 1978.

Riesz, F.; Sz.-Nagy, B.

1. *Leçons d'analyse fonctionnelle*, Akad. Kiadó, Budapest, 1952.

Rosenblum, M.

1. Perturbations of continuous spectrum and unitary equivalence, *Pacific J. Math.*, **7**(1957), 997-1010.

Ryan, R.

1. Boundary values of analytic vector-valued functions, *Proc. Nederlandes Acad. Van Wetenschappen Ser. A*, **65**(1952), 558-572.

Saitô, T.

1. *Hyponormal operators and related topics*, Lecture Notes in Math., vol.**247**, Springer-Verlag, Berlin-Heidelberg-New York, 1972.

Salinas, N.

1. The Grassmann manifold of a C^*-algebra and Hermitian holomorphic bundles, in *Special classes of linear operators and other topics*, Birkhäuser, Basel--Boston-Stuttgart, 1988.

Schmüdgen, K.

1. On commuting unbounded self-adjoint operators. I, *Acta Sci. Math. (Szeged)*, **47**(1984), 131-146.

2. On commuting unbounded self-adjoint operators. III, *Manuscripta Math.*, **54**(1985), 221-247.

3. On commuting unbounded self-adjoint operators. IV, *Math. Nauch.*, **125**(1986), 83-102.

Schmüdgen, K.; Friedrich, J.

1. On commuting unbounded self-adjoint operators. II, *Integral Equations Operator Theory*, **7**(1984), 815-867.

Schwarz, L.

1. Théorie des distributions à valeurs vectorielles. I, *Ann. Inst. Fourier*, **7**(1957), 1--141; II, Ibid., **8**(1958), 1-209.

Shields, A.L.

1. Weighted shift operators and analytic function theory, in *Topics in operator theory*, Math. Surveys vol.**13**, Amer. Math. Soc., Providence, R.I., 1974.

Stampfli, J.

1. Hyponormal operators, *Pacific J. Math.*, **12**(1962), 1453-1458.

2. Hyponormal operators and spectral density, *Trans. Amer. Math. Soc.*, **117**(1965), 469-476.

3. Which weighted shifts are subnormal?, *Pacific J. Math.*, **17**(1966), 367-379.

4. On hyponormal and Toeplitz operators, *Math. Ann.*, **183**(1969), 328-336.

5. A local spectral theory for operators. I, *J. Funct. Anal.*, **4**(1969), 1-10.

6. A local spectral theory for operators. II, *Bull. Amer. Math. Soc.*, **75**(1969), 803--806.

7. A local spectral theory for operators. III: Resolvents, spectral sets and similarity, *Trans. Amer. Math. Soc.*, **168**(1972), 133-151.

8. A local spectral theory for operators. IV: Invariant subspaces, *Indiana Univ. Math. J.*, **22**(1972), 159-167.

9. A local spectral theory for operators. V: Spectral subspaces for hyponormal operators, *Trans. Amer. Math. Soc.*, **217**(1976), 285-296.

Stampfli, J.; Wadhwa, B.L.

1. On dominant operators, *Monatsh. Math.*, **84**(1978), 143-153.

Stein, E.M.

1. *Singular integrals and differentiable properties of functions*, Princeton University Press, Princeton, N.J., 1970.

Stochel, J.

1. Subnormal contractions and seminormal composition operators, preprint, 1987.

Stochel, J.; Szafraniec, F.H.

1. A characterization of subnormal operators, in *Spectral theory of linear operators and related topics*, Birkhäuser-Verlag, Basel-Boston-Stuttgart, 1984, pp. 261-263.

2. On normal extensions of unbounded operators. I, *J. Operator Theory*, **14**(1985), 31-55.

3. On normal extensions of unbounded operators. II, *Acta Sci. Math. (Szeged)*, to appear.

4. On normal extensions of unbounded operators. III, preprint, 1988.

Stone, M.H.

1. *Linear transformations in Hilbert space and their applications to analysis*, Amer. Math. Soc. Colloq. Bull., vol. **15**, Providence, R.I., 1932.

Sun, S.

1. Bergman shift is not unitarily equivalent to a Toeplitz operator, *Kexue Tongbao*, **28**(1983), 1027-1030.

Szegö, G.

1. On certain Hermitian forms associated with the Fourier series of a positive

function, *Comm. Sem. Math. Univ. Lund., tome suppl. dedié a Marcel Riesz* (1952), 228-238.

Sz.-Nagy, B.

1. *Unitary dilations of Hilbert space operators and related topics*, in Reg. Conference Series in Math., vol.**19**, Amer. Math. Soc., Providence, R.I., 1974.

Sz.-Nagy, B.; Foiaş, C.

1. *Harmonic analysis of operators on Hilbert space*, Akad. Kiadó, Budapest, 1970.
2. An application of dilation theory to hyponormal operators, *Acta Sci. Math. (Szeged)*, **37**(1975), 155-159.
3. Toeplitz type operators and hyponormality, in *Dilation theory, Toeplitz operators and other topics*, Birkhäuser-Verlag, Basel-Boston-Stuttgart, 1983, pp. 371-388.

Szymanski, W.

1. Positive forms and dilations, *Trans. Amer. Math. Soc.*, **301**(1987), 761-780.
2. Dilations and subnormality, *Proc. Amer. Math. Soc.*, **101**(1987), 251-259.

Thomson, J.E.

1. Invariant subspaces for algebras of subnormal operators, *Proc. Amer. Math. Soc.*, **3**(1986), 462-464.

Vasilescu, F.-H.

1. *Analytic functional calculus and spectral decompositions*, Ed. Academiei and D. Reidel Co., Bucharest and Dordrecht, 1982.

Verblunsky, S.

1. Two moment problems for bounded functions, *Proc. Cambridge Philos. Soc.*, **42**(1946), 189-196.
2. On the initial moments of a bounded function, *Proc. Cambridge Philos. Soc.*, **43**(1947), 275-279.

Voiculescu, D.

1. Some extensions of quasitriangularity, *Rev. Roumaine Math. Pures Appl.*, **18**(1973), 1303-1320.
2. Some results on norm-ideal perturbations of Hilbert space operators, *J. Operator Theory*, 2(1979), 3-37.
3. A note on quasitriangularity and trace-class self-commutators, *Acat Sci. Math. (Szeged)*, **42**(1980), 195-199.
4. Some results on norm-ideal perturbations of Hilbert space operators. II, *J. Operator Theory*, 5(1981), 77-100.
5. Remarks on Hilbert-Schmidt perturbations of almost normal operators, in *Topics in modern operator theory*, Birkhäuser-Verlag, Basel-Boston-Stuttgart, 1981, pp. 311-318.
6. Hilbert space operators modulo normed ideals, in *Proc. Int. Congress Math. Warszawa*, PWN-Polish Scientific Publishers, North-Holland, Warszawa and

Amsterdam, 1984, pp. 1041-1047.

7. On a trace formula of M.G. Krein, in *Operators in indefinite metric spaces, scattering theory and other topics*, Birkhäuser-Verlag, Basel-Boston-Stuttgart, 1987, pp. 329-332.

Wadhwa, B.L.

1. A hyponormal operator whose spectrum is not a spectral set, *Proc. Amer. Math. Soc.*, **38**(1973), 83-85.

Widom, H.

1. Asymptotic behaviour of block Toeplitz matrices and determinants. II, *Adv. in Math.*, **21**(1976), 1-29.

2. *Asymptotic expansions for pseudo-differential operators on bounded domains*, Lecture Notes in Math., vol.**1152**, Springer-Verlag, Berlin-Heidelberg-New York, 1985.

Williams, L.R.

1. Quasisimilarity and hyponormal operators, *J. Operator Theory*, **5**(1981), 127--139.

Wolniewicz, T.M.

1. The Hilbert transform in weighted spaces of integrable vector-valued functions, *Colloq. Math.*, **43**(1987), 103-108.

Xia, D.

1. On non-normal operators. I, *Chinese J. Math.*, **3**(1963), 232-246; II, *Acta Math. Sinica*, **21**(1978), 187-189.

2. On the spectrum of hyponormal or semi-hyponormal operators, *J. Operator Theory*, **5**(1981), 257-266.

3. *Spectral theory of hyponormal operators*, Birkhäuser-Verlag, Basel-Boston--Stuttgart, 1983.

4. On the analytic model of a class of hyponormal operators, *Integral Equations Operator Theory*, **6**(1983), 134-157.

5. On the kernels associated with a class of hyponormal operators, *Integral Equations Operator Theory*, **6**(1983), 444-452.

6. On the semi-hyponormal n-tuple of operators, *Integral Equations Operator Theory*, **6**(1983), 879-898.

7. The analytic model of a subnormal operator, *Integral Equations Operator Theory*, **10**(1987), 258-289.

8. Analytic theory of subnormal operators, *Integral Equations Operator Theory*, **10**(1987), 880-903.

Yoshihiro, N.

1. Principal functions and invariant subspaces of hyponormal operators, *Hokkaido Math. J.*, **1**(1983), 1-9.

NOTATION AND SYMBOLS

For the convenience of the reader, we assemble below some standard notations which are used constantly throughout the book.

General symbols

N, Z, Q, R, C : natural, integer, rational, real, and respectively, complex numbers.

intσ, σ^-, $\partial\sigma$: interior, closure, and respectively, boundary of a set σ.

spanσ : linear subspace generated by a set σ.

$\bigvee\sigma$ = (**span**σ)$^-$.

coσ, **aco**σ : convex hull, and respectively, absolute convex hull of a set σ.

χ_σ : characteristic function of a set σ.

$D(\omega,r)$: open disk centered at ω and with radius r.

D = $D(0,1)$: the open unit disk in **C**.

T = ∂**D** : the one dimensional torus.

dim E : dimension of a complex vector space E.

Hilbert space and operator symbols

H, K, ... : Hilbert spaces; <u>all Hilbert spaces are assumed to be complex and separable</u>.

$\langle\zeta,\eta\rangle_H$: inner product of two vectors $\zeta,\eta \in H$; sometimes denoted simply by $\langle\zeta,\eta\rangle$.

$S^\perp$: subspace orthogonal to a subset S.

$L(H,K)$: linear bounded operators from H into K.

$L(H) = L(H,H)$.

Ker T, **Ran** T : algebraic kernel, and respectively, range of an operator T.

T^* : adjoint of an operator T.

$\sigma(T)$: spectrum of an operator T.

$\sigma_{ess}(T)$: essential spectrum of an operator T.

$\rho(T) = \mathbf{C} \setminus \sigma(T)$: resolvent set of T.

$\rho_{ess}(T) = \mathbf{C} \setminus \sigma_{ess}(T)$: essential resolvent set of T.

ind : Fredholm index.

Tr : trace functional.

$C_1(H)$, $|\cdot|_1$: trace class operators on H and trace norm.

$C_2(H)$, $|\cdot|_2$: Hilbert-Schmidt operators on H and Hilbert-Schmidt norm.

$[A, B] = AB - BA$: commutator of two operators A and B.

$[T^*, T]$: self-commutator of an operator T.

P_K^H : orthogonal projection from H onto a closed subspace K.

$T|K$: restriction of an operator T to a closed invariant subspace K.

Function spaces

$\ell^2(\mathbf{N})$, $\ell^2(\mathbf{Z})$: the Hilbert spaces of square summable complex sequences of the form $(\xi_n)_{n \in \mathbf{N}}$, and respectively, $(\xi_n)_{n \in \mathbf{Z}}$.

$L^p(\Omega)$, $1 \leq p \leq \infty$: Lebesgue space of complex-valued p-summable functions on a Borel subset Ω of $\mathbf{R}^n (n \geq 1)$, with respect to the usual Lebesgue measure.

$\|\cdot\|_{p,\Omega}$: the norm of $L^p(\Omega)$.

$L^p_{comp}(\Omega)$: subspace of those functions in $L^p(\Omega)$ with compact essential support.

$L^1_{loc}(\Omega)$: locally integrable functions on Ω.

$L^p(\Omega, X)$: X-valued p-summable functions on Ω, where X is a Banach space.

$O(\Omega)$: analytic complex-valued functions on an open subset Ω of $\mathbf{C}$.

$O(\sigma)$: germs of complex analytic functions defined in open neighbourhoods of a closed subset σ of $\mathbf{C}$.

$O(\Omega, X)$, $O(\sigma, X)$: spaces of X-valued analytic functions, where X is a Banach space.

$E(M)$: complex-valued smooth (C^∞) functions on a smooth manifold M.

Symbols defined in the text

INDEX